CURRENT ORNITHOLOGY

VOLUME 8

A Continuation Order Plan is available for this series. A continuation order will bring delivery of each new volume immediately upon publication. Volumes are billed only upon actual shipment. For further information please contact the publisher.

CURRENT ORNITHOLOGY

VOLUME 8

Edited by
DENNIS M. POWER
Santa Barbara Museum of Natural History
Santa Barbara, California

PLENUM PRESS • NEW YORK AND LONDON

The Library of Congress cataloged the first volume of this title as follows:

Current ornithology.—Vol. 1–
New York: Plenum Press, c1983–
v.: ill.; 24 cm.
Annual.
Editor: Richard F. Johnston.
ISSN 0742-390X = Current ornithology.
1. Ornithology—Periodicals. I. Johnston, Richard F.
QL671.C87 598′.05—dc19 84-640616
[8509] AACR 2 MARC-S

ISBN 0-306-43640-X

A Division of Plenum Publishing Corporation
233 Spring Street, New York, N.Y. 10013

Printed in the United States of America

CONTRIBUTORS

ALLAN J. BAKER, Department of Ornithology, Royal Ontario Museum, Toronto, Ontario M5S 2C6, Canada; and Department of Zoology, University of Toronto, Toronto, Ontario M5S 1A1, Canada

I. LEHR BRISBIN, JR., Savannah River Ecology Laboratory, Aiken, South Carolina 29801

PATRICIA ADAIR GOWATY, Department of Biological Sciences, Clemson University, Clemson, South Carolina 29634-1903

HUGH I. JONES, Department of Zoology, University of Western Australia, Nedlands, Western Australia 6009, Australia. *Present address:* Gwynedd Health Authority, Coed Mawr, Bangor, Gwynedd LL57 4TP, Wales

IAN G. McLEAN, Department of Zoology, University of Canterbury, Christchurch, New Zealand

MELINDA A. PRUETT-JONES, Natural Reserve System, Scripps Institution of Oceanography, University of California at San Diego, La Jolla, California 92093. *Present address:* Field Museum of Natural History, Chicago, Illinois 60605

STEPHEN G. PRUETT-JONES, Department of Biology, University of California at San Diego, La Jolla, California 92093. *Present ad-*

dress: Department of Ecology and Evolution, University of Chicago, Chicago, Illinois 60637

GILLIAN RHODES, Department of Psychology, University of Canterbury, Christchurch, New Zealand

HANS TEMRIN, Division of Ethology, Department of Zoology, University of Stockholm, S-106 91 Stockholm, Sweden

JOSEPH M. WUNDERLE, Jr., Institute of Tropical Forestry, Southern Forest Experiment Station, United States Department of Agriculture Forest Service, Rio Piedras, Puerto Rico 00928-2500; and Department of Biology, University of Puerto Rico, Cayey, Puerto Rico 00633

PREFACE

Within the cycle of ornithological events, the occurrence of an International Ornithological Congress ranks high. It brings focus on the work and the scientists in the host country. The Twentieth IOC in New Zealand held in December 1990 is no exception. The editorial board suggested that this issue of *Current Ornithology* emphasize a bit more than usual a specific geographic region and, of course, the appropriate region would be New Zealand and the southern Pacific Ocean.

Allan J. Baker, a New Zealander now based at the Royal Ontario Museum, Toronto, has contributed a summary of ornithological work in New Zealand. He highlights some of the important contributions to ornithology in general, identifying certain key, long-term studies, and discusses some of the unique conservation issues involving the avifauna there.

Ian G. McLean and Gillian Rhodes of the Departments of Zoology and Psychology, respectively, University of Canterbury, review enemy recognition and response in birds in light of their own work on cowbird parasitism of certain New Zealand species. They emphasize cognitive processes and challenge the optimality paradigm that focuses on costs and benefits of a trait in terms of fitness.

Stephen G. Pruett-Jones, now at the University of Chicago, Melinda A. Pruett-Jones, now at the Field Museum of Natural History, Chicago, and Hugh I. Jones, University of Western Australia, now at the Gwynedd Health Authority, look at parasitism and sexual selection in New Guinea. Their work addresses the correlation observed by others between parasite loads and plumage showiness in males. They find that the relationship in a New Guinea avifauna is not a simple one and that using the sexual selection hypotheses to explain this correlation may not yet be warranted.

Elsewhere in the world, our attention was brought to Chernobyl in the Soviet Union when in April 1986 there was an explosion and fire in a nuclear power plant that resulted in significant environmental contamination. I. Lehr Brisbin, Jr. reviews studies in avian radioecology and summarizes years of work at the Savannah River Ecology Laboratory sponsored by the United States Department of Energy. He addresses how risks to birds and man may be measured.

Patricia Adair Gowaty of Clemson University provides a provocative analysis of sex ratios in birds. She suggests there is statistical evidence that birds adjust sex ratio through relative allocation of resources to sons and daughters. More may be happening with progeny sex ratios than was previously suspected.

Hans Temrin, University of Stockholm, summarizes his own work on certain European passerines and that of other behavioral ecologists on mating status in passerines and the role of deception, if any, in polyterritorial species. Fledging success of primary and secondary broods, timing of breeding, female mate choice, and female behavior (whether "coy" or "fast") all add to the complexity and challenge of studying mating strategies.

Joseph M. Wunderle, Jr., of the University of Puerto Rico, reviews the mechanics of juvenile foraging proficiency. Juveniles generally are less proficient than adults, but studies have shown that in some species there are interesting ways young birds may compensate.

We continue to solicit reviews, syntheses, and position papers addressing current topics and active areas of research in avian biology. Interested authors may contact any member of the editorial board or send a letter and brief prospectus to me at the Santa Barbara Museum of Natural History, Santa Barbara, California 93105.

I am grateful to the editorial board for suggesting potential contributors and for their advice on the suitability of papers. Special thanks and appreciation are due the authors who worked diligently on the chapters in this volume.

Dennis M. Power

Santa Barbara, California

CONTENTS

CHAPTER 1

A Review of New Zealand Ornithology

Allan J. Baker

1. Introduction 1
2. Origin of the New Zealand Avifauna 4
   2.1. Biogeographic Elements 4
   2.2. Routes of Colonization 9
   2.3. Double Invasions 10
3. Geographic Variation, Island Subspecies, and Speciation .. 12
   3.1. Tomtits and New Zealand Robins 13
   3.2. New Zealand Bellbirds 20
   3.3. Variable Oystercatchers 22
   3.4. Silver Gulls 24
   3.5. Speciation in Parrots and Bush Wrens 24
4. Flightlessness and Melanism 27
   4.1. Flightlessness 28
   4.2. Melanism 28
5. Extinct Birds 30
6. Rare and Endangered Species 35
7. Conservation of Birds 38
   7.1. Scope of Conservation Work 38
   7.2. Island Refuges and Forest Reserves on the Mainland 42

7.3. Examples of Conservation of Birds 43
7.4. Conservation Genetics 44
8. Introduced Birds 45
9. History of Ornithological Discovery in New Zealand 49
9.1. Mainland New Zealand 49
9.2. Offshore Islands 51
10. Ornithological Organizations in New Zealand 52
11. Exemplary Long-Term Population Studies 53
11.1. Yellow-eyed Penguins 54
11.2. Silver Gulls 55
11.3. Southern Great Skuas 56
11.4. Pukekos 58
11.5. North Island Saddlebacks 60
12. Future Prospects 61
References 62

CHAPTER 2

AVIAN RADIOECOLOGY

I. LEHR BRISBIN, JR.

1. Introduction 69
2. Some Basic Principles of Ionizing Radiation 72
2.1. Types of Ionizing Radiation 73
2.2. Some Important Radionuclides and Their Characteristics 74
2.3. Radionuclide Cycling and Turnover 76
2.4. Radionuclide Uptake and Concentration 80
2.5. Detection and Measurement of Radiation 81
2.6. Units of Radiation Measurement 83
3. Radiation Effects Studies 85
3.1. Mortality Responses to Radiation 86
3.2. Growth Responses to Radiation 90
3.3. Responses of Reproduction to Radiation 93
3.4. Radiation Effects and Hormesis 98
4. Cycling of Radioactive Contaminants 100
4.1. The Savannah River Plant and Its Avifauna: A Case Study 101
4.2. Radionuclide Distribution in the SRP Avifauna 105
4.3. Radionuclide Uptake and Dispersal by SRP Waterfowl 112
4.4. Long-Term Changes in Radionuclide Levels of SRP Waterfowl 119

5. Applied Avian Radioecology: The Chernobyl Accident ... 121
5.1. Estimation of Radionuclide Transport by Waterfowl ... 124
5.2. Conclusions: The Chernobyl Nuclear Accident and Risks to Man and Bird ... 128
6. Summary ... 132
References ... 134

CHAPTER 3

FACULTATIVE MANIPULATION OF SEX RATIOS IN BIRDS: RARE OR RARELY OBSERVED?

PATRICIA ADAIR GOWATY

1. Introduction ... 141
2. What is Facultative Manipulation of Sex Ratios? ... 143
2.1. Definition ... 143
2.2. Mechanisms of Facultative Manipulation ... 144
2.3. Ecological Correlates of Facultative Manipulation ... 145
2.4. How Large Should Facultative Skews in Sex Ratios Be? ... 146
2.5. Tests for Facultative Manipulation of Sex Ratios ... 147
3. The Binomial Test ... 148
4. What Data Are Necessary and Sufficient to Show Facultative Manipulation? ... 157
4.1. Zebra Finches ... 158
4.2. Yellow-headed Blackbirds ... 160
4.3. Red-winged Blackbirds ... 161
4.4. Red-cockaded Woodpeckers ... 162
4.5. Harris' Hawks ... 165
4.6. Bald Eagles ... 166
5. Are Facultative Sex Ratios Rare or Rarely Observed in Birds? ... 168
References ... 169

CHAPTER 4

ENEMY RECOGNITION AND RESPONSE IN BIRDS

IAN G. MCLEAN AND GILLIAN RHODES

1. Introduction ... 173
2. A Cognitive Model of Recognition and Response ... 176

3. Mobbing Behavior 180
3.1. Predictions Derived from Optimality Theory 182
3.2. An Alternative Scenario: The Feedback Hypothesis . 185
3.3. Evidence for the Feedback Hypothesis 187
3.4. Other Factors Influencing Feedback, the Response Curve, and Variance in Response 190
3.5. Do Mobbing Birds Exhibit Reciprocal Altruism? 192
3.6. The Significance of Cultural Transmission 195
4. A Case Study of Recognition: the Gerygone and the Cuckoo 197
4.1. Recognition of Adult Cuckoos as an Enemy by Grey Gerygones 198
4.2. Vocal Mimicry of Begging Calls by Shining Bronze Cuckoo Chicks 201
5. General Discussion 203
References 205

CHAPTER 5

PARASITES AND SEXUAL SELECTION IN A NEW GUINEA AVIFAUNA

STEPHEN G. PRUETT-JONES, MELINDA A. PRUETT-JONES, AND HUGH I. JONES

1. Introduction 213
2. Study Area 216
3. Field Methods 217
3.1. Collection and Examination of Blood Smears 217
3.2. Measures of Parasitemia 217
3.3. Host Plumage Brightness and Showiness 218
3.4. Host Ecology 219
3.5. Statistical Analysis 219
4. Results 220
4.1. Distribution of Parasites across Species, Altitude, and Seasons 220
4.2. Distribution of Parasites across Individuals 223
4.3. Plumage Brightness versus Showiness 225
4.4. Parasitism and Showiness 226
4.5. Parasites and Ecology 229
4.6. Parasites and Phylogenetic Order 231
5. Discussion 232
6. Summary and Conclusions 237

7. Appendix: Parasite Prevalences and Intensity Values for Species and Families of Papua New Guinea Birds 239
References 243

CHAPTER 6

DECEIT OF MATING STATUS IN PASSERINE BIRDS: AN EVALUATION OF THE DECEPTION HYPOTHESIS

HANS TEMRIN

1. Introduction 247
2. Polyterritorial Behavior 248
3. Testing the Deception Hypothesis 250
4. Factors Affecting Deception of Mating Status 256
4.1. The Relationship between Paternal Care and Breeding Success 257
4.2. The Probability of Becoming a Secondary Female and Losing in Fitness 260
4.3. The Cost of Assessing Male Mating Status 263
5. Conclusions 264
6. Appendixes 267
6.1. Calculations of Reproductive Success for Uninformed Females (F_u) versus Informed Females (F_i)[a] 267
6.2. Calculations of the Maximum Number of Days a Female Could Use to Inform Herself about Mating Status and Find an Unmated Male, and Still Do Better Than Uninformed Females 268
6.3. Examples of the Relationship between the Probability of Becoming a Secondary Female, the Fledging Success (N_2) of Secondary Females, and the Costs of Assessing the Mating Status of Males .. 268
References 269

CHAPTER 7

AGE-SPECIFIC FORAGING PROFICIENCY IN BIRDS

JOSEPH M. WUNDERLE, JR.

1. Introduction 273
2. The Nature of the Literature 276

3. Foraging Components 278
3.1. Foraging Site 278
3.2. Search Methods and Patterns 283
3.3. Food Recognition and Selection 286
3.4. Prey Capture 291
3.5. Food Handling Time and Techniques 297
4. Discussion 303
5. Summary 307
6. Appendix: Summary of Species in Which Age-Related Differences in Foraging Proficiency Have Been Documented 308
References 315

INDEX 325

CURRENT ORNITHOLOGY

VOLUME 8

CHAPTER 1

A REVIEW OF NEW ZEALAND ORNITHOLOGY

ALLAN J. BAKER

1. INTRODUCTION

The New Zealand Avifauna

Ornithologists visiting New Zealand for the first time will quickly discern two characteristics of the avifauna that are direct consequences of its long period of isolation. First, the nonmarine avifauna has low species diversity, with only 65 native species of land and freshwater birds now extant (Bull and Whitaker, 1976). If seabirds are included, however, there is a total of 149 native and endemic species breeding in New Zealand and its outlying islands in the north (Kermadecs), east (Chathams), and subantarctic (Antipodes, Auckland, Bounty, Campbell, Macquarie, and Snares; see Fig. 1). A further 34 species introduced by humans in the last 100–130 years make up the 183 total list of breeding species (Table I). The remainder of the 276 species in the new Zealand recent avifauna (excluding 9 species from the Ross Dependency, a region of Antarctica administered by New Zealand) comprises 24 nonbreeding migrants (16 of which are shorebirds) and 69 extralimital stragglers (Kinsky, 1970; Williams, 1973). Another 47 species are known only from subfossil or fossil remains.

ALLAN J. BAKER • Department of Ornithology, Royal Ontario Museum, Toronto, Ontario, M5S 2C6,Canada; and Department of Zoology, University of Toronto, Toronto, Ontario M5S 1A1, Canada.

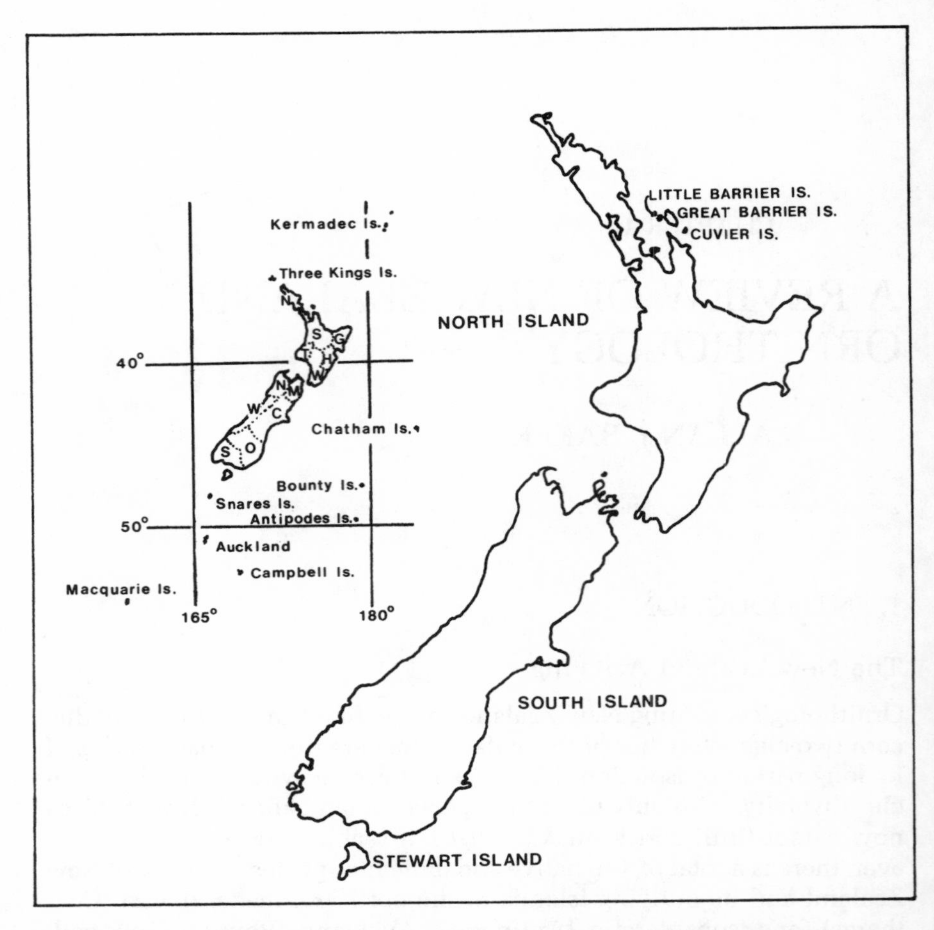

FIGURE 1. Map of the New Zealand region showing the major outlying island groups around the New Zealand mainland (after Kinsky, 1970). Provinces of the North and South Islands are indicated with letters on inset.

Although New Zealand is depauperate in native landbird species, its nonmarine avifauna harbors some unique lineages of birds that are a "must see" for visiting ornithologists. The famous and bizarre kiwis, along with the now extinct moas, comprise two endemic orders of flightless landbirds (Apterygiformes and Dinornithiformes, respectively). Additionally, three endemic families of passerines occur, the New Zealand wrens (Acanthisittidae), the New Zealand wattlebirds (Callaeaidae), and the probably extinct New Zealand thrushes (Turn-

TABLE I
The Number of Indigenous, Introduced, and Endemic Breeders in the Recent Avifauna of New Zealand

Common name	Family	Usual habitat	A[a]	B[b]	C[c]	D[d]
Kiwis	Apterygidae	Terrestrial	3	—	—	3
Penguins	Spheniscidae	Marine	11	2	—	3
Grebes	Podicipedidae	Freshwater	3	1	—	1
Albatrosses	Diomedeidae	Marine	9	5	—	2
Petrels	Procellariidae Hydrobatidae Pelecanoididae	Marine	48	25	—	9
Tropicbirds	Phaetontidae	Marine	1	1	—	—
Pelicans	Pelecanidae	Marine	1	—	—	—
Gannets	Sulidae	Marine	3	2	—	—
Cormorants	Phalacrocoracidae	Freshwater	3	2	—	1
		Marine	5	1	—	4
Darters	Anhingidae	Marine	1	—	—	—
Frigatebirds	Fregatidae	Marine	2	—	—	—
Herons, bitterns	Ardeidae	Freshwater	8	3	—	—
		Marine	1	1	—	—
Ibises, spoonbills	Threskiornithidae	Freshwater	3	1	—	—
Waterfowl	Anatidae	Freshwater	16	3	4	4
		Marine	1	—	—	1
Hawks, falcons	Accipitridae	Terrestrial	3	1	—	1
Pheasants, partridges, quail	Phasianidae	Terrestrial	8	1	7	—
Cranes, rails	Gruidae Rallidae	Terrestrial & Freshwater	12	6	—	3
Waders	S.O. Charadrii	Freshwater	12	5	—	3
		Marine	36	5	—	3
Skuas, gulls, terns	Stercorariidae	Freshwater	3	—	—	1
	Laridae Sternidae	Marine	19	11	—	1
Pigeons, doves	Columbidae	Terrestrial	3	—	2	1
Parrots	Cacatuidae Nestoridae Platycercidae	Terrestrial	10	1	3	6
Cuckoos	Cuculidae	Terrestrial	6	2	—	—
Owls	Strigidae	Terrestrial	4	1	1	1
Swifts	Apodidae	Terrestrial	2	—	—	—
Kingfishers, rollers	Alcedinidae Coraciidae	Terrestrial	3	1	1	—
Songbirds	S.O. Tyranni S.O. Passeres	Terrestrial	45	3	16	21
			285	79	34	70

[a]Total number of species in Kinsky (1970).
[b]Indigenous breeders, excluding endemics.
[c]Introduced breeders.
[d]Endemic breeders.

agridae). The degree of endemicity is, of course, much higher in land and freshwater birds than in the seabirds that often traverse the southern oceans. For example, 37 (57%) of the 65 land and freshwater species of birds are endemic, and this climbs to 70% endemicity if subfossil and recently extinct species are included. In contradistinction, the degree of endemicity falls to only about 30% in the species of seabirds and shorebirds inhabiting the New Zealand region, although many local subspecies have been described (Bull and Whitaker, 1976).

2. ORIGIN OF THE NEW ZEALAND AVIFAUNA

2.1. Biogeographic Elements

Biogeographic analysis of the New Zealand avifauna has been based on perceived affinities, rather than on more rigorous assessment of phylogenetic relationships of taxa, area cladograms, and generalized tracks. Thus the traditional view is that the New Zealand avifauna is composed largely of elements that have dispersed to New Zealand long after the breakup of Gondwanaland. This view is supported by the lack of fossils (other than of penguins) from the lower Cenozoic and upper Cretaceous, the clear similarities of many extant New Zealand and Australian taxa (implying sister-group relationships), and the prevailing westerly wind patterns that regularly deposit Australian species in New Zealand. For example, eight species of Australian birds have successfully colonized New Zealand in the past 150 years (Table II), whereas at least three others have failed to persist: Red-necked Avocet (*Recurvirostra novaehollandiae*), White-eyed Duck (*Aythya australis*), and

TABLE II
Species That Successfully Colonized New Zealand in the Last 150 Years

Species	Country of origin	Approximate date of arrival
Grey-backed White eye (*Zosterops lateralis*)	Australia	1856
Grey Teal (*Anas gibberifrons*)	Australia	1916?
White-faced Heron (*Ardea novaehollandiae*)	Australia	1940
Masked Plover (*Vanellus miles*)	Australia	1940
Royal Spoonbill (*Platalea regia*)	Australia	1940's
Common Coot (*Fulica atra*)	Australia	1954
Black-fronted Dotterel (*Charadrius melanops*)	Australia	1954
Pacific Swallow (*Hirundo tahitica*)	Australia	1958

Dusky Wood Swallow (*Artamus cyanopterus*). Not surprisingly, then, the Australian element is a dominant one in the New Zealand avifauna, although such an apparently high rate of successful colonization would have given rise to a much larger avifauna than currently exists unless it was accompanied by historically high rates of extinction (Fleming, 1962a). Alternatively, the man-modified habitat in New Zealand may presently be much more favorable for colonization than it was in the more distant past.

Three other major biogeographic elements are distinguishable in the New Zealand avifauna, referred to as the Holarctic, Austral, and Archaic elements. The Holarctic element is small, with only three species of obvious northern hemisphere origin. The South Island Pied Oystercatcher (*Haematopus ostralegus finschi*) is very clearly related to Holarctic races of *H. ostralegus* (Baker, 1974). The New Zealand Scaup (*Aythya novaeseelandiae*) is closely related to the Tufted Duck (*A. fuligula*) of the Palaearctic and the Greater and Lesser scaups (*A. marila* and *A. affinis*) of the Holarctic region (Johnsgard, 1965). The closest relative of the extinct Auckland Island Merganser (*Mergus australis*) is the Chinese Merganser (*M. squamatus*) in China (Kear and Scarlett, 1970). The extinct Chatham Island Eagle (*Haliaeetus australis*) is most closely related to sea-eagles in the northern hemisphere, even though the congeneric White-bellied Sea Eagle (*H. leucogaster*) occurs in Australia (Olson, 1984). Thus the Chatham Island Eagle appears to be another Holarctic element that established itself by chance colonization in the New Zealand region.

The rationale for recognizing Holarctic elements in the New Zealand avifauna is that they did not colonize by way of Australia or elsewhere, but this can lead to problems in interpretation when sister-group relationships are not known. For example Falla (1953) and Fleming (1962a,b) have assumed that the endemic Black-billed Gull (*Larus bulleri*) is a Holarctic immigrant because it is similar in wing tip pattern to the Black-headed Gull *Larus ridibundus*. Yet it is very similar morphologically to the Silver (Red-billed) Gull (*L. novaehollandiae*), with which it readily hydridizes and produces viable F_1 progeny (Gurr, 1967). Mitochondrial DNA studies indicate that Black-billed and Silver Gulls are sister-species (Fig. 2), and thus the former is more properly referred to as the Australian element.

As might be expected, the Austral element in the New Zealand avifauna is composed exclusively of seabirds, most notably penguins, albatrosses, and petrels (Bull and Whitaker, 1976). The penguins are an archaic element, with a fossil record from the Eocene to Pliocene. They are today represented by eight species in five genera that breed in New

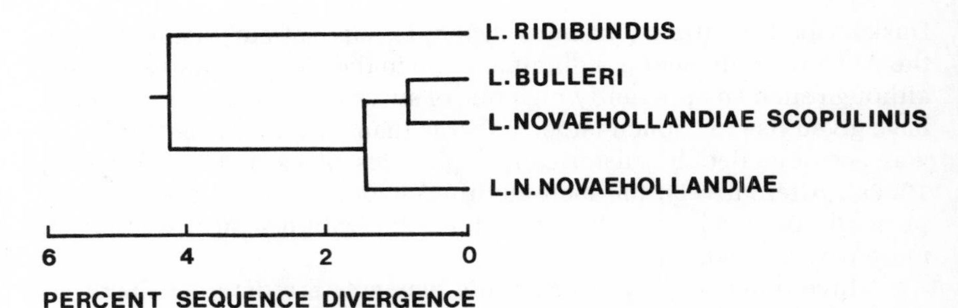

FIGURE 2. Relationships of the Black-billed Gull (*L. bulleri*), New Zealand (Red-billed Gull) (*L. novaehollandiae scopulinus*), Australian Silver Gull (*L. n. novaehollandiae*), and the European Black-headed Gull (*L. ridibundus*), as indicated by UPGMA cluster of percent sequence divergence in mitochondrial DNA. Sequence divergence was estimated based on restriction analysis of mtDNA with 16 endonucleases (A. J. Baker *et al.*, unpublished manuscript).

Zealand or the subantarctic islands. Because of their ocean-traversing prowess, many of the albatrosses and petrels in the New Zealand region presently have a circumpolar distribution, although a few, such as the Royal Albatross (*Diomedea epomophora*) and the Common Diving Petrel (*Pelecanoides urinatrix*), have evolved different subspecies on various islands (Harper and Kinsky, 1978). The Southern Black-backed Gull (*Larus dominicanus*) is another Austral element, occurring in New Zealand, South Africa (where it has a different iris color), and South America. Bull and Whitaker (1976) identified certain marine cormorants as possible Austral elements because the distinct races in the New Zealand region have their nearest relatives in South America.

The Archaic element is made up of endemic taxa with such a long evolutionary history in New Zealand that no close relatives can be identified elsewhere. The kiwis and extinct moas make up two endemic orders, and the New Zealand wrens, wattlebirds, and thrushes three endemic families. Endemic genera usually included in this element are parrots *Nestor* and *Strigops*, the snipe *Coenocorypha*, the plovers *Thinornis* and *Anarhynchus*, the Blue Duck *Hymenolaimus*, the giant extinct rail-like birds *Diaphorapteryx* and *Aptornis*, and the rails *Gallirallus* and *Notornis*. To some degree the Archaic element is an artificial construct because it overlaps to an unknown extent with other elements. Future phylogenetic analyses will undoubtedly determine sister-groups with taxa in other biogeographic regions, and thus the archaic element will be eroded. For example, *Notornis* is most likely a sister-group to *Porphyrio*, and thus is simply an early-colonizing Australian element, and *Thinornis* is likely referable to *Charadrius*.

The Australian element has been described in detail by Falla (1953). It includes members of at least 18 families, and is particularly well represented in the land and freshwater birds. In addition to the eight Australian species mentioned above whose arrival and spread has been documented by Europeans, 11 more species are still clearly referable to Australian subspecies, implying that they are also recent colonists. By my count, another 14 species of Australian origin have been in New Zealand long enough to evolve distinctive subspecies, although this number is clearly approximate because it depends on historically unstable intraspecific taxonomy.

A rough temporal framework for the evolution of this intraspecific diversity has been constructed by Fleming (1962a). He noted that well differentiated subspecies have evolved in the North and South Islands since Cook Strait last became a barrier to gene flow about 15,000 years ago, and in the formerly glaciated Auckland Islands following the growth of forests about 10,000 years ago. Using these calibrations, Fleming argued that the New Zealand subspecies of Australian species are not much older than the late Pleistocene (about 20,000 years). This scenario is also consistent with the evolution of subspecies of northern hemisphere birds such as Canada Geese (*Branta canadensis*) following retreat of Pleistocene ice sheets some 15,000–10,000 years ago (Van Wagner and Baker, 1990).The colonization of New Zealand from Australia has undoubtedly been occurring for much longer than the late Pleistocene, and is reflected at higher taxonomic levels. For example, Falla (1953) noted that endemic New Zealand species such as the New Zealand Shelduck (*Tadorna variegata*), Brown Teal (*Anas aucklandica*), Black Stilt (*Himantopus novaezealandiae*), Variable Oystercatcher (*Haematopus unicolor*), and the Tomtit (*Petroica macrocephala*) appear to be most closely related (i.e., sister-species) to Australian species. Using a gradualistic neo-Darwinian philosophy, Fleming (1962a) extrapolated his subspecies divergence estimate backwards in time, and concluded that endemic species could have arisen in the New Zealand archipelago in the last million years or so. While these estimates were made with appropriate cautions as to their speculative nature, they are not out of line with modern biochemical evidence. For example, speciation of the Chaffinch and Blue Chaffinch (*Fringilla coelebs* and *F. teydea*, respectively) seems to have occurred within the last million years in the Atlantic islands, judging from divergence in protein-encoding genes and mtDNA (Baker *et al.*, 1989).

At the level of New Zealand endemic genera, relationships with sister-taxa in Australia are much more difficult to discern, implying more ancient colonizations than for endemic species. However, three

New Zealand birds seem to have clear affinities to Australian genera, and most probably are sister groups: *Anthornis* (New Zealand Bellbird, *A. melanura*) and *Meliphaga* (Australian honeyeaters), *Notiomystis* (Stitchbird, *N. cincta*) and *Phylidonyris* (Australian honeyeaters), and *Cnemiornis* (extinct New Zealand Goose) and *Cereopsis* (Cape Barren Goose, *C. novaehollandiae*). Based on the degree of divergence between these pairs of genera, plus the uncertain affinities of other endemic genera such as *Hymenolaimus* (Blue Duck), *Thinornis* (Shore Plover), *Anarhynchus* (Wrybill Plover, *A. frontalis*), *Coenocorypha* (Sub-Antarctic Snipe, *C. aucklandica*), *Notornis* (Takahe), *Hemiphaga* (New Zealand Pigeon, *H. novaeseelandiae*), *Sceloglaux* (Laughing Owl, *S. albifacies*), *Bowdleria* (Fernbird, *B. punctata*), *Mohoua* (Whitehead, *M. albicilla* and Yellowhead, *M. ochrocephala*), *Finschia* (New Zealand Brown Creeper, *F. novaeseelandiae*), and *Prosthemadera* (Tui, *P. novaeseelandiae*), Fleming (1962a) suggested that they have resulted from colonizations during the 25 million years of the late Tertiary, or possibly even into the Eocene for some genera.

The three endemic families of birds, Acanthisittidae, Callaeaidae, and Turnagridae, because of their distinctness, were thus thought to have colonized New Zealand even earlier in the Eogene. Fleming thought it was most unlikely that any of these colonizations was as old as the Upper Cretaceous (70 mya). Finally, the colonization of New Zealand by kiwis and moas, two endemic orders, was attributed to the Upper Cretaceous (Fleming, 1962b). DNA hybridization studies of the Acanthisittidae and Ratites have provided invaluable tests of these speculative biogeographic scenarios, and the results are the opposite of Fleming's predictions. The New Zealand wrens diverged from the Tyrannides (Pittas, broadbills, and New World suboscines) about 84 mya (Sibley *et al.*, 1982), and thus must be a vicariant element because the Tasman Sea between Australia and New Zealand opened up about 80 mya (Fleming, 1975). They may well be the oldest living group of endemic New Zealand birds because similar DNA derived dates indicate that ancestral New Zealand ratites diverged from the Australo-Papuan ratites only about 40–50 mya, and thus must have dispersed to New Zealand across a northern archipelagic land bridge via Norfolk Island (Sibley and Ahlquist, 1981).

This hypothesis of ratite origins from a flightless ancestor widely distributed in Gondwanaland has been challenged recently by Houde (1986). He argued instead that the *Lithornis*-cohort of palaeognathous carinates from the Palaeocene and Eocene deposits in North America and Europe are the ancestors of ostriches, and because they most likely had the power of flight it is possible that kiwis could be descendents of a palaeognath that flew to New Zealand and became secondarily flightless. It is clear that the biogeographic history and systematics of New Zealand birds could benefit enormously from further studies utilizing biochemical genetic, DNA, and fossil data, particularly in relation to sister taxa in the Australian region. The potential of such a multifaceted approach is well illustrated by recent studies of the adaptive radiation and phylogenetic relationships of Hawaiian passerines by Freed *et al.* (1987) and Johnson *et al.* (1989).

2.2. Routes of Colonization

The dominance of the Australian element in the New Zealand avifauna points to post-Gondwanaland dispersal across the Tasman Sea via west wind drift as the principal route of colonization. Because the Tasman Sea began forming about 80 mya, vicariant elements from an ancestral Gondwanaland avifauna are rare, with only the Acanthisittidae appearing to be old enough to have originated this way. Yet the traditional view of dispersal from the south (Antarctica, South America, and southern Africa), west (Australia), and north (Holarctic) is almost certainly too simplistic, and is beginning to yield to more detailed analysis with the generalized track method of Croizat (1952, 1958, 1964). For example, cladistic analysis of platycercine parrots establishes that New Zealand and New Caledonian taxa are sister-groups, and that their sister-group is in southeast Australia (Croizat, 1958; Smith, 1975). As pointed out by Craw (1982), this pattern of relationships is congruent with the fragmentation of Gondwanaland. However, as McDowell (1978) has noted, this pattern could equally arise by dispersal (or some combination of vicariance and dispersal), and dating of divergence times using DNA hybridization or other molecular data would be useful in attempting to distinguish between these two alternatives.

As noted by Bull and Whitaker (1976), it is possible that some northern hemisphere species have reached New Zealand via Australia where they have subsequently become extinct, although there is at present no fossil or other evidence supporting this possibility. The direct route to New Zealand is used regularly by migrant shorebirds that breed in the arctic or subarctic, including the Asiatic Golden Plover (*Pluvialis dominica fulva*), Eastern Bar-tailed Godwit (*Limosa lapponica baueri*), Ruddy Turnstone (*Arenaria interpres*), Knot (*Calidris canutus*), and Parasitic Jaeger (*Stercorarius parasiticus*). Another 32 species of Holarctic shorebirds are either regular visitors or occasional stragglers, thus attesting to their ability for trans-hemispheric dispersal. It is conceivable that the endemic South Island Pied Oystercatcher arose from a once migrant Holarctic stock that became sedentary in New Zealand. Dispersal of Holarctic birds to New Zealand is not limited to marine species; the Northern Shoveler (*Anas clypeata*) has recently been recorded as an occasional visitor.

2.3. Double Invasions

Given the proximity of Australia (1200 miles at present) and the prolonged period over which dispersal to New Zealand could have occurred, it would be surprising if some species did not make more than one successful colonization. The chances of detecting multiple invasions are enhanced when they occur far enough apart in time to allow differentiation of the descendents of the first colonizers from those of subsequent ones. Polymorphism and/or speciation in some pairs of New Zealand birds appear to be attributable to double (or more) invasions of the same ancestral stock. The endemic Black Stilt is likely the product of an earlier invasion of the Pied Stilt (*H. leucocephalus*) (Falla, 1953). Stilts now occurring in New Zealand are descendents of a more recent invasion, possibly as recently as the early 19th century (Pierce, 1984). The two species occasionally hybridize and produce phenotypically intermediate F_1 progeny (referred to as 'smudgies' by local ornithologists), indicating that the time between successive colonizations has been too short for the evolution of complete reproductive isolation. Introgressive hybridization has resulted in the Pied Stilts in New Zealand becoming distinguishable from the Australian parental stock; the former possess shorter tarsi, longer tails, and variable plumage markings (Pierce, 1984).

A more complex example, again involving melanism, is provided by the Variable Oystercatcher (*H. unicolor*). On Stewart Island and most of the South Island the birds are wholly melanistic, but in the North Island the frequency of pied and mixed plumage birds increases dramatically towards the far north. To explain this pattern of differentiation within New Zealand, as well as the occurrence of morphologically and ecologically well differentiated pied (*Haematopus longirostris*) and black (*H. fuliginosus*) forms of oystercatchers in Australia, Mayr (in Falla, 1953) hypothesized that an ancestral Australian pied stock originally invaded New Zealand. As it spread southward it evolved increasing melanism, and later reinvaded southern Australia to give rise to the Sooty Oystercatcher. Reanalysis of the relationships of these oystercatcher taxa by Baker (1972) using multiple datasets indicates that the New Zealand species most likely have arisen from a double invasion of

H. longirostris stock. Descendents of the first invasion became melanic in New Zealand, and these birds now hybridize extensively with pied birds that have recently invaded the North Island from Australia (Fig. 3). Hybridization is so common now in northern New Zealand that very few pure pied forms are seen, and there is no evidence of selection against hybrids (Baker, 1973). As in the stilts, the time between successive colonizations has apparently been too short to allow the evolution of reproductive barriers between the parental morphs.

Pied Oystercatchers resident in the Chatham Islands 645 kilometers to the east of New Zealand are very likely a remnant of the first invasion of Australian pied birds, because they comprise a distinctive subspecies adapted to foraging on rocky shores (Baker, 1974). The Chatham Island birds have a restricted patch of white on the rump, as do the Variable and Australian pied oystercatchers. Phylogenetic analysis based on biochemical genetic data supports a sister-group relationship for *H. unicolor* and *H. longirostris,* and their sister-group is probably *H. fuliginosus,* consistent with the double invasion hypothesis (Fig. 4).

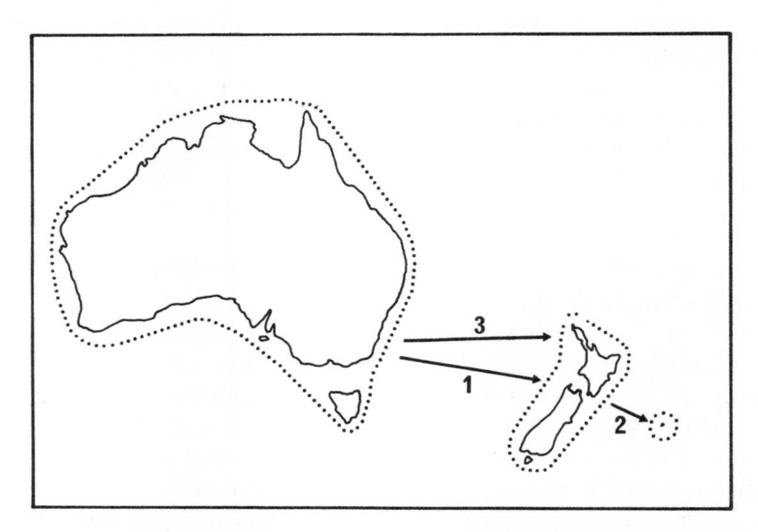

FIGURE 3. Hypothesized double invasion of New Zealand by Australian pied oystercatchers to account for related New Zealand forms. The first invasion (1) gave rise to the Variable Oystercatcher in mainland New Zealand, which later became secondarily melanistic. Not long after the first invasion, pied birds colonized the Chatham Islands (2). A recent invasion (3) has resulted in hybridization of black and pied morphs, especially in the northern North Island.

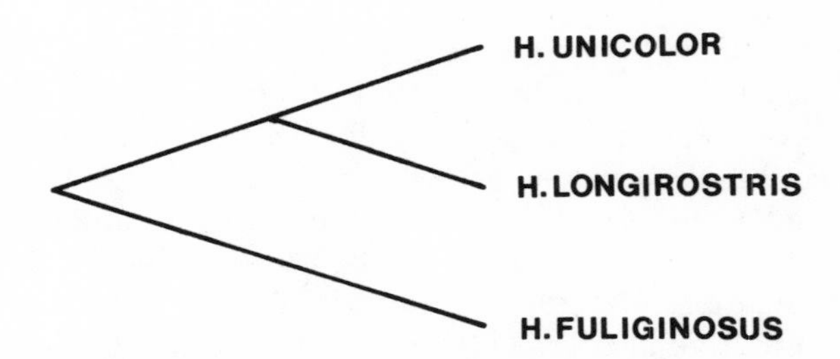

FIGURE 4. Sister-group relationships of the Variable (*H. unicolor*), Australian pied (*H. longirostris*), and Sooty (*H. fuliginosus*) oystercatchers, based on cladistic analysis of protein-encoding nuclear genes at 37 loci (A. J. Baker, unpublished manuscript).

Another example of the phenomenon of double invasion, this time probably of considerably greater antiquity than the two previous examples, is provided in the muscicapid flycatcher genus *Petroica*. The Tomtit (*P. macrocephala*) of New Zealand is clearly allied to the Scarlet Robin (*P. multicolor*) of Australia and certain Pacific Islands, whereas the larger New Zealand Robin (*P. australis*) with its markedly longer tarsus is classified in a separate subgenus *Miro* (Fleming, 1950a,b). To account for this speciation, Fleming hypothesized that the robins arose from an earlier invasion of *Petroica* stock and the tit from a second, much later invasion. The two invasions were thought to have occurred in the Pliocene and the early Pleistocene, respectively, and in the relatively long intervening period of isolation in New Zealand the robins speciated.

3. GEOGRAPHIC VARIATION, ISLAND SUBSPECIES, AND SPECIATION

Although the New Zealand landmass only extends about 1100 miles on its major north-south axis, fragmentation of the avifauna is promoted by its mountainous terrain, by water barriers between the three major islands, and the occurrence of numerous offshore islands of various sizes and ecological complexity. Additionally, the Chatham Islands to the east and the subantarctic islands to the south are well isolated by substantial ocean barriers to dispersal. In such a natural "laboratory" it might be expected that intraspecific variation of native land birds would have been intensively studied, especially because of its relevance to current controversies on the roles of homogenizing gene flow and founder effects in speciation theory (Carson and Templeton,

1984; Barton and Charlesworth, 1984). Although the broad outlines of geographic variation have been sketched for most species, there is a dearth of modern multivariate and genetic studies to complement the limited morphological data available from relatively small museum collections (principally skins).

3.1. Tomtits and New Zealand Robins

The geography of intraspecific variation has been analyzed in detail for only a handful of native birds in New Zealand. A classic study is that of Fleming (1950a,b), who described geographic variation in relation to subspeciation and speciation in Tomtits and New Zealand Robins. Fleming's study portends modern systematic methodology in that he employed sample sizes adequate for statistical analysis where possible, he integrated phylogeny with historical reconstructions of fragmentation of geographic ranges and climates, and he compared putative ancestral and descendent populations (see Fig. 5).

The Tomtits (or New Zealand Tits as they are sometimes called) show a wide range of plumage variation including melanism, as well as geographic variation in body size and proportions. The plumage differences among the different subspecies are depicted in Fig. 6, and are portrayed in color in Plate 42 of Falla *et al.* (1982). Briefly, males of the North Island subspecies (*P. macrocephala toitoi*) have white breasts and are smaller than the yellow-breasted males of the South Island or Chatham Island subspecies (*P. m. macrocephala* and *P. m. chathamensis*, respectively. The latter two subspecies differ in that the Chatham birds have longer tarsi and the brown upperparts of the female. The two subantarctic subspecies are more strongly differentiated than the others. Both sexes of the Auckland Island form (*P. m. marrineri*) are alike and resemble the South Island subspecies in breast color, but in the Snares Island form (*P. m. dannefaerdi*) both sexes are wholly black. The subantarctic birds are also larger.

Fleming (1950a) found a strong association between mean temperature of the center of the geographic ranges of subspecies of Tomtits and their mean size (as measured by wing length), with larger birds occurring in cooler southern climates (Fig. 7). Because this cline in body size is in accordance with Bergmann's rule, he interpreted it as an adaptive response to the temperature gradient. The only exception to this smooth clinal trend was provided by the Snares Islands subspecies, which appears to have a longer mean wing length than their more southerly counterparts from the Auckland Islands (although this could also arise from sampling error). To explain this apparent anoma-

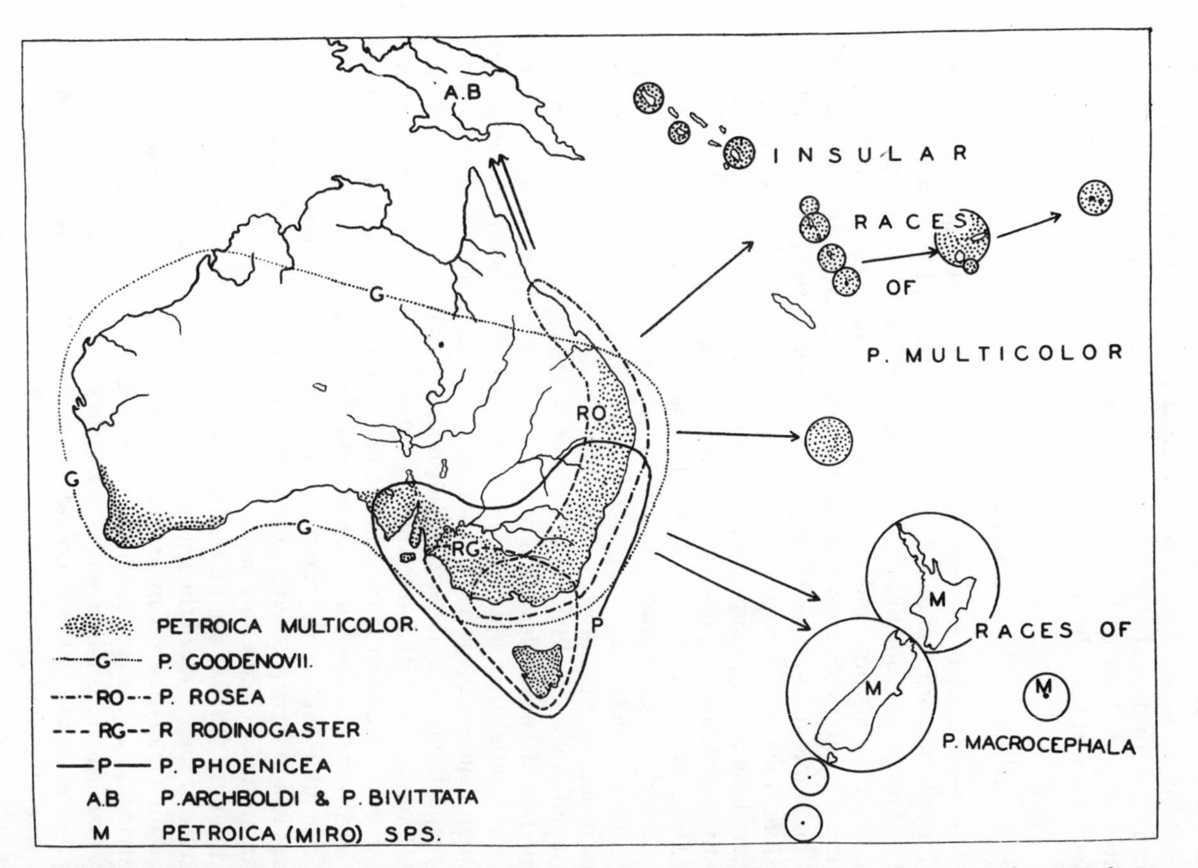

FIGURE 5. Map showing the geographic distribution and inferred colonization routes of forms in the Australasian flycatcher genus *Petroica* (from Fleming, 1950a).

FIGURE 6. Plumage differences among different subspecies of the Tomtit (redrawn from Fleming 1950a, by Marianne Collins of the Royal Ontario Museum). Subspecies are identified as follows: row one, *Petroica macrocephala toitoi;* row two, *P. m. dannefaerdi*. Adult and juvenile males are depicted in columns one and two, respectively, and adult and juvenile females in columns three and four, respectively.

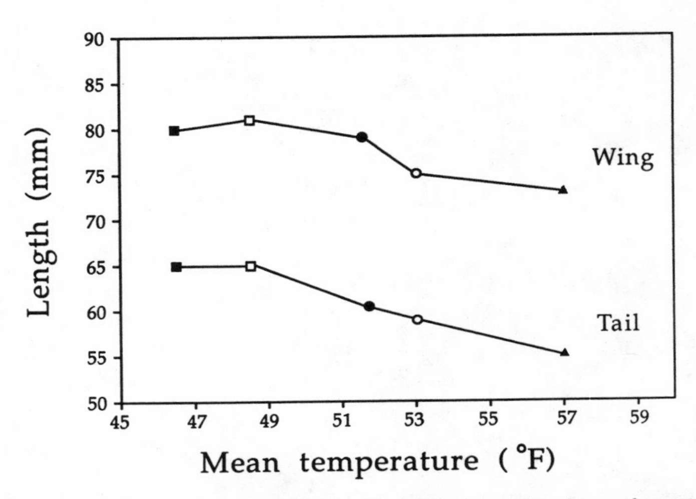

FIGURE 7. Plot of mean wing and tail lengths of males of the five subspecies of the Tomtit against mean temperature at the center of their respective ranges (redrawn from Fleming, 1950a).

ly, Fleming (1950a) invoked the Sewall Wright effect of genetic drift operating in a small population of about 500 breeding pairs, and thus the deviation from a smooth cline was interpreted as nonadaptive.

As further evidence for this conclusion, he noted that coefficients of variation were higher in the three main islands of New Zealand than in the other more isolated island subspecies, suggesting that the latter have restricted genetic variability relative to the widely ranging mainland populations. This interpretation is suspect, however, because it confuses within-population variation in the isolated island populations with among-population variation in the pooled New Zealand mainland populations. Nevertheless, recent theoretical work has established that within-population character variances scale with effective population size in neutral traits, and thus the ad hoc invocation of selection as the primary agent of the clinal pattern of differentiation in Tomtits needs re-examination with modern biochemical and statistical techniques.

The evolution of much more distinctive subspecies of Tomtits in the outlying islands than in mainland New Zealand is exemplified in other species of birds such as the Fernbird (*Bowdleria punctata*), New Zealand Bellbird, New Zealand Gerygone (*Gerygone igata*), and in Yellow-fronted Parakeets (*Cyanorhamphus auriceps*) (Fleming, 1950a). Such a generalized trend points to the role of isolation in the formation of these strong subspecies. Although the Chatham Islands were not glaciated during the Pleistocene and thus were available for colonization during this period, the Snares and Auckland Islands cannot have been colonized until near the end of the Pleistocene or early Recent because suitable habitat was not present before this time (Fleming, 1950a). Yet the North and South Island populations were separated by the Cook Strait water barrier earlier in the Pleistocene, so the degree of differentiation of subspecies does not obviously relate to the antiquity of isolation. Instead, the degree of isolation seems most relevant, which argues strongly for the homogenizing role of gene flow in slowing the divergence of the mainland populations. For example, the slight differentiation of the Stewart Island population of the Tomtit from the neighboring South Island one is presumably due to gene flow across the Foveaux Strait water barrier between them, and this in turn is facilitated by the chain of islands in the Strait acting as bridge between the two populations.

New Zealand Robins are sufficiently distinctive to be classified as a separate subgenus (*Miro*) within the genus *Petroica*, and provide an interesting contrast in geographic variation to the Tomtits. Three subspecies are currently recognized in the New Zealand region, one each in the North Island, South Island, and Stewart Island. Males of the North Island race (*P. australis longipes*) differ from females in having the lower breast and belly white, whereas males in the other two mainland New Zealand are yellow-breasted. Males in the Stewart Island subspecies (*P. a. rakiura*) differ from those in the South Island subspecies (*P. a. australis*) in being darker dorsally and in having white breasts, but the former is only a weakly differentiated subspecies. The wholly melanistic Chatham Island Robin (*P. traversi*) is regarded as a separate species.

Although data are sparse for the robins, the trend in geographic variation is the direct reverse to that in Tomtits, except that the North Island subspecies seems to be smaller than the South Island one. Within subspecies, however, clines in size are opposite to those predicted by Bergmann's rule, with larger birds in the north (Fig. 8). There is a major discontinuity in morphometrics either side of Cook Strait, attesting to its effectiveness as a barrier to gene flow between the North and South islands. Fleming (1950b) was unable to advance any explanation for this pattern of geographic variation since it has no clear adaptive basis, but it may relate to isolation in different glacial refugia with very different climates, and subsequent expansion into newer habitats. It is noteworthy that passerines such as House Sparrows (*Passer domesticus*) and Common Mynas (*Acridotheres tristis*) introduced by man to New Zealand in the last century also have developed a similar pattern of

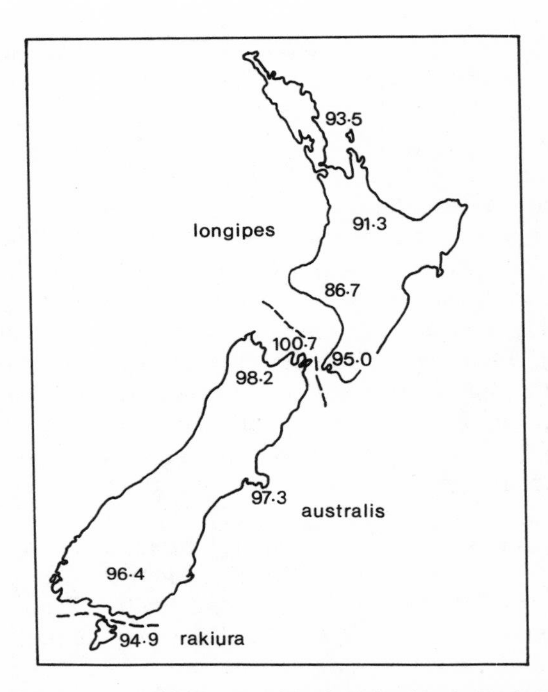

FIGURE 8. Geographic variation in mean wing length of New Zealand Robins in the South Island, North Island, and Stewart Island (redrawn from Fleming 1950b).

geographic variation in size (Baker, 1980; Baker and Moeed, 1979). The tendency for body size to increase in warmer northern localities (Fig. 9) may simply reflect an increased nongenetic component in more productive environments, or it may result from selection for factors such as larger food items in the north (Baker and Moeed, 1979).

Comparison of wing shape in Australasian *Petroica* revealed that the New Zealand Tomtits have shorter and rounder wings than their Australian congeners, the New Zealand Robins have even rounder wings, and the Chatham Island Robin has the shortest and roundest (most degenerate) wing of all. Fleming (1950b) hypothesized that this rounding of the wing was a result of relaxed selection pressure in predator-free insular environments. Similarly, he argued that the evolution of longer tarsi in the New Zealand Robins, the Chatham Island race of the New Zealand Tomtit, and the nominate race of the Scarlet Robin (*P. multicolor multicolor*) on Norfolk Island was an adaptive response to forest-floor foraging in almost predator-free insular forests.

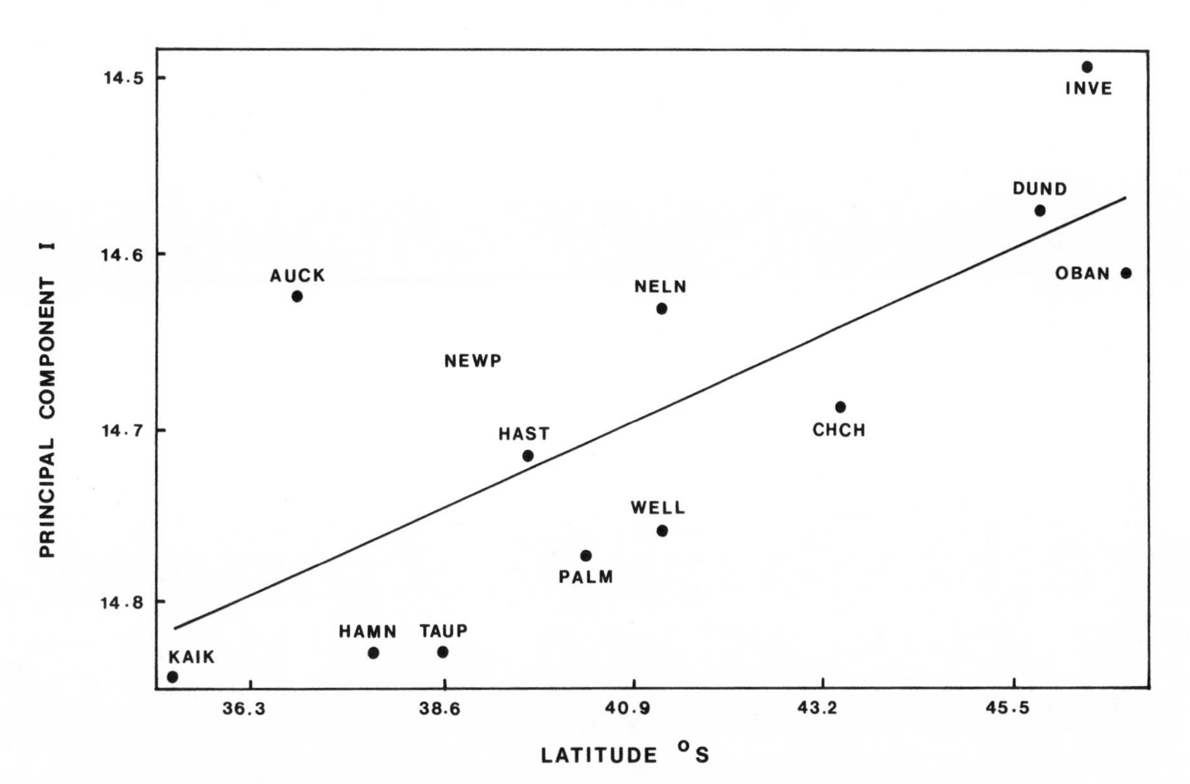

FIGURE 9. Plot of principal component I (size) of 16 skeletal measurements against latitude for New Zealand samples of male House Sparrows (from Baker, 1980). The regression equation is $Y = 0.021X - 17.754$ ($p = 0.023$).

3.2. New Zealand Bellbirds

Another detailed study of intraspecific variation is that of Bartle and Sagar (1987) who sampled a wide range of populations of the New Zealand Bellbird (Fig. 10). Despite having to use external measurements taken from a mixture of museum skins and live-caught birds, as well as having measurements made by several people, two clear patterns emerged from their analysis. First, both body size and plumage coloration vary clinally among populations in mainland New Zealand,

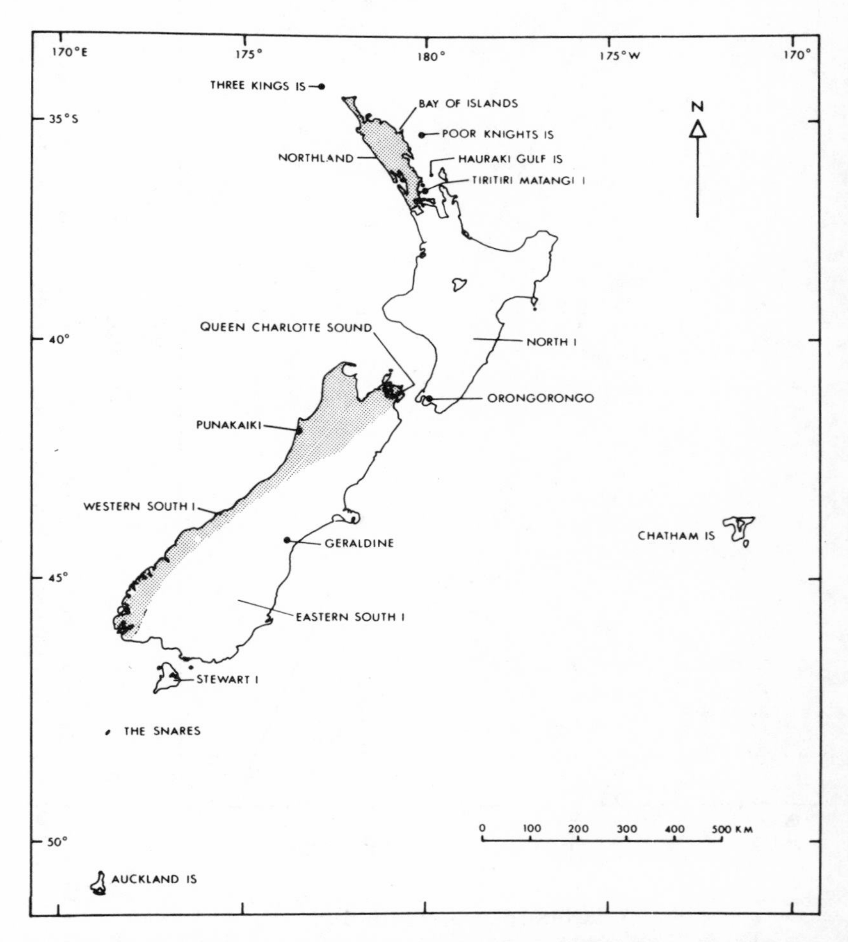

FIGURE 10. Map of the region showing where New Zealand Bellbirds have occurred, and location of sample sites (from Bartle and Sagar, 1987). The western South Island and Northland are stippled.

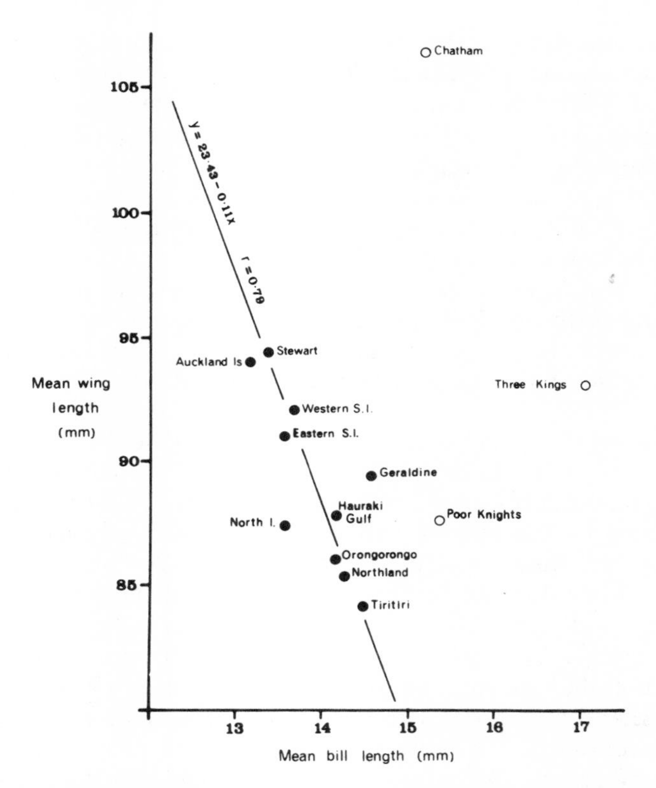

FIGURE 11. Plot of mean bill length against mean wing length of adult male New Zealand Bellbirds (from Bartle and Sagar, 1987). Localities with open circles were not used in calculating the regression equation.

nearshore islands, and Auckland Island in the subantarctic. Bellbirds are larger and more brightly colored in southern latitudes, but bill length displays a reverse cline (Fig. 11). The degree of melanism, which determines the intensity and hue of the iridescent head and olive plumage coloration, increases southwards and on all isolated islands except Three Kings Islands in the north of their range. Increased melanism is correlated with increased rainfall, especially in wet areas such as the western South Island, Stewart Island, Chatham Island, and Auckland Island.

Second, the clinal pattern of variation is disrupted by a distinctive insular trend in populations on some isolated Islands (Chathams, Three Kings, and Poor Knights) that were colonized relatively early, probably during the Pleistocene. Both sexes, but males in particular, are larger than their mainland conspecifics at similar latitudes, a phenomenon common in island birds (Grant, 1965). Because these populations occur on islands with fewer competitors and high densities of bellbirds, Bartle and Sagar (1987) argued that the insular trend for increased body size was an adaptive response to enable the sexes to exploit wider niches and to reduce intraspecific competition, as proposed earlier for island birds by Selander (1966). Somewhat surprisingly, this trend is not evident in the extremely isolated population on Auckland Island, but Bartle and Sagar suggest that this area may have been colonized recently by bellbirds. Perhaps the most spectacular example of divergence in morphometrics is provided by the now extinct Chatham Island Bellbird, which has a mean tarsus length about 20% larger than elsewhere.

Based on their analysis, Bartle and Sagar (1987) recognized four subspecies of bellbirds in the New Zealand region, one each on Three Kings (*A. m. obscura*), Poor Knights (*A. m. oneho*), and Chatham (*A. m. melanocephala*) islands, as well as the nominate subspecies ranging throughout the New Zealand mainland and also Auckland Island. The clinal trend in color and size was attributed by them to environmental induction and possibly adaptive differentiation conforming to Bergmann's rule. Similarly, they interpreted the divergence of the more isolated populations as a product of a long period of isolation from their mainland ancestors, with only the Poor Knights population (the least differentiated and most recently separated race) likely to have experienced any significant gene flow. Although these views have intuitive appeal, they remain speculative without supporting genetic data and long-term population studies.

3.3. Variable Oystercatchers

Modern geographic variation studies have also been conducted on two species of shorebirds, the black morph of the Variable Oystercatcher and Silver Gulls. In both cases, geographic patterns are roughly clinal. Two major contrasting trends are evident in oystercatchers. First, bill, wing, tail, and tarsus lengths all increase clinally from north to south, as does body weight (with the exception of the Northland sample (Fig. 12). Second, variation in middle toe length shows a reverse cline, except that the sample from Stewart Island in the far south has the largest mean toe length of all. This latter exception correlates with the primarily rock-dwelling habits of the Stewart Island birds, in line with the general observation that all species of oystercatchers that for-

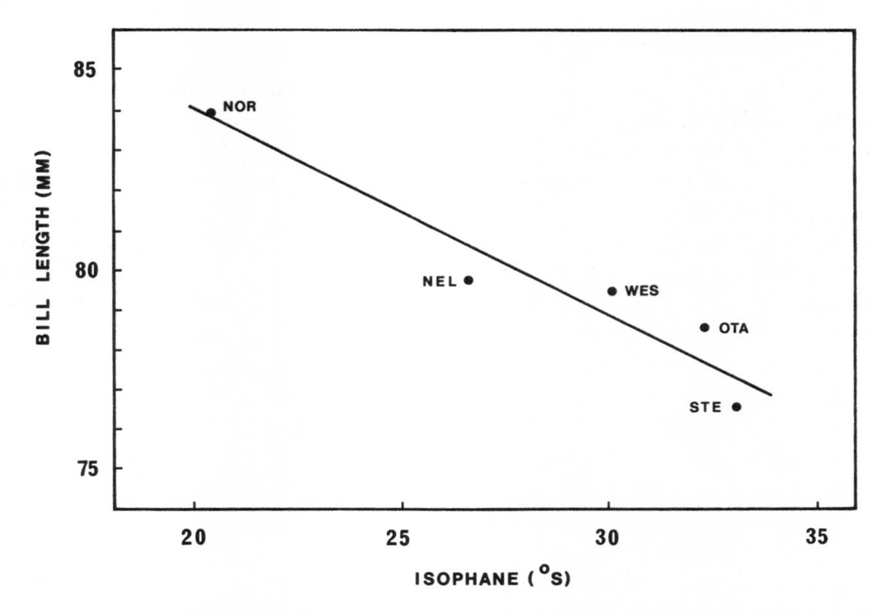

FIGURE 12. Plot of mean bill length of adult male Variable Oystercatchers against isophane (a composite of latitude, longitude, and altitude). Sample localities are identified as follows: Northland (NOR), Nelson (NEL), Westland (WES), Otago (OTA), and Stewart Island (STE). The regression equation is $Y = 93.974 - 0.503X$ ($p < 0.01$).

age on slippery wave-washed rocky shores have elongated toes, presumably an adaptation for balance and agility for food-getting in these environments (Fleming, 1939).

In attempting to explain these trends, Baker (1975) computed a regression analysis of locality means on various environmental variables. Because New Zealand has an essentially north–south gradient in temperature, the positive association of the size of body extremities with decreasing temperature is predicted by Allen's rule, although these associations may simply be correlates of the clinal trend for increased body size in cooler southern climates. The latter is in accord with Bergmann's rule, implying thermoregulatory advantage for bigger-bodied birds in cooler southern latitudes. One problem with this adaptationist narrative is that it does not exclude nongenetic environmental induction, and genetic data would be invaluable in providing information on levels of gene flow, selection, and drift as potential determinants of geographic variation in Variable Oystercatchers.

3.4. Silver Gulls

A similar but much weaker cline in size is apparent among Silver Gulls sampled from breeding colonies in New Zealand (Fig. 13), but the degree of geographic variation in external morphological characters is generally very low. This result is expected from recoveries of banded birds, which demonstrates that dispersal occurs among colonies (J. A. Mills, personal communication). Protein electrophoresis resolving genetic variation at 43 loci revealed that the New Zealand populations are essentially panmictic (F_{st} = 0.055), and thus gene flow is extensive enough to prevent geographic differentiation (Baker *et al.*, unpublished manuscript). A concomitant effect is that geographically adjacent localities are most similar genetically because they experience greater levels of gene flow than do localities further apart (Fig. 14). Thus in Silver Gulls homogenizing gene flow prevents the evolution of subspecies in the New Zealand mainland, and local selective forces are clearly not strong enough to overcome the de-differentiating effects of gene flow. Comparison of New Zealand and southeastern Australian populations, representing two morphometrically distinct subspecies, indicates that the Tasman Sea is an effective barrier to gene flow (F_{st} = 0.155). In seabirds in general there has been very little differentiation and subspeciation in the New Zealand region, and where it occurs it mostly involves very isolated populations with restricted ranges in the subantarctic islands. As noted earlier by Falla (1953), this lack of geographic structuring in seabird populations points up their enhanced dispersal ability relative to landbirds, and this commonly translates into realized gene flow.

3.5. Speciation in Parrots and Bush Wrens

Under the neo-Darwinian model of evolution, processes of population differentiation can be extrapolated through time to account for the gradual adaptive divergence of new species (Gould, 1980). Geographic variation should therefore grade into subspecies differences and eventually into species differences under this gradualistic model. Perhaps the clearest example of allopatric divergence across different taxonomic levels in New Zealand birds is that of the species-pair of endemic parrots, the Kea (*Nestor notabilis*) and the Kaka (*N. meridionalis*). The Kaka is subdivided into two subspecies, one each in the North and South Islands. The North Island subspecies (*N. m. septentrionalis*) is smaller and has duller plumage than the South Island subspecies (*N. m. meridionalis*). The Kea is restricted to the alpine zone and high moun-

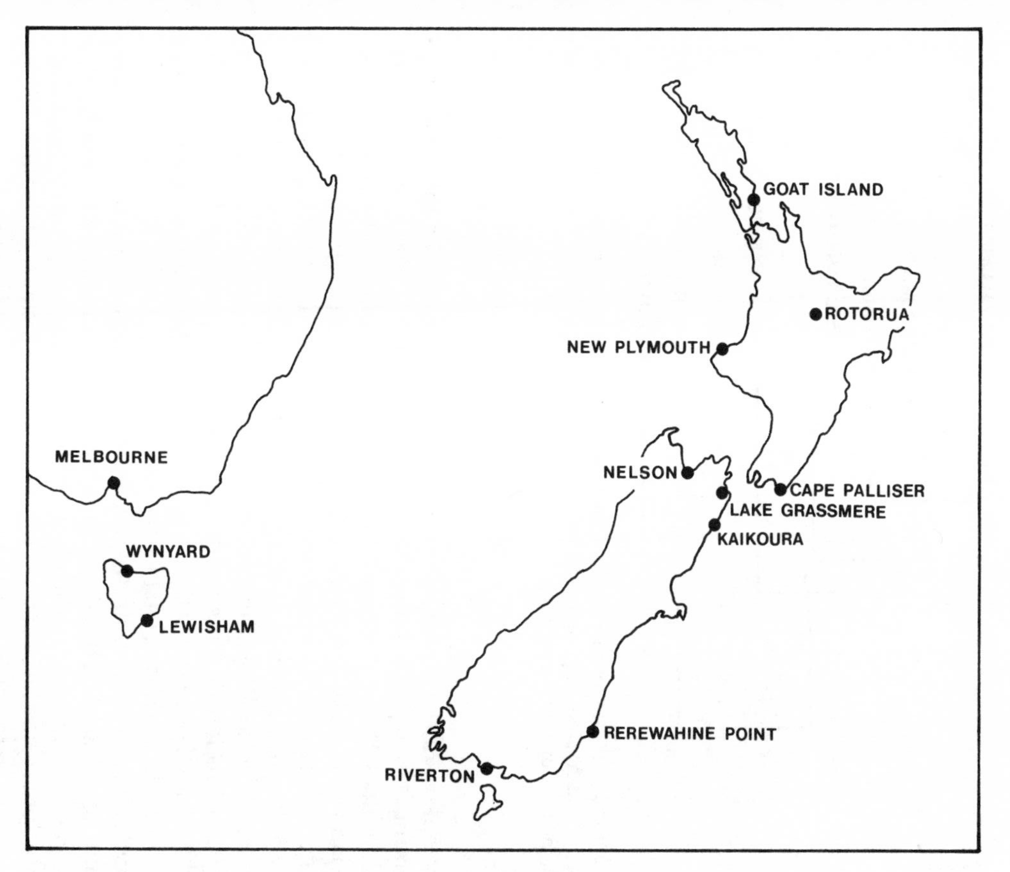

FIGURE 13. Locations in New Zealand and Australia where samples of Silver Gulls were obtained for protein electrophoresis. The map is not drawn to scale.

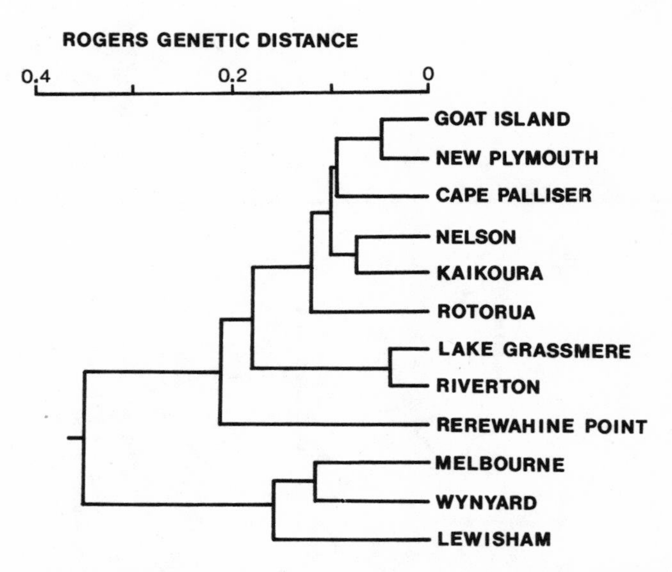

FIGURE 14. UPGMA cluster analysis of Rogers' (1972) genetic distances among New Zealand and Australian samples of Silver Gulls, based on allele frequencies at 43 loci of protein-encoding nuclear genes.

tain valleys of the South Island. To account for this geographic pattern, Fleming (1975) posited that the common ancestor to Keas and Kakas was widely distributed throughout the forests covering the continous New Zealand landmass that existed in the Tertiary (Fig. 15). When the Manawatu Strait formed and segmented the landmass into two major islands, the southern birds had to adapt to life in the alpine zone as a consequence of the cooling climate and associated loss of forests at the onset of the Ross Glaciation. This resulted in ecological specialization of North and South Island populations, and gradual acquisition of reproductive isolation. Later post-Pleistocene expansion by Kakas into the South Island when new forests were available for colonization lead to the evolution of the South Island subspecies, presumably mediated by reduced gene flow across the Cook Strait water barrier. Exactly the same evolutionary scenario can account for subspeciation in the Bush Wren (*Xenicus longipes*) in the North and South Islands, as well as for the origin of its alpine sister species, the Rock Wren (*X. gilviventris*). The latter species is confined to alpine and subalpine habitats in the South Island.

The general conclusion that emerges from these studies is that they

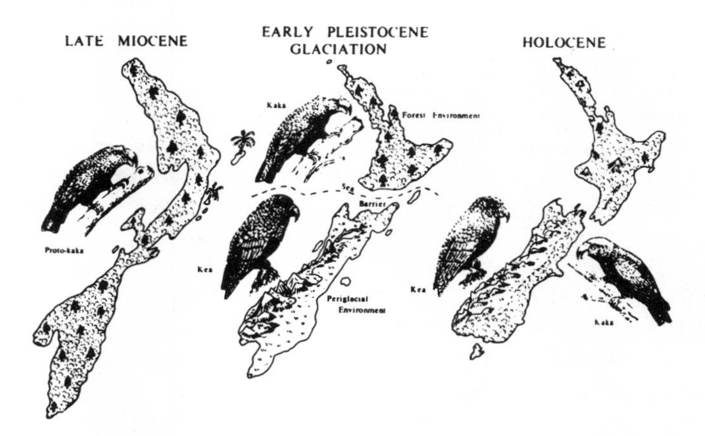

FIGURE 15. Hypothetical scenario for speciation and subspeciation in the Kea and Kaka in New Zealand. The Tertiary forest-dwelling proto-Kaka is thought to have subspeciated either side of the Manawatu Strait in the early Pleistocene. The South Island population then speciated in the alpine zone during the Ross Glaciation, and the Kaka later reinvaded the South Island and subspeciated again (from Fleming, 1975).

are consistent with gradual divergence of populations in allopatry. Patterns of geographic variation, subspeciation, and speciation in the New Zealand region are coincident with population fragmentation by water barriers and glacial events during the Pleistocene. This implies that the disruption of homogenizing gene flow between populations, coupled with selection for ecological specialization in changing environments, has been crucial in the generation of avian diversity in this remote part of the world. Just what role founder-induced speciation has played (if any) must await future population genetic studies.

4. FLIGHTLESSNESS AND MELANISM

The New Zealand avifauna exhibits a relatively high incidence of flightlessness and melanism, and the common cause for this is assumed to be the lack of selection pressure in an essentially predator-free environment (Bull and Whitaker, 1976). There were no mammalian land predators in New Zealand until Polynesian man introduced rats and dogs a few hundred years ago, although the impact of mammalian predation was likely to have been felt most severely last century when Europeans introduced mustelids, and other species of rats, cats, and dogs.

4.1. Flightlessness

The most extreme examples of flightlessness in New Zealand birds is provided by the now extinct moas. All currently recognized 11 species were not only large heavy-boned birds, but some were among the largest birds ever to have lived. They do not have any trace of even vestigial wing bones, so it is assumed that they have secondarily lost their wings (Pycraft, 1900; De Beer, 1956) because their sister group (Tinamous) is volant (Cracraft, 1974). The closely related kiwis are also flightless, though they do have tiny vestigial wings. As in many other parts of the world (Ripley, 1977) several species of rails are flightless. Other New Zealand birds that are flightless or nearly so are the Kakapo (*Strigops habroptilus*), the extinct goose (*Cnemiornis*), the extinct Owlet-nightjar (*Megaegotheles*), and the Auckland Island and Campbell Island subspecies of the Brown Teal (*Anas aucklandica aucklandica* and *A. a. nesiotis*, respectively) (Bull and Whitaker, 1976; Rich and Scarlett, 1977). In addition to these flightless and near flightless species, at least another seven species of native land birds have weak powers of flight (Table III).

4.2. Melanism

Birds with totally black plumage are quite common in New Zealand, and are often southern morphs in a polymorphic species. For example, the Grey Fantail (*Rhipidura fuliginosa*) has both black and "pied" morphs, though the black morph is rare in the North Island. In native hardwood forests in eastern Otago Province in the South Island, Craig (1972) recorded the highest frequency of melanics (21%), but the average in all vegetation types was about 12%.

As pointed out by Craig (1972), the distribution of melanics and more heavily melanized races of New Zealand forest birds suggests that climatic factors such as temperature and precipitation are determinants of this phenomenon. The Chatham Island Robin is melanic, in line with the cool, wet climate in this archipelago. Similarly, the darker melanic phase of the Weka (*Gallirallus australis*) is found in the high rainfall and cooler regions of the Western South Island and Fiordland. In the Tomtit, races occurring in places with cooler and wetter climates have totally black plumage (the Snares Islands race), or are darker (South Island race). Yet this does not explain why the most southerly race on Auckland Island is not melanic since it has a cool and wet climate.

Melanism also occurs in the shorebirds and cormorants. The black morph of the Variable Oystercatcher increases in frequency south-

TABLE III
Flightless and Weakly-Flying Species of New Zealand Birds

Family	Species	Distribution
Flightless:		
Apterygidae	Little Spotted Kiwi (*Apteryx oweni*)	South Island
	Common Kiwi (*A. australis*)	North, South, and Stewart Islands
	Great Spotted Kiwi (*A. haasti*)	South Island
Rallidae	Weka (*Gallirallus australis*)	North, South, Stewart, and Chatham Islands
	Takahe (*Notornis mantelli*)	South Island
Anatidae	Brown Teal (*Anas aucklandica*)	Races on Auckland Island and Campbell Island
Weak Flyers:		
Psittacidae	Kakapo (*Strigops habroptilus*)	South and Stewart Islands
Acanthisittidae	Bush Wren (*Xenicus longipes*)	South Island
	South Island Rock Wren (*X. gilviventris*)	South Island
	Stephen Island Wren (*X. lyalli*)	Stephen Island
Muscicapidae	Fernbird (*Bowdleria punctata*)	North, South, Stewart, Codfish, Snares Islands, Chatham Island[a]
Callaeidae	Saddleback (*Creadion carunculatus*)	Hauraki Gulf and offshore from Stewart Island
	Kokako (*Callaeas cinerea*)	North and possibly South and Stewart Island
	Huia (*Heteralocha acutirostris*)*	North Island

[a]Extinct.

wards, and all birds in the southern South Island and Stewart Island are black. Similarly, the Black Stilt has wholly melanic plumage. Two color phases occur in the Little Pied Cormorant (*Halietor melanoleucus*), some of which are pied but the majority are black with white throats. Although the pied phase is most abundant in the far north of New Zealand, it still constitutes only about one bird in three or four there (Falla *et al.*, 1982). The Stewart Island race of the Rough-faced Cormorant (*P. carunculatus chalconotus*) is also dimorphic in plumage, and approximately half the populations are composed entirely of the darker bronze-green birds lacking the white ventral plumage of the pied phase. The general explanation for melanism in New Zealand birds may reside with the persistence of random mutations for melanic alleles that would be selected against because of the conspicuousness they encode phenotypically in other predator-rich environments. Thus able to persist, these alleles may actually convey thermoregulatory advantage associated with cooler environments in southern parts of New Zealand, but this hypothesis needs testing with precise measures of heat economy of color phases in a dimorphic species.

5. EXTINCT BIRDS

New Zealand has a surprisingly rich avian fossil and subfossil record, particularly of large flightless birds that became extinct in relatively recent times. The most spectacular material is that of the gigantic flightless moas, which is mostly well preserved because many birds died in caves, limestone sink-holes, sandhills, and swamps. Additionally, they are well known from Polynesian burial sites and kitchen middens. Earlier studies by Archey (1941) and Oliver (1949) had led to the recognition of about 27 species, but this number was later reduced to 13 by Cracraft (1976a) who noted that two species (*Anomalopteryx oweni* and *Megalapteryx benhami*) were based on dubious material and may not be worthy of recognition. This proved to be an accurate prediction because Millener (1982) and Worthy (1988) showed with further material that they can be referred to *A. didiformis* and *M. didinus* respectively, leaving a total of 11 species in six genera. They were confined to the three main islands of New Zealand, and North Island forms of comparable taxa are generally smaller than their South Island counterparts. The moas ranged in size from the giant *Dinornis maximus* standing about 3.5 m tall to the small *Euryapteryx curtus* and *Megalapteryx didinus* about 1 m (Williams, 1976). They had very thick and heavy leg bones, especially in *Pachyornis elephantopus,* although *Anomalopteryx didiformis* was more gracile than other species.

The insular adaptive radiation of the moas is every bit as spectacular as the famous examples provided by Darwin's finches (Geospizinae) in the Galapagos and the Hawaiian honeycreepers (Drepanididae), and they deserve more attention in the literature of avian evolutionary biology than heretofore has been the case. Cracraft (1976a) has rightly emphasized that this radiation is unique because in addition to the changes in feeding mechanisms associated with the passerine radiations, it also involved body size and proportions. Part of the reason that the adaptive radiation of the moas is poorly known is the lack of comprehensive analyses of morphometric variation using multivariate statistical methods; only Cracraft (1976b,c) has attempted such analyses, and he used predominantly or exclusively variables measured on leg bones.

To remedy this situation, we measured 12 variables chosen to represent major trends in skeletal variation in the 10 species of moas for which reasonable sample sizes were available in museum collections (Baker and McGowan, unpublished manuscript). For comparative purposes we also measured skeletons of the three extant species of kiwis as well as Ostrich (*Struthio camelus*), Emu (*Dromaius novaehollandiae*), Rhea (*Rhea americana*) and Cassowary (*Casuarius casuarius*) skeletons. Principal component analysis based on the variance-covariance matrix of logarithmically-transformed data is presented in Table IV and Fig. 16). PC I is a general size axis because all variables load positively and highly (with the exception of mandible length), and PC II represents shape changes involving body proportions (notably mandible length, cranium length, skull depth, and synsacrum width). The tremendous radiation in size and shape in moas relative to extant ratites is clearly depicted in their scatter in Fig. 16. The smaller kiwis are very distinct morphometrically from either the moas or the other living ratities.

The morphometric scale of the moa radiation can be gauged by comparison with the Geospizinae, based on morphometric data for the latter from Lack (1947). Morphometric differentiation, as measured by

TABLE IV
Loadings of Characters of Ratites on
Principal Components I (PCI) and II (PCII)

Character	PCI	PCII
Cranium length	0.922	0.346
Cranium depth	0.947	0.242
Mandible length	0.192	0.781
Sternum width	0.959	−0.084
Synsacrum length	0.991	0.070
Synsacrum width	0.971	−0.232
Femur length	0.988	0.072
Femur width	0.991	−0.119
Tibiotarsus length	0.991	0.080
Tibiotarsus width	0.991	−0.104
Tarsometatarsus length	0.879	0.451
Tarsometatarsus width	0.990	−0.104

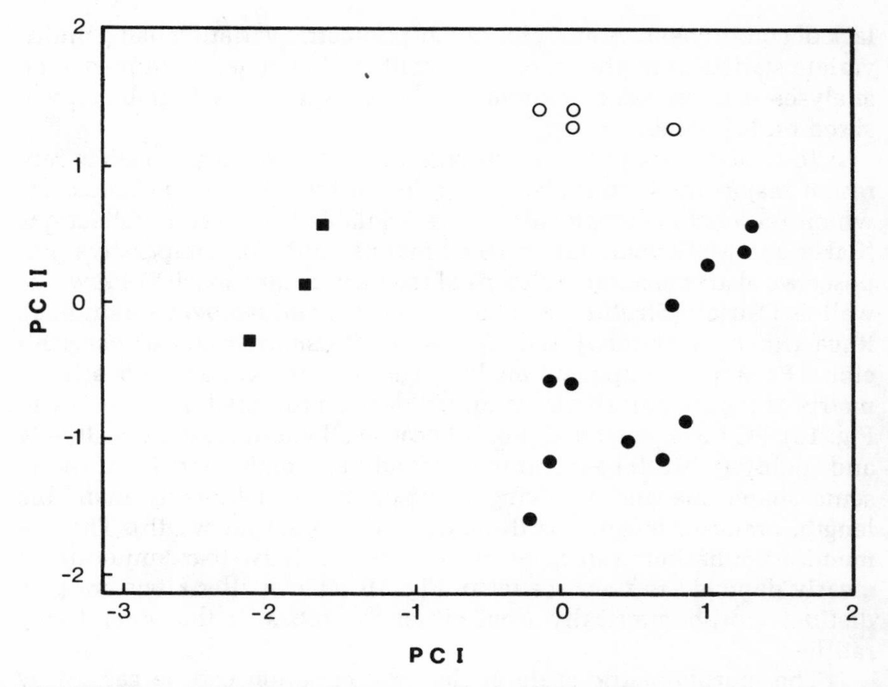

FIGURE 16. Plot of PCI against PCII based on 12 skeletal measurements of species of moas (●), kiwis (■), and other extant ratites (Cassowary, Emu, Rhea, and Ostrich—○).

average taxonomic distances (± standard error) among species, is comparable in magnitude in moas and Darwin's finches (d = 0.159 ± 0.0129 and 0.144 ± 0.0091, respectively). However, if bill measurements are omitted, morphometric differentiation is greater in the moas (d = 0.158 ± 0.0131 and 0.119 ± 0.0090, respectively), indicating that a significant part of the Geospizine radiation has involved changes in bill size and shape. This confirms Cracraft's (1976a) contention that the moa adaptive radiation is more broadly based morphometrically, and involves both size and shape changes as well as modification of the trophic apparatus.

Analysis of the contents of *Dinornis* gizzards indicates that they ate predominately twigs of a wide range of woody shrubs and trees that were sheared off with their presumably sharp-edged bills, but they also ate seeds, fruits, and leaves (Burrows, 1980; Burrows *et al.*, 1981). Gizzards of *Emeus* and *Euryapteryx* contain mostly twigs of a smaller range of plant species (Falla, 1941; Gregg, 1972), but these species may have fed more on seeds and leaves than did *Dinornis* (Burrows, 1980). Although no dietary information is available for *Pachyornis, Anomalopteryx,* and *Megalapteryx,* it seems highly likely that the generic differences in bill size and shape (see Oliver, 1949) were integral to niche diversification in moas. As in Darwin's finches (Grant, 1986) evidence in moas indicates that larger forms had a broader diet than smaller ones. Nevertheless, it seems certain that moas were birds of forests and forest margins where they were the ecological equivalents of browsing mammals.

Although moas are one of the few ancient elements in the New Zealand avifauna, with an origin in the Cretaceous, their adaptive radiation is probably not just a function of their long period of isolation. Instead, it appears that the changing climates and geographies of the New Zealand archipelago through time, coupled with Pleistocene-induced modifications of habitats, have cumulatively accounted for species diversity via allopatric divergence. The opportunities for allopatric speciation in an old group can best be appreciated in terms of the changing geography of the New Zealand archipelago through time (Fig. 17). Speciation would thus be promoted by cycles of invasion, geographic isolation, and later re-invasion of islands, much as has been envisioned for Darwin's finches in the Galapagos Islands. The speciation of moas is thought to have begun in the Tertiary when long intervals of warm temperate climate imply that New Zealand was covered in forest. Because the oldest fossil bones are Upper Miocene and most species have been found in subfossil deposits in the Holocene, however, some workers have speculated that there may have been cycles of speciation in the Pleistocene when populations were isolated in forest refugia during glacial maxima (e.g., Cracraft, 1976a; Trotter and McCulloch, 1984).

The discovery of very large caches of bones (and sometimes eggs) in middens, particularly at mass butchering sites in the eastern South Island, is convincing evidence of the role that Polynesian humans played in the extinction of these giant birds (Anderson, 1984; Trotter and McCulloch, 1984). Archaeologists have established that moas survived climatic changes in the Cenozoic, and probably existed in large numbers before Polynesians colonized New Zealand about 1000 years ago. Furthermore, these populations were hunted and slaughtered for food in large numbers, and their eggs were eaten too. This overexploitive phase in moa hunting probably occurred about a century after the arrival of Maoris, coincident with rapid growth of the human population. Unlike most extant ratities, moas appear to have had low fecundity, and probably laid only one egg. Such low reproductive rates

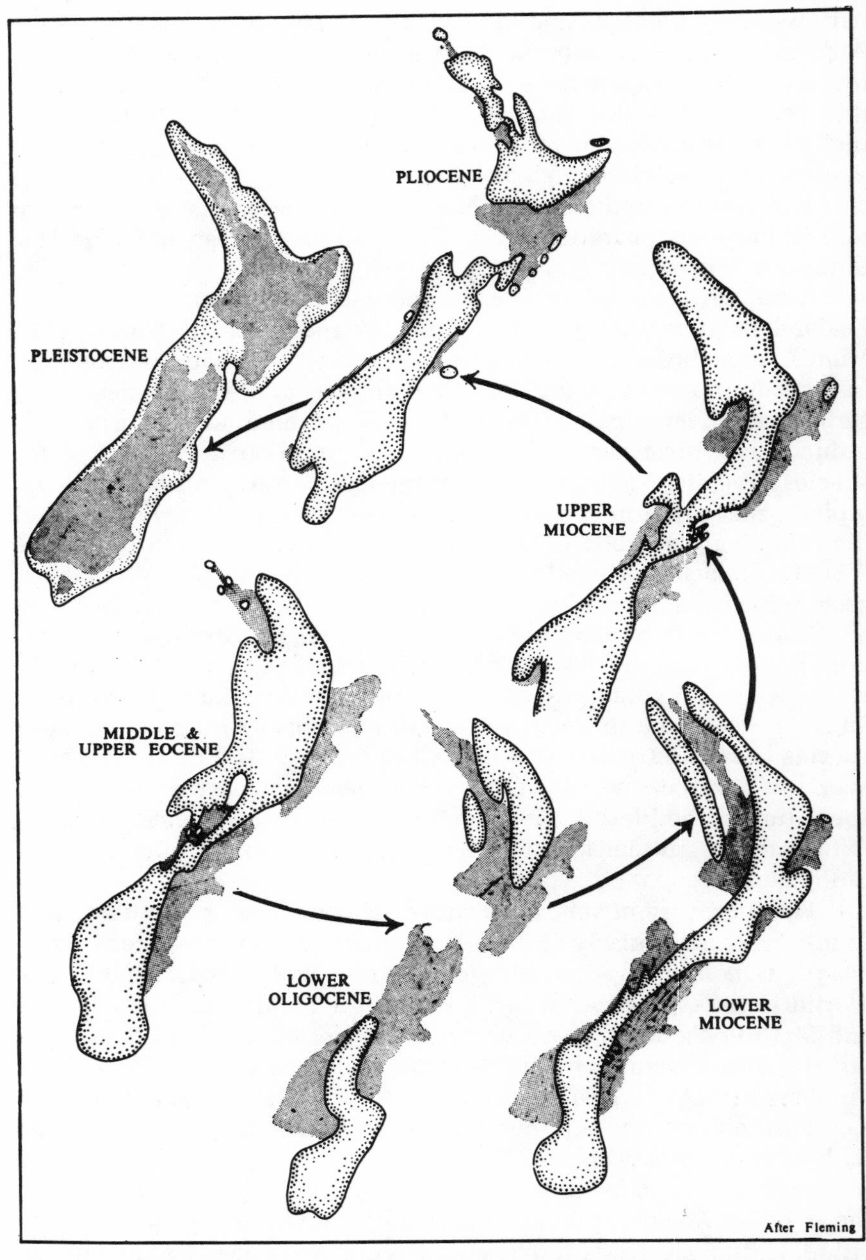

FIGURE 17. The changing outline of the New Zealand archipelago during the Cenozoic (from Fleming, 1975).

would have exacerbated the voracious predation by Maoris. Extensive burning of forests in the next few centuries led to the destruction of much of their natural habitat, thus accelerating their decline until they had become so rare by about 400 B. P. that they were no longer regularly hunted. It is thought that the last of the moas became extinct just before European settlement last century.

Trotter and McCulloch (1984) have drawn a parallel between the extinction of moas in New Zealand with the extinction of the large ratites and lemurs on Madagascar at the hands of humans. They also noted that smaller nocturnal species of lemurs survived human predation better (presumably because they were harder to hunt and had lower food value), as did the small nocturnal kiwis in New Zealand.

Other birds thought to have been extirpated by Polynesian predation include three flightless species; a goose (*Cnemiornis calcitrans*), a swan (*Cygnus sumnerensis*), and a giant rail (*Aptornis otidiformis*). Additionally, the Giant Eagle (*Harpagornis moorei*) was an indirect victim of moa extermination because the latter were its main food supply (Trotter and McCulloch, 1984). Of 26 species of land and freshwater species of birds with fossil records in the Quaternary that are known to have become extinct before European settlement, all but two have been found in association with Maori occupation sites (Millener, 1981). It is not known, however, whether Polynesian predation was the cause of extinction in all cases (e.g., Fleming, 1962c; Millener, 1981; Diamond, 1982; Anderson, 1983), or whether climatic and vegetational changes were responsible (Archey, 1941; Oliver, 1949; Williams, 1960), or both. Williams (1962) argued that because the rates of extinction before 1800 were similar in the North, South, and Chatham Islands (which had very different-sized Polynesian populations), it is difficult to unambiguously ascribe extinctions to the Polynesians. However, this argument misses the point that sustained but lower rates of hunting could have the same effect in extinction as short-term slaughter.

As well as these recent extinctions, considerably older fossil penguins have been described from the New Zealand region. Eight species in six genera are listed in Oliver (1955), occurring in marine sediments varying in age from lower Oligocene to upper Eocene.

6. RARE AND ENDANGERED SPECIES

Mills and Williams (1979) reported that of the 318 species and subspecies of rare or endangered birds in the Red Data Book, about 11% are from the New Zealand region. They attributed this “unenviable record" to the vulnerability of species-depauperate avifaunas on long isolated oceanic islands, coupled with the activities of humans, especially Europeans in the last two centuries. Apart from the destruction wrought by Polynesians chronicled above, European settlement after 1800 has rendered even greater modifications to habitats previously used by birds. For example, in the last 200 years or so, the proportion of New Zealand covered by forest has been reduced from about 66% to just 22%. More importantly, higher proportions of species-rich lowland podocarp forests have been lost, and introduced grazing mammals such as goats, deer, and possums have severely altered the remaining forests. In this same time period seven species, seven subspecies, and many populations of birds have become extinct (Table V), and another 24 species or subspecies have become rare or endangered (Table VI). These authors point out that the impact on the Chatham Islands avifauna has been even more severe, with populations reduced to 100 or fewer birds for five species or subspecies (Table V).

There is good evidence that the spread of the black rat (*Rattus rattus*) has coincided with major declines in native forest birds (Atkinson, 1973). A striking illustration of the impact of this species on birds was provided in 1964 when some came ashore from fishing boats and became established on three small islands west of Stewart Island. The rat populations erupted to plague level within three years, and eliminated four endemic races [Bush Wren (*Xenicus longipes variabilis*), New Zealand Robin (*Petroica australis rakiura*), Fernbird (*Bowdleria punctata stewartiana*), and Sub-Antarctic Snipe (*Coencorypha aucklandica iredalei*)], and severely reduced the populations of four more subspecies [Red-fronted and Yellow-fronted Parakeets (*Cyanoramphus n. novaezelandiae* and *C. a. auriceps*, respectively), Saddleback (*Creadion c. carunculatus*), and the New Zealand Bellbird (*Anthornis m. melanura*)]. The Saddlebacks then disappeared within the next five years (Mills and Williams, 1979). The rats eat the eggs and young of ground-nesting and forest-dwelling species.

In summarizing the status of New Zealand birds, Lockley (1980) noted that of the approximately 280 species and subspecies known to have bred regularly in the past since European contact, nearly 25% are endemics, and nearly all of these are in danger. Fragmented island distributions, habitat destruction and depredations by humans, and the introduction of mammalian ground predators in the last 1000 years, have led to precipitous declines in effective population sizes, concomitant inbreeding depression, and extinction or near-extinction. An excellent example is that of the Takahe, an endangered rail that was once widely distributed through New Zealand. Analysis of the distribution

TABLE V
Extinctions of the Avifauna in Different Parts of New Zealand in European Times[a]

Species	Date	Authority
North Island		
New Zealand Quail (*Coturnix n. novaezealandiae*)	1860–70	Buller, 1883, 1905
Sub-Antarctic Snipe (*Coenocorypha aucklandica barrierensis*)	ca.1870	Oliver, 1955
Kakapo (*Strigops habroptilus*)	ca. 1930	Williams, 1956
N.I.Laughing Owl (*Sceloglaux albifacies rufifacies*)	ca. 1890	Williams and Harrison, 1972
Bush Wren (*Xenicus longipes stokesi*)[a]		Kinsky, 1970
Huia (*Heteralocha acutirostris*)	ca.1907	Oliver, 1955
Stitchbird (*Notiomystis cincta*)[c]	ca. 1885	Kinsky, 1970
Little Spotted Kiwi (*Apteryx oweni*)[c]		
New Zealand Thrush (*Turnagra capensis tanagra*)[b]		
South Island		
New Zealand Quail (*Coturnix n. novaezealandiae*)	ca. 1875	Buller, 1883
Weka (*Gallirallus australis hectori*)[c]	ca.1920	Oliver, 1955
Laughing Owl (*Sceloglaux a. albifacies*)		
Stephen Island Wren (*Xenicus lyalli*)	ca.1894	Buller, 1905
Bush Wren (*X. l. longipes*)[b]		
Saddleback (*Creadion c. carunculatus*)[a]	ca. 1925	Williams, 1962
New Zealand Thrush (*Turnagra c. capensis*)[b]		
Kokako (*Callaeas c. cinerea*)[b]		
Stewart Island		
Sub-Antarctic Snipe (*Coenocorypha aucklandica iredalei*)	1964	Blackburn, 1965
Bush Wren (*Xenicus longipes variabilis*)	1967	Merton, 1975a
Brown Teal (*Anas aucklandica chlorotis*)	1972	Hayes and Williams, 1982
Chatham Islands		
Brown Bittern (*Botaurus poiciloptilus*)	1910	Fleming, 1939
Brown Teal (*Anas aucklandica chlorotis*)	1915	Fleming, 1939

(*continued*)

TABLE V
(*Continued*)

Species	Date	Authority
Chatham Island Rail (*Rallus modestus*)	1900	Fleming, 1939
Banded Rail (*Rallus philippensis dieffenbachi*)	1840	Fleming, 1939
Fernbird (*Bowdleria punctata rufescens*)	1900	Fleming, 1939
Bellbird (*Anthornis melanura melanocephala*)	1906	Fleming, 1939
Auckland Island		
Auckland Island Merganser (*Mergus australis*)	1905	Kinsky, 1970

[a]From Mills and Williams, 1979.
[b]Presumed extinct.
[c]Extant on offshore island.

and age of subfossil remains of the birds, plus their specialized alpine tussock feeding ecology, led Mills *et al.* (1984) to conclude they were widespread during the glacial periods of the Pleistocene when this grassland covered most of New Zealand (Fig. 18). When the climate warmed at the end of the Pleistocene, forests replaced much of this grassland, and thus the Takahe were restricted to fewer localities where they were much more vulnerable to hunting by Polynesians in the last millenium. They were thought to be extinct this century, but in 1948 a small population was discovered in a remote alpine valley in Fiordland. About 120 birds remain today, and attempts to breed them in captivity have had poor success, possibly due in part to inbreeding depression. One of the major factors limiting their success in natural grassland habitats appears to be competition with introduced red deer for high quality tussocks (Mills *et al.*, 1984). Thus in this one species we see all postulated factors operating to severely limit population numbers and dramatically enhance its chances of extinction.

7. CONSERVATION OF BIRDS

7.1. Scope of Conservation Work

Because of the large proportion of rare and endangered taxa of birds in New Zealand relative to the total number in the avifauna, as

TABLE VI
Native New Zealand Avifauna in Danger of Becoming Extinct[a]

Species/subspecies	Numbers	Trend in last decade	Distribution	Requirements for survival[b,c]
Probably Extinct				
North Island Bush Wren				
South Island Bush Wren				
Stewart Island Bush Wren				
North Island Laughing Owl				
South Island Laughing Owl				
South Island Kokako				
North Island New Zealand Thrush				
South Island New Zealand Thrush				
Endangered				
Chatham Island Taiko	Very low	Declining?	Very restricted	M?
Chatham Island Robin	11	Declining	Very restricted	M
Orange-fronted Parakeet	Very low	Declining?	Very restricted	M?
Kakapo	At least 50	Declining	Very restricted	M
Takahe	ca. 120	Declining	Very restricted	M
Black Stilt	ca. 70	Declining	Very restricted	M
Little Spotted Kiwi	Very low	Declining	Very restricted	M
Chatham Island New Zealand Pigeon	ca. 50	Declining	Very restricted	H
Rare				
Forbes' Parakeet	ca. 40	Recently increased	Very restricted	M
North Island Kokako	Low	Declining	Scattered	H
North Island Saddleback	Low	Increased	Restricted	M

(continued)

TABLE VI
(Continued)

Species/subspecies	Numbers	Trend in last decade	Distribution	Requirements for survival[b,c]
South Island Saddleback	Low	Recently increased	Restricted	M
Stitchbird	Low	Stable	Very restricted	HM
Codfish Fernbird	Low	Declining	Very restricted	H
Brown Teal	Low	Declining	Restricted	MH
Blue Duck	Low	Declining?	Locally	H
King Cormorant	200–300	Stable	Restricted	H
Chatham Island Petrel	Low	Declining	Very restricted	H?
Kermadec Storm Petrel	Very low	Unknown	Very restricted	H?
Auckland Island Rail	Low	Stable	Very restricted	H
Chatham Island Oystercatcher	ca. 60	Increased	Restricted	H
Shore Plover	120	Stable	Very restricted	H
Auckland Island Banded Dotterel	ca. 200	Declining?	Very restricted	H?
Chatham Island Sub-Antarctic Snipe	ca. 1000	Increased	Restricted	H
Vulnerable				
Westland Petrel	3–9000	Increased	Very restricted	H
Hutton's Shearwater	5–20,000	Stable	Very restricted	H

[a]From Mills and Williams, 1979.
[b]Active management necessary to increase the chance of survival.
[c]Protection of the habitat necessary for continued survival.

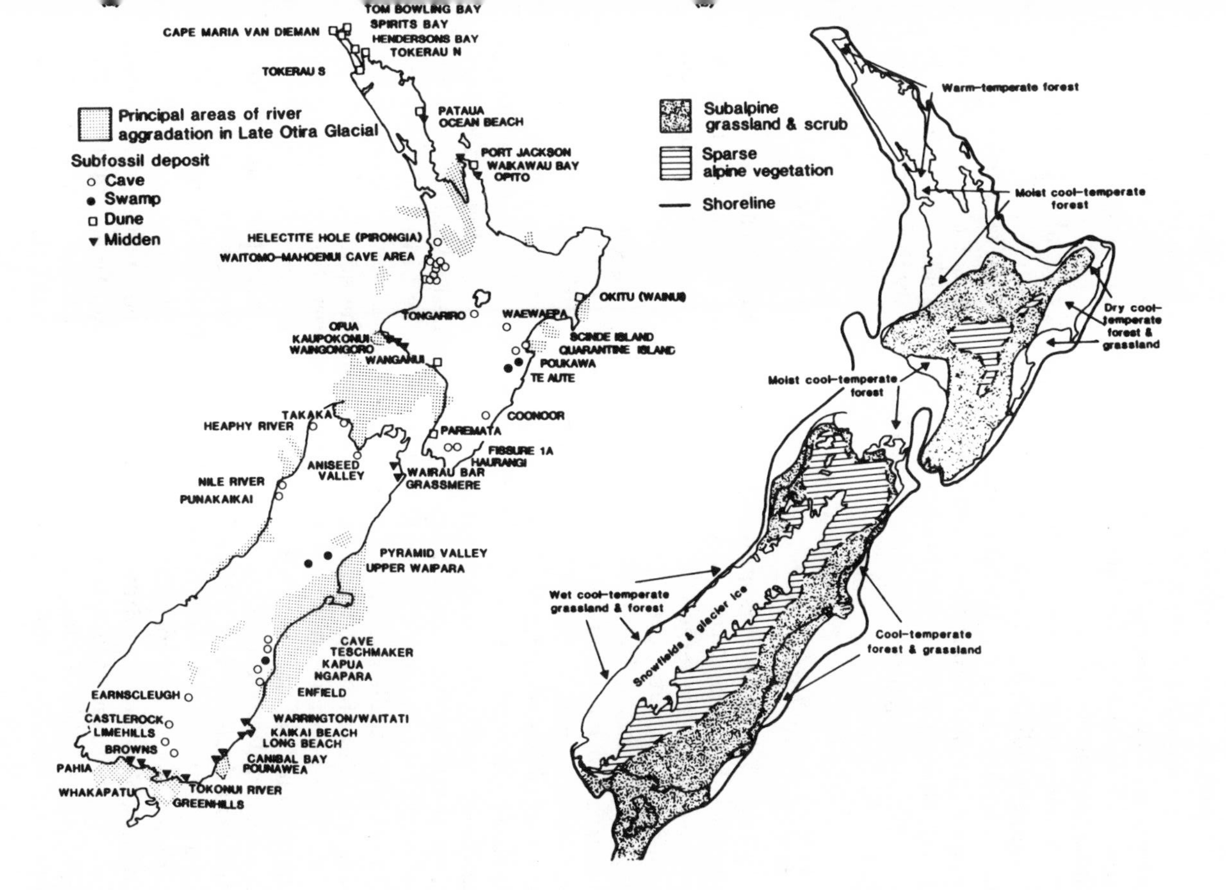

FIGURE 18. Distribution of Takahe subfossil remains (a) in relation to late Quaternary paleogeography (b) (from Mills *et al.*, 1984).

well as their vanishing habitats, much effort is now expended in conservation. Primary responsibility for conservation of birds rests with the Department of Conservation, a New Zealand Government agency whose headquarters are in Wellington. Recently, these responsibilities have been decentralized to regional offices around New Zealand to promote greater local decision-making in areas of immediate relevance for conservation.

The broad spectrum of conservation efforts undertaken by the new Zealand Wildlife Service (a precursor to the Department of Conservation) is outlined in Mills and Williams (1979). All the species listed as endangered or vulnerable in Table VI are the subject of conservation efforts, though limited financial resources and staff have resulted in most effort being directed towards the most severely threatened species (Bell, 1975).

7.2. Island Refuges and Forest Reserves on the Mainland

The offshore islands around the New Zealand coast are playing a pivotal role in conservation of rare and endangered species of birds because some are largely umodified by humans and lack introduced mammalian ground predators. The Wildlife Service has had notable success in transferring small numbers of threatened taxa to such island refuges, and have thereby established self-sustaining populations for subsequent transfer elsewhere (Merton, 1975; Williams, 1977). In an analysis of the occurrence of indigenous forest-dwelling birds on New Zealand offshore islands, East and Williams (1984) showed a general tendency for species in greatest need of conservation to be those with greatest area requirements. Thus large unmodified islands such as Little Barrier and Kapiti are valuable refuges for the Little Spotted Kiwi, Stitchbird, Saddleback, and Kakapo, species that are unlikely to survive predation from introduced mammals on the mainland.

It is important to remember that the preservation of taxa on offshore island refuges is not a good alternative to their conservation in their natural environments on the mainland where this can be done successfully (Crawley, 1982). Large reserves of tens or hundreds of square kilometers of forest are probably required on the mainland to enhance survival prospects of endemic species such as the Weka, Kokako, New Zealand Brown Creeper, Long-tailed Cuckoo, Whitehead, Yellowhead, New Zealand Robin, Rifleman, Kaka, Yellow-fronted Parakeet, and the kiwis (Dawson, 1984; East and Williams, 1984). Reserves adequate for these more vulnerable species should also preserve species diversity in New Zealand forests because they will also be havens for less area-dependent birds.

7.3. Examples of Conservation of Birds

Three recent examples of conservation work on endangered species of birds deserve special mention because they underscore the problems faced in saving perilously small populations. Perhaps the most famous is that of the Chatham Island Robin. This species was once distributed over most of the islands in the group, but forest destruction and predation by feral cats caused a catastrophic decline in numbers to the point where only 20 or so birds remained in a tiny patch of forest on Little Mangere Island (Flack, 1975). Natural damage to this forest reduced the population to seven birds (two of which were females) in 1975, so in 1977 all birds were transferred to another forest remnant on neighboring Mangere Island. When rockfalls partially destroyed this habitat, birds were then moved to South East Island which has a much larger forest on it. Since 1980, when only five birds and one effective breeding pair existed, the population has rebounded from the brink of extinction, thanks to an innovative recovery program masterminded by Don Merton of the Department of Conservation. Merton cross-fostered robin clutches to New Zealand Tomtits, but to prevent imprinting on the foster species he transferred eggs back to robins just before they hatched. Nest boxes were used to protect against losses from petrels crashing through the forest canopy at night en route to their burrows. Survival of robins has improved dramatically on South East because of the abundance of suitable habitat, and the total population now exceeds 100 birds.

The Chatham Island Taiko was thought to have been extinct for the about the last 100 years. It was known from only one specimen collected in the south Pacific ocean in 1867 by the Italian research vessel Magenta, from which it derives its scientific name, *Pterodroma magentae*. Reports of strange calls of birds flying over Chatham Island at night prompted David Crockett to mount repeated expeditions to the Tuku Valley area, and on the night of 1 January 1978 three Taikos were caught using spotlights and nets. A total of 14 birds have been caught and banded until now, but only one burrow containing breeding birds has been located, despite the use of radio-tracking techniques. It seems that the few breeding pairs that exist are scattered over an unknown area.

The third example is provided by the flightless Owl-Parrot or Kakapo, which once occurred in North, South, and Stewart Islands. It is now restricted to remote rainforests of Fiordland in the South Island, and intensive searches from 1973–1981 yielded only 18 birds, all males. Fortunately, in 1977 Kakapos were rediscovered on Stewart Island, where 58 birds were caught and banded between 1977 and 1983 (Mills and Williams, 1979). Karl and Best (1982) found that predation by cats was seriously threatening the Stewart Island population, so the Wildlife Service decided to move birds to Little Barrier Island because it is free of cats and is large enough to support an expanding population. In early 1983, a total of 21 birds had been transferred to Little Barrier, and at that time it was estimated that 30–50 Kakapos remained on Stewart Island.

7.4. Conservation Genetics

The Wildlife Service and now the Department of Conservation deserve great credit for their pioneering work in the conservation of rare and endangered species of birds. One of the most significant recent developments has been the establishment of a modest program in conservation genetics within this Government agency, with the realization that the maintenance of genetic diversity in threatened populations and potential problems with inbreeding depression are integral components of any conservation strategy. Genetic studies from this program and others are beginning to yield results of immediate relevance to conservation. For example, Stewart Island and Fiordland populations of Kakapos are not genetically differentiated in their protein-encoding genes, suggesting that it it reasonable to combine birds from both of these populations in the Little Barrier Island refuge. However, Fiordland birds are genetically depauperate relative to their Stewart Island conspecifics, consistent with recent bottlenecks in effective population size (Triggs *et al.*, 1989). Thus the latter population alone is a poor candidate for any breeding program where the goal should be to maintain genetic diversity and thus minimize any possible deleterious effects from inbreeding depression.

Genetic studies also have great potential to identify different populations and taxa that are at risk of extinction. For example, Meredith and Sin (1988) have demonstrated with protein electrophoresis that gene frequencies in Little Blue Penguins (*Eudyptula minor*) vary clinally along the east coast of New Zealand, corresponding to a similar cline in morphometric variation. This argues against the recognition of three mainland subspecies as in Kinsky (1970) because levels of gene flow between populations are apparently high, and they therefore are effectively one panmictic unit. Thus in mainland New Zealand we are not only dealing with conservation of one subspecies, but the increased census population size decreases concern with its conservation status.

Relationships and taxonomic status of parakeets that hybridize in the New Zealand region are clearly fundamental to any conservation strategy for this group of threatened or declining populations. C. H. Daugherty and S. J. Triggs (personal communication) have conducted important genetic studies of these birds, and their conclusions have ramifications for conservation. The genetic distance between Yellow-fronted and Red-fronted parakeets is typical of species-level divergence in other birds, arguing that they are good species despite known hybridization on Little Barrier Island. Yellow-fronted Parakeet populations are structured genetically into two groups that meet in the central South Island, indicating that two subspecies should be recognized where only one was known before. Lastly, Forbes' Parakeet is genetically distinct from the Chatham Island Red-fronted Parakeet with which it hybridizes. Because it is clearly derived from the Red-fronted rather than the Yellow-fronted Parakeet, Daugherty and Triggs (personal communication) recommend removing Forbes' Parakeet as a subspecies of the latter and raising it to full species status.

Daughtery and Triggs also have studies in progress on Brown and Little Spotted kiwis, Blue Ducks, New Zealand Teal, Grey Duck, Mallard and Orange-fronted Parakeets (*Cyanoramphus malherbi*). This work is urgently needed in the New Zealand region and deserves priority funding from conservation agencies. Given the fragmented distributions and small population sizes of many threatened species of endemic birds in New Zealand, it is imperative to integrate conservation genetics research in any recovery plans. Informed decisions can then be made about the choice of new breeding stocks in refuge populations, breeding strategies can be designed in captive flocks to equalize contributions from as many founders as possible (and thus maintain allelic diversity), and the potential effects of inbreeding depression can be assessed and minimized.

8. INTRODUCED BIRDS

For the visitor to New Zealand, one of the most conspicuous features of the avifauna, particularly in cities, farms, exotic forests, and the edges of native forests, is the great abundance of European passerines introduced by humans late last century. House Sparrows, Chaffinches, European Starlings, Blackbirds, and Song Thrushes are among the most abundant birds in the country. At least 143 exotic species were re-

TABLE VII
The Introduced Avifauna of New Zealand[a]

Family	Species	Country/area of origin	Date of first introduciton
Anatidae	Mute Swan (*Cygnus olor*)	United Kingdom	1866
	Black Swan (*Cygnus atratus*)	Australia	Before 1864
	Canada Goose (*Branta canadensis*)	U.S.A.	1876
	Mallard (*Anas platyrhynchos*)	U.S.A. and U.K.	1867
Phasianidae	Chukar (*Alectoris chukar*)	Asia	
	Grey Patridge (*Perdix perdix*)	Denmark	1962
	Brown Quail (*Synoicus ypsilophorus*)	Australia	Before 1870
	Northern Bobwhite (*Colinus virginianus*)	U.S.A.	1898
	California Quail (*Lophortyx californica*)	U.S.A.	1865
	Ring-necked Pheasant (*Phasianus colchicus*)	Asia	1842
	Common Peafowl (*Pavo cristatus*)	Asia	1843
Columbidae	Rock Dove (*Columbia livia*)	United Kingdom	?
	Spotted Dove (*Streptopelia chinensis*)	Asia	1950?
Cacatuidae	Sulphur-crested Cockatoo (*Cacatua galerita*)	Australia	1920
Platycercidae	Crimson Rosella (*Platycercus elegans*)	Australia	1910
	Eastern Rosella (*P. eximius*)	Australia	1910
Strigidae	Little Owl (*Athene noctua*)	Germany	1906
Alcedinidae	Kookaburra (*Dacelo gigas*)	Australia	1866
Alaudidae	Skylark (*Alauda arvensis*)	United Kingdom	1964
Pycnonotidae	Red-vented Bubul (*Pycnonotus cafer*)[b]	Fiji?	1952
Prunellidae	Dunnock (*Prunella modularis*)	United Kingdom	1867
Muscicapidae	Song Thrush (*Turdus philomelos*)	United Kingdom	1986
	Blackbird (*Turdus merula*)	United Kingdom	1862
Emberizidae	Yellowhammer (*Emberiza citrinella*)	United Kingdom	1862
	Cirl Bunting (*E. cirlus*)	United Kingdom	1871
Fringillidae	Chaffinch (*Fringilla coelebs*)	United Kingdom	1862
	Greenfinch (*Carduelis chloris*)	United Kingdom	1862
	Goldfinch (*C. carduelis*)	United Kingdom	1862
	Redpoll (*Acanthis flammea*)	United Kingdom	1862
Ploceidae	House Sparrow (*Passer domesticus*)	United Kingdom	1862

(*continued*)

TABLE VII
(*Continued*)

Family	Species	Country/area of origin	Date of first introduciton
Sturnidae	European Starling (*Sturnus vulgaris*)	United Kingdom	1862
	Indian Myna (*Acridotheres tristis*)	India	1870
Cracticidae	Black-backed Magpie (*Gymnorhina t. tibicen*)	Australia	1864?
	White-backed Magpie (*Gymnorhina tibicen hypoleuca*)	Australia	1864
Corvidae	Rook (*Corvus frugilegus*)	United Kingdom	1862

[a]Adapted from Williams (1973).
[b]Believed exterminated.

leased, of which 34 species have colonized successfully, with 18 species from Europe (mainly England), seven from Australia, five from Europe, and four from North America (Table VII). These birds were imported to control plagues of insects that accompanied agriculturization of the land, and for sentimental reasons. Importations were made primarily by Acclimatisation Societies in the major provinces, especially the Auckland, Wellington, Nelson, Canterbury, and Otago Societies. Once birds were successfully established, they were quickly spread to other localities by the Societies, bird fanciers, and interested amateurs. Many have subsequently spread widely in the New Zealand region (Table VIII), and are self-introduced to most offshore and outlying islands including the Chathams (Williams, 1953) and the subantarctic islands (Williams, 1973). The history of the liberations of introduced birds in New Zealand is chronicled by Thomson (1922) and Long (1981).

The impact of introduced species on the native avifauna and their possible role in extinction of native species has been debated because it is difficult to control for confounding factors such as mammalian predation, forest loss, depletion of food plants by introduced browsing mammals, and diseases. Diamond and Veitch (1981) were able to largely unravel these effects by examining forest bird communities on some island in the Hauraki Gulf (in the northern North Island) which lacked mammalian browsers and had very few mammalian predators. They found introduced bird species that were abundant in modified

TABLE VIII
Dispersal of Introduced European Passerines from New Zealand to Offshore Islands[a]

Species	Island group[b,c,d]							
	Three Kings (55)	Kermadecs (805)	Chathams (645)	Antipodes (805)	Campbell (565)	Aucklands (320)	Snares (95)	Macquarie (885)
European Starling	Xb	Xb	Xb	Xb	Xb	Xb	X	Xb
House Sparrow	X	—	Xb	—	Xb	Xb	Xb	—
Redpoll	X	X	Xb	Xb	Xb	Xb	Xb	Xb?
Goldfinch	X	X	Xb	Xb	Xb	Xb	Xb	X
Greenfinch	—	X	Xb	—	Xb	X	X	—
Chaffinch	Xb	X	Xb	Xb	Xb	Xb	X	—
Yellowhammer	X	Xb	Xb	—	X	—	X	—
Blackbird	Xb	Xb	Xb	—	Xb	Xb	Xb	—
Song Thrush	Xb	Xb	Xb	X	Xb	Xb	Xb	X
Dunnock	X	—	Xb	Xb	Xb	X	X	—
Skylark	—	X	Xb	—	X	Xb	—	—
Total number of species in group	9	9	11	6	11	10	10	4

[a]From Williams (1973).
[b]Distance in km from mainland New Zealand in parentheses.
[c]Xb, breeding on this island.
[d]X, present on this island.

mainland forests were almost absent from unmodified islands on the islands. Their conclusion was that competition from introduced species could only invade forests after native species were decimated and browsing mammals had altered forest structure. In keeping with this conclusion is the finding by Abbott and Grant (1976) that the number of passerine species on New Zealand offshore islands is not at equilibrium, but is generally increasing. This implies that introductions of exotic species has not been matched by extinctions of native species, but instead the former have mostly occupied niches that were formerly vacant.

9. HISTORY OF ORNITHOLOGICAL DISCOVERY IN NEW ZEALAND

9.1. Mainland New Zealand

A comprehensive history of ornithological discovery in New Zealand to about 1950 has been written by Oliver (1955), but a brief summary is presented here to highlight events of ornithological importance. Although birds were noted as plentiful and some were shot for food on Captain Cook's first voyage of discovery to New Zealand in 1769–1770, it was not until his second visit in 1773–1774 that the first serious ornithological work was done on this completely unknown avifauna. This transpired because Cook had two naturalists on board, the father and son combination of John and George Forster. The Forsters collected and preserved 38 species of birds in the region of New Zealand, 35 of which were also drawn. This material was subsequently featured in Latham's *General Synopsis of Birds* (1785). On Cook's third voyage (1776–1777) 26 species of birds were mentioned by the ship's surgeon (W. Anderson), and seven were drawn.

Nothing more of ornithological significance occurred in the next 50 years with the notable exception of the capture of the first kiwi in 1812. Then followed expeditions by French, American, and British explorers. Familiar names on the French visits in 1824 and 1827 were Dumont D'Urville and R. P. Lesson, both avid collectors, as well as the zoologists J. R. C. Quoy and P. Gaimard. Between them they discovered six new taxa, including the North Island race of the New Zealand Robin, the New Zealand Tomtit, the Grey Gerygone, New Zealand Quail, Wrybill, and Fernbird. The first specimens of the bronze phase of the Stewart Island race of the Rough-faced Cormorant was collected on D'Urville's third expedition in 1839. Captain Charles Wilkes, in command of the American Exploring Expedition, visited the Bay of Islands in the North Island in 1840. Twelve species of birds including the Common Cormorant (*P. carbo*) were described by T. Peale, the ornithologist to the expedition. In 1841 the British ships *Erebus* and *Terror* visited New Zealand after stops at Auckland and Campbell Islands. Birds collected on this expedition along with other material sent to the British Museum were the basis of the first full description by G. R. Gray of all New Zealand birds (99 species) known at that time. Around the same time (1837) the first moa bone to reach England was examined by Richard Owen, who immediately realized that it represented a new struthious bird unlike extant ratites. The Takahe was discovered on Resolution Island in Fiordland in 1849.

The avifauna of New Zealand did not receive intensive scrutiny until resident ornithologists began their work, and the last half of the 19th century saw a great advance in knowledge of the species present and their natural history. T. H. Potts was an excellent naturalist who arrived in New Zealand in 1853. He traveled extensively in Bank's Peninsula, Westland, and Canterbury, concentrating on the nesting habits of native birds. Potts described the Great Spotted Kiwi and the Black-billed Gull. Sir Walter Buller was born in New Zealand in 1838, and quickly developed into the premier ornithologist in the Dominion. From 1865 to 1905 he published 74 papers as well as four monumental works, as follows: *History of the Birds of New Zealand* (1873), *Manual of the Birds of New Zealand* (1882), *History of the Birds of New Zealand, Second edition* (1888), and *Supplement to a History of Birds of New Zealand* (1905). In the same period three geologists, J. von Haast, Sir James Hector, and F. W. Hutton added substantially to ornithological knowledge specifically and to zoological knowledge generally with their field observations, as is readily apparent from the number of New Zealand taxa named after them.

The first comprehensive life history studies of New Zealand birds were carried out by H. Guthrie-Smith and E. Stead, both of whom used photography to augment their accounts. Guthrie-Smith published five books entitled *Birds of the Water, Wood, and Waste* (1910), *Mutton Birds and other Birds* (1914), *Tutira* (1921), *Bird Life on Island and Shore* (1925), and *Sorrows and Joys of a New Zealand Naturalist* (1936), and Stead one, *Life Histories of New Zealand Birds* (1932). The whole of Guthrie-Smith's second book and much of his fourth was on the birds of Stewart Island and numerous small islands surrounding it. His final book recorded observations on the birds of the Kermadec Islands.

W. R. B. Oliver published the first complete list of Stewart Island birds in 1926, followed by his tome *New Zealand Birds* in 1930.

The many and varied contributions of R. A. Falla and C. A. Fleming, including important systematic studies, highlighted the 1940s and 1950s, eventually culminating in Knighthoods for both men in recognition of their services to ornithology. One of the most exciting moments in New Zealand ornithology came in 1948 when G. B. Orbell rediscovered living Takahes in Fiordland, and this has been followed by equally important rediscoveries of Kakapo on Stewart Island in 1977 by the New Zealand Wildlife Service, and of the Chatham Island Taiko in 1978 by a field party led by David Crockett.

9.2. Offshore Islands

The first ornithological information on the Chathams dates to a visit by E. Dieffenbach in 1840, as recorded in the appendix (by G. R. Gray) to his *Travels in New Zealand*. Visits by Buller in 1855, Travers in 1871, and Forbes in 1892 added specimens and observations of Chatham species. A list of birds observed was published by G. Archey and C. J. Lindsay following their stay in the Chathams in 1923. C. A. Fleming and E. G. Turbott visited in 1937, and results of their more detailed studies were published by Fleming (1939).

The Auckland Islands were visited in 1839 by the French expedition, where the Yellow-eyed Penguin and now extinct Merganser were discovered. The following year the Sub-Antarctic Snipe was collected by the American expedition. A general account of the birds of the Auckland and Campbell Islands by E. R. White appeared in *The Subantarctic Islands of New Zealand* (1939) following the Canterbury Philosophical Institute expedition in 1937. Watching stations established on these islands during the second world war led to extensive observations of birds and collection of valuable specimens. Ornithological discovery on the Snares Islands has been more recent. R. C. Murphy, E. F. Stead, R. A. Falla, and C. A. Fleming collected material and made observations in 1947, and this was followed by a party composed of L. E. Richdale and W. M. C. Denham which resulted in detailed study of Buller's Albatross (*Diomedea bulleri*). Falla was also instrumental in early surveys of birds on Macquarie Island in 1930 (as a member of the British, Australian, and New Zealand Antarctic Research Expedition) and the Antipodes and Bounty Islands (with E. G. Turbott and R. K. Dell) in 1950. At the other end of the New Zealand mainland, the Three Kings Islands were visited several times from 1934 onwards, and the birds were reported on by Turbott in 1948.

10. ORNITHOLOGICAL ORGANIZATIONS IN NEW ZEALAND

The Ornithological Society of New Zealand (OSNZ) was founded in 1939 with the objective to encourage, organize, and carry out field work on a national scale. The first president, quite fittingly, was Sir Robert Falla. In 1949 the quarterly journal of the society, *New Zealand Bird Notes* (now called *Notornis*), was first published, and at the time of writing this review it has reached its 50th anniversary. There is no doubt that it has been a major stimulus to ornithological research in New Zealand, and has provided a valuable outlet for dissemination of results of work performed by amateur and professional ornithologists alike.

The OSNZ has been extremely effective in stimulating an interest in birds by conducting annual field camps in different regions of the country, by organizing annual surveys of selected species, by censusing all species in a wide variety of bird haunts through the vehicle of regional branches (published in *Notornis* as Summarized Classified Notes roughly equivalent to Christmas bird counts in *American Birds*, for example), by running the beach patrol scheme to monitor wrecks of petrels and other seabirds following major storms, by keeping a repository of the nest records, and by maintaining a library of the world's top ornithological journals. The bird-banding scheme started by the OSNZ in 1950 was transferred to the Wildlife Service in 1966. In 1953 the OSNZ published its *Checklist of New Zealand Birds*, and demand was so great that it published an enlarged version in 1970 entitled *Annotated Checklist of New Zealand Birds*. A new edition is currently being prepared. In collaboration with the New Zealand Wildlife Service and the Ecology Division of the Department of Scientific and Industrial Research (DSIR), the OSNZ implemented a national mapping scheme designed to outline the present distribution of bird species in New Zealand. This initiative grew from concerns about plans to destroy large areas of *Nothofagus* forest (or to replace them with exotic forests) in the South Island, and the increasing exploitation of avian habitats by man. This effort culminated in the publication of *The Atlas of New Zealand Birds* (Bull *et al.*, 1985). This joint effort thus involved the three major ornithological organizations in New Zealand.

In 1945 the Wildlife Branch of the Department of Internal Affairs in Wellington was established to study primarily introduced forest and high country mammals, but in 1956 this mandate was transferred to the New Zealand Forest Service. This freed the Wildlife Branch to focus on birds, though initially waterfowl and upland game birds were a priority. The dramatic rediscovery of the Takahe in 1948 provided the stimulus for much needed studies of the native avifauna (Williams, 1973). This evolution continued with increasing emphasis on the conservation of rare and endangered species, aided by the transformation of the Branch into the Wildlife Service and recently into a separate Department of Conservation.

The DSIR established a research program in 1948 to study animals of agricultural importance, from which emerged the current Ecology Division. This group has a strong tradition of applied research on birds as agricultural pests and as intregral components of New Zealand ecosystems. Ornithologists employed by the Forest Research Institute, the Department of Lands and Survey, and the Royal Forest and Bird Protection Society have concentrated on forest birds. The community of professional ornithologists is enlarged and enriched by staff of Zoology Departments in New Zealand universities, with a particularly strong group at the University of Auckland. A much smaller number are employed in major museums at Auckland, Wellington, Christchurch, and Dunedin. Failure by universities to provide effective training in systematics in the past three decades has led to the decline of ornithological research in museums.

11. EXEMPLARY LONG-TERM POPULATION STUDIES

New Zealand ornithology has long been heavily influenced by its British heritage gained through the hiring of British ornithologists in professional organizations, especially universities, training of graduate students in British universities, and respect for the excellence of ecological work conducted at the Edward Grey Institute at Oxford University. As a graduate student in New Zealand in the late 1960s and early 1970s I well remember the dominating impact of David Lack's work there, and the concomitant enthusiasm for long-term studies of marked populations of birds. With apologies to all those ornithologists whose work I have not included, the following is a selection of long-term studies which have made major contributions to our knowledge of the population biology of birds, or have the potential to do so.

11.1. Yellow-eyed Penguins

The first long-term study of a sea bird is that of L. E. Richdale, who studied a marked population of Yellow-eyed Penguins on the Otago Peninsula near Dunedin in the South Island (Richdale, 1957). The study was remarkable in several aspects. First, Richdale was an amateur ornithologist who worked alone. Second, he continued his studies in his spare time for 18 consecutive breeding seasons from August 1936 to March 1954. Third, he marked many of the birds in his study with individually recognizable bands and footmarks. Fourth, he focused on the effects of age on reproductive rate, breeding success, and population turnover.

Richdale established that a sizable nonbreeding population occurred each year on the breeding colony he studied, representing 38% of all birds averaged over all years. Most of these nonbreeders were subadults from one to four years of age, and others were unmated adult males. He found that the penguins do not breed until they are at least two years old and that males have delayed maturity, presumably as a result of the greater long-term survival of males and the consequent unbalanced sex ratio in adults. Most females first attempt breeding when they are two or three years old, whereas males delay until they are three, four, or five.

Site fidelity in breeding birds was found to be high, as was mate fidelity, with the divorce rate being about 14% when both birds returned to the colony. Richdale further showed that the average laying date each year was remarkably constant, quite unlike small passerines in the northern hemisphere studied at that time by Lack and his colleagues (Lack, 1966). Because individual females (and their daughters) lay at the same date relative to others each year, Richdale deduced that the differences among females were heritable.

Perhaps his most important contribution was to demonstrate the effects of age on breeding performance. Richdale showed that younger birds had much lower breeding success. He found that almost 40% of the females that bred at two years old laid only one egg, and this declined to 6% in three year olds, and 1.5% in older penguins. The usual clutch size in mature birds is two eggs. Considering all pairs in which one bird was a two year old, he showed that they had smaller clutch sizes and much higher losses of eggs and possibly young than older birds, and thus argued that delayed maturity would be selected for in most birds. Annual mortality among breeding birds three years and older banded by Richdale was estimated at 13%, and was balanced by recruitment. Interestingly, he encountered one poor year (1938–39) when breeding success was very low, presumably due to a failure in the food supply like that encountered around New Zealand in 1987–88, when warm water inshore led to food shortage and breeding failure for Silver Gull colonies. When Richdale ended his study in 1954 the colony he studied had grown to 82 breeding pairs from a low of 21 pairs in 1940. He did not enter the debate at that time about factors regulating population size, but as predation was apparently very low Lack (1966), argued that food supply was likely the principal determinant of breeding success and winter survival in these penguins. As a postscript to Richdale's monumental study, the Yellow-eyed Penguin is now regarded as one of the rarest penguins in the world, and current research by J. Darby of the Otago Museum has identified predation by feral cats as a major cause of the decline in the population.

11.2. Silver Gulls

One of the most important long-term population studies of birds anywhere in the world is that conducted by J. A. Mills on Silver Gulls at the large Kaikoura colony on the east coast of the South Island. Since 1964 Mills has visited the colony annually, and about 70,000 young have been banded by him and others from 1958. Since 1967 he has followed individuals identified with unique combinations of color bands, and some idea of the massive scale of his work can be appreciated from the 4,408 adults and 1078 young he has color-banded. Mills has accumulated information on 32,000 birds in a computer file, and now has 105,000 recoveries. Much of this huge database is being prepared for publication in the next two years, beginning with a chapter that is in press in the book *Lifetime Reproduction in Birds* edited by Ian Newton (Academic Press). Studies of sexing, band loss, the influence of age and pair-bond on breeding biology, and egg size have previously been published (Mills, 1971, 1972, 1973, 1979).

By following a sample of 81 males and 66 females through their lives, Mills has been able assess their lifetime contributions to future generations. He found that the mortality of young birds between fledging and breeding was so high that most did not contribute offspring to the next generation (Mills, 1990), a similar finding to that reported by Newton (1985) for the European Sparrowhawk (*Accipiter nisus*). Among the birds that survived to breed, another 36% of males and 39% of females did not even fledge any young in their lifetimes. Overall, 20% of the males produced 58% of the fledglings, and 15% of the females produced 52% of the fledglings. However, only 17% of males and 24% of females that bred actually produced offspring that were later recruited into the population as breeding adults.

These obvious disparities in the lifetime reproductive outputs of the sexes is related to the female-biased sex ratio in Silver Gulls. Mills found that females had greater longevity but had delayed maturity and bred less frequently than males. A staggering 49% of adult females did not breed each year, whereas only 14% of males did not attempt to breed annually. Parent-offspring data collected over five years indicates that the Kaikoura population is being maintained by a small number of highly productive birds. However, their productivity is apparently not strongly heritable, and neither is their laying date or frequency of breeding attempts in different years. These last two factors account for the greatest variability in lifetime fledging success.

Such asymmetries between the sexes and low lifetime contributions by most gulls in a colony has important evolutionary consequences. First, it reduces the long-term average effective population size (N_e), which in turn affects the amount of genetic variability maintained in populations. In this regard, recent studies have revealed low average heterozygosity at allozyme loci in the Kaikoura colony (A. J. Baker *et al.*, unpublished manuscript). Second, it cautions against adaptationist hypotheses that seek to extrapolate variance in reproductive investment (particularly in terms of clutch size) among individuals as an index of fitness (see, for example Nur, 1986).

11.3. Southern Great Skuas

After many years of research on South Polar Skuas (*Catharacta maccormicki*) in Antarctica, E. C. Young began his study in 1974 of Southern Great Skuas (*C. skua lonnbergi*) breeding on the Chatham Islands. He was surprised to discover that birds bred not only in pairs but also quite commonly in trios, and more rarely in larger groups of up to seven birds. However, more than 50 years earlier, communal breeding in Southern Great Skuas had been reported on the temperate outlying islands in the New Zealand region, particularly Stewart Island (and its surrounding islets), the Snares, and the Chathams (Guthrie-Smith, 1925; Stead, 1932). These observations were ahead of their time and thus were largely dismissed as anecdotal or were ignored; Richdale (1965) noted that ornithologists in the northern hemisphere still reacted to reports of trios with skepticism. The burgeoning of interest in communal breeding in birds and in sociobioloical theories to explain the phenomenon has raised awareness to nonmonogamous associations of breeding adults.

Young and his associates have confirmed that Southern Great Skuas are exceptional among seabirds in the high incidence of communal breeding they display in the New Zealand region, the northern extreme of their circumpolar range in the southern hemisphere. Long-term studies have thus been conducted on the small populations breeding on South East Island and Mangere Island in the Chatham group to attempt to ascertain the factors that have led to the evolution of communal breeding in this large predatory seabird. On these islands Young has established that of 62 breeding groups 48 (77%) are in pairs, 12 (19%) are in trios, one (2%) is composed of four birds, and one (2%) now has five birds (at one time this group had seven birds). These groups are stable in size and composition through time, especially on South East Island where only three changes occurred among 90 birds in the 1988–89 breeding season (E. C. Young, personal communication). One trio, for example, has been intact since they were banded in 1977.

Since the degree of relatedness of birds in trios and other communal groups is central to theories attempting to explain communal breeding, much effort has been put into developing a banded population of known parentage. This was initially attempted by banding all birds in groups and each year's production of chicks. The skuas prey on small petrels which abound on the islands, and therefore breeding success and chick survival is quite high. Age of first breeding is delayed relative to South Polar Skuas, however, so that only 14 individually identified chicks out of a total of 727 banded have as yet begun to breed as adults. The mean age of these 14 birds at their first breeding attempt was 7.6 ± 0.6 (± S.E.) years as opposed to 3–4 years in South Polar Skuas, attesting to the difficulty of working with long-lived birds. None of these 14 have returned to their natal territories or have joined groups with their parents or siblings.

DNA fingerprinting is now in progress to determine the parentage of young produced by groups and possibly relationships within groups. Preliminary analysis of four trios has shown that the beta bird (as determined from behavioral data) is not the offspring of the other two birds. Though it is remotely conceivable that the beta bird is the offspring of a previous partner in the trio, this seems an unlikely explanation for all four trios, especially given the stability of the group through time. Current work with larger sample sizes will permit stronger inferences about sibling and offspring membership in groups.

A. D. Hemmings has compared the reproductive performance of pairs, trios, and larger groups. He found that there were no significant differences between pairs or trios in egg size, chick growth and condition, and chick production per territory. However, trios are less productive than pairs on a per-adult basis. Breeding success is high in the Chathams relative to other populations studied, but there is no evidence in the short or longer term that communal breeding in the Southern Great Skua results in enhanced production of young (Hemmings, 1989; Young, personal communication).

The value of such a long term study is graphically illustrated by earlier observations that trios on South East Island were predominantly situated on the northern side of the island where most of the petrels breed. The topsoil there is an excellent substrate in which the petrels can dig burrows for breeding attempts. Thus the obvious hypothesis emerged that trios were necessary to defend the higher quality territories on the food-rich northern side. Radio-tracking of birds at night, however, showed that the male of a pair without petrel burrows on their territories was able to capture petrels inland off the territory, and it is clear that many other pairs do so successfully (Young *et al.*, 1988). Recently, the number of trios on the rockier southern side of the island has increased to match the frequency of those on the northern side. Breeding success is comparable in both areas, and more apparently suitable space for territories is available, thus arguing against ecological constraints (Emlen, 1982) as the cause of communal breeding.

11.4. Pukekos

Some of the most thought-provoking and original work on communal breeding in birds in recent years has been conducted on Pukekos by J. L. Craig at the University of Auckland. An initial three year study in the Manawatu region in the southern part of the North Island described the behavior, social organization, breeding success, and communal breeding of the species (Craig, 1976, 1977, 1979, 1980a, 1980b). Since that time a longer-term intensive study was carried out from 1979–1986 near Auckland city, the prime objective of which was to document levels of dispersal, inbreeding, and possible inbreeding avoidance in this gallinule. The Pukeko is ideal for this kind of investigation because it has a polygynandrous mating system, it is large and easy to observe, and it is possible to determine matings because copulations are frequent and conspicuous events.

The social organization of the Pukeko is complex, varying from pairs through groups of promiscously breeding adults to groups with nonbreeding yearlings and pairs and groups assisted by juveniles from first broods of the year. The groups thus consist of one to two breeding females, from one to seven breeding males, and up to an additional seven nonbreeding helpers. All nonbreeding helpers are young birds from previous breeding seasons, and range in age from one to four years. Although birds are capable of breeding when they are one year old, this rarely happens and breeding is typically delayed for both sexes (Craig, 1980a). Territories are defended throughout the year. Females in groups lay their eggs in the same nest, and incubation is performed by all birds that participated in copulations. Feeding and brooding of young, as well as territory defense, is carried out by all members of each group. Linear dominance hierarchies exist within breeding groups, but dominant males do not apparently monopolize copulations. Although this is consistent with male–male cooperation to enhance survival of young when promiscuity is advantageous to females, Jamieson and Craig (1985) argued persuasively that males were indifferent to mating competitors. Consequently, they predicted that the incidence of multiple paternity should be high in the breeding groups of Pukekos.

By tracking individually recognizable birds over the course of the seven year study, Craig and Jamieson established that most Pukekos remain to breed in their natal territories, and only males dispersed to breed in other territories. They commonly observed incestuous matings, and inbreeding appeared to be more common than outbreeding (Craig and Jamieson, 1988). Dispersal of subadult males was thought to be linked to parental dominance in which they were prevented from copulating by aggressive behavior of their mothers. Matings occurred between daughters and fathers as well as between sisters and brothers, thus demonstrating that incest is not avoided where the usual system of male dominance prevails.

Because the level of incestuous matings is higher in Pukekos than is predicted from present theories on the evolution of vertebrate mating systems, Craig and Jamieson (1988) questioned whether inbreeding in this species now has deleterious genetic effects. Instead, they noted that a population with a long history of close inbreeding would be purged of recessive lethals, and thus incest would not necessarily be a problem. Additionally, if nonrandom fusion of gametes occurred, this could also minimize the deleterious effects of inbreeding. They point out that when few disadvantages accrue in a population with a long history of inbreeding due to low dispersal, then males should simply attempt to breed with any females irrespective of their genetic relatedness. Craig and Jamieson conclude provocatively that the existing orthodoxy in mating system theory results from the unquestioning assumption by many researchers that the detrimental effects of inbreeding have led to selection for inbreeding avoidance in natural populations.

11.5. North Island Saddlebacks

In 1968 the New Zealand Wildflife Service transferred 29 Saddlebacks from the sole remaining population on Hen Island to nearby Cuvier Island as a conservation measure. P. F. Jenkins realized the potential of this new population for studies of the development of song traditions and their cultural transmission across generations, and two years later he commenced his classic study. Because Saddlebacks are almost flightless, and they have varied repertoires, permanent pair-bonds, life-long territories, and are unafraid of man, they proved ideal subjects for study despite the rugged terrain on Cuvier. Almost all adults and young were made individually recognizable by color-banding, and thus Jenkins was able to study spatial and temporal aspects of groups of birds of known identity in a population not subject to the confounding influence of immigration.

Major findings of the study were not only novel in their detail of cultural transmission in ancestral-descendent lineages, but also were often at variance with the existing orthodoxy on mechanisms of song learning. Spatially discrete song groups composed of up to 20 birds occupying contiguous territories sang the same song patterns, with each group having a distinctively different pattern (Jenkins, 1978). Each song group was centered around one or two original colonizers, and thus song traditions had clearly been transmitted culturally to the island-bred birds in the group. Although Jenkins referred to these spatial clusters of songs as dialects, he also found that where the territories of song groups abutted they sang both traditions. All five sons whose parentage was known happened to settle in different songs groups than their fathers, and learned to sing the song of their neighbors rather than singing their fathers' songs. Birds switched songs when they switched groups and also modified their songs later in life when new neighbors sang different songs, indicating that they do not have a restricted sensitive period. This led him to hypothesize that dialects in Saddlebacks functioned as cultural markers for inbreeding avoidance (Jenkins, 1978).

The inbreeding avoidance hypothesis has recently been tested by an ingenious field experiment in which a new population of Saddlebacks was established on Tiritiri Matangi Island near Auckland. In 1984, 24 birds whose song patterns were known were transferred from Cuvier to Tiritiri. The birds were released in selected remnants of forest to mimic the early pattern of settlement on Cuvier. Each forest patch had different song patterns, but within each patch one to three pairs of birds shared one or more patterns (Craig and Jenkins, unpublished manuscript). Contrary to Jenkins (1978), they found no evidence of inbreeding avoidance in this population. They followed color-banded birds of both sexes, and discovered that females dispersed more than males and settled randomly with respect to their natal dialects. Males did not disperse if territorial vacancies occurred in their natal dialect areas, and close inbreeding was observed in the two instances where relatives were unmated at the same time. Craig and Jenkins suggest that, as in the Pukeko, the opportunity to breed in Saddlebacks may have more impact on fitness than any possible deleterious effects of inbreeding, especially where the turnover of the breeding population is low. Song matching with neighbors is therefore implicated as the cause of local song groups, and philopatry of males to natal dialects would benefit them in acquiring vacant territories there by familarity with local songs.

12. FUTURE PROSPECTS

Given the tremendous awareness and concern for environmental problems and the fragile nature of its endemic avifauna, current emphasis on conservation is to be expected. However, the level of funding from Government and other agencies for ornithological research in New Zealand is woefully inadequate. It should also be pointed out that the quality of applied research in conservation is inextricably interwoven with the fabric of good quality basic research so often neglected by funding agencies. The single most pressing need is for the institution of a competitive grants programs to foster research excellence, with emphasis on areas of national priority such as conservation genetics and the evolution and ecology of island biotas.

New Zealand has an opportunity be a world leader in management and conservation genetics of endemic species, especially birds. Numerous offshore islands provide refuges of various sizes and ecological complexity for preserving genetic diversity in endangered species. There is a clear need to integrate ecological, physiological, and behavioral research with genetic and management studies through a nationally-funded research program. One possible solution to maximize the limited financial and human resources available in New Zealand is to create research teams composed of government, university, museum, and possibly amateur groups, but this would have to be encouraged by program funding rather than by legislative fiat. Finally, the deplorable state of funding for researchers and for the preservation of natural history collections in museums is little short of a national disaster, and should be rectified as soon as possible. Systematics is basic to other disciplines in the life sciences, and the legacy left to New Zealand ornithology by Walter Buller, Robert Falla, and Charles Fleming is fitting testimony to this truism.

ACKNOWLEDGMENTS. I would like to thank Michael Dennison for expert assistance with literature searches, data analysis, and selection of topics, all of which facilitated the preparation of this chapter. For permission to cite their unpublished work as well as for providing reprints of their major papers, I thank John Craig, Charles Daugherty, Alan Hemmings, Ian Jamieson, Peter Jenkins, Jim Mills, Sue Triggs, and Euan Young. Financial support for my own research was provided by grants over the years from the Natural Sciences and Engineering Research Council of Canada, which I gratefully acknowledge.

REFERENCES

Abbott, I., and Grant, P. R., 1976, Nonequilibrial bird faunas on islands, *Am. Natural.* **110**:507–528.

Anderson, A., 1983, Faunal depletion and subsistence change in the early prehistory of southern New Zealand, *Archaeology in Oceania* **18**:1–10.

Anderson, A., 1984, The extinction of moa in southern New Zealand, in: *Quaternary Extinctions* (P. S. Martin and R. G. Klein, eds.), University of Arizona Press, Tuscon, Arizona, pp. 728–740.

Archey, G., 1941, The moa. A study of the Dinornithiformes, *Auck. Inst. Mus. Bull.* **1**:1–145.

Atkinson, I. A. E., 1973, Spread of the ship rat (*Rattus r. rattus* L.) in New Zealand, *J. Roy. Soc. N. Z.* **3**:457–472.

Baker, A. J., 1972, Systematics and affinities of New Zealand oystercatchers, Ph.D. dissertation, University of Canterbury, Christchurch, New Zealand.

Baker, A. J., 1973, Genetics of polymorphism in the variable oystercatcher, (*Haematopus unicolor*), *Notornis* **20**:330–345.

Baker, A. J., 1974, Ecological and behavioural evidence for the systematic status of New Zealand oystercatchers, *Life Sciences Contributions, Royal Ontario Museum* **96**:1–34.

Baker, A. J., 1975, Morphological variation, hybridization and systematics of New Zealand oystercatchers (Charadriiformes: Haematopodidae), *J. Zool., Lond.* **175**:357–390.

Baker, A. J., 1980, Morphometric differentiation in New Zealand populations of the house sparrow (*Passer domesticus*), *Evolution* **34**:638–653.

Baker, A. J., and Moeed, A., 1979, Evolution in the introduced New Zealand populations of the common myna, *Acridotheres tristis, Can. J. Zool.* **57**:570–584.

Baker, A. J., Dennison, M. D., Lynch, A., and Le Grand, G., 1990, Genetic divergence in peripherally isolated populations of chaffinches in the Atlantic islands. *Evolution,* **44**:981–999.

Bartle, J. A., and Sagar, P. M., 1987, Intraspecific variation in the New Zealand bellbird *Anthornis melanura, Notornis* **34**:253–306.

Barton, N. H., and Charlesworth, B., 1984, Genetic revolutions, founder effects and speciation. *Ann. Rev. Ecol. Syst.* **15**:133–164.

Bell, B. D., 1975, The rare and endangered species of the New Zealand region and the policies that exist for their management, *XII Bull. Int. Council Bird Preservation*, pp. 165–172.

Blackburn, A., 1965, Muttonbird islands diary, *Notornis* **12**:191–207.

Bull, P. C., and Whitaker, A. H., 1976, The amphibians, reptiles, birds and mammals, in: *Biogeography and Ecology of New Zealand*, (G. Kuschel, ed.), W. H. Junk, The Hague, pp. 231–276.

Bull, P. C., Gaze, P. D., and Robertson, C. J. R., 1985, *The Atlas of Bird Distribution in New Zealand*, New Zealand Government Printer, Wellington.

Buller, W. L., 1883, On some rare species of New Zealand birds, *Trans. N. Z. Inst.* **16**:308–318.

Buller, W. L., 1905, *Supplement to A History of the Birds of New Zealand*, published by the author, London.

Burrows, C. J., 1980, Some empirical information concerning the diet of moas, *N. Z. J. Ecol.* **3**:125–130.

Burrows, C. J., McCulloch, B., and Trotter, M. M., 1981, The diet of moas based on gizzard contents samples from Pyramid Valley, North Canterbury, and Scaifes Lagoon, Lake Wanaka, Otago, *Rec. Cant. Mus.* **9**:309–336.

Carson, H. L., and Templeton, A. R., 1984, Genetic revolutions in relation to speciation phenomena: The founding of new populations, *Ann. Rev. Ecol. Syst.* **15**:97–131.

Cracraft, J., 1974, Phylogeny and evolution of ratite birds, *Ibis* **116**:494–521.

Cracraft, J., 1976a, The species of moas (Aves, Dinornithidae), *Smithson. Contrib. Paleobiol.* **27**:189–205.

Cracraft, J., 1976b, Covariation patterns in the postcranial skeleton of moas (Aves, Dinornithidae): A factor analytic study, *Paleobiology* **2**:166–173.

Cracraft, J., 1976c, The hindlimb elements of the moas (Aves, Dinornithidae): A multivariate assessment of size and shape, *J. Morph.* **150**:495–526.

Craig, J. L., 1972, Investigation of the mechanism maintaining polymorphism in the New Zealand fantail, *Rhipidura fuliginosa* (Sparrman), *Notornis* **19**:42–55.

Craig, J. L., 1976, An interterritorial hierarchy: An advantage for a subordinate in a communal territory, *Ethology* **42**:200–205.

Craig, J. L., 1977, The behaviour of the pukeko, *Porphyrio porphyrio melanotus*, *N. Z. J. Zool.* **4**:413–433.

Craig, J. L., 1979, Habitat variation in the social organization of a communal gallinule, the pukeko, *Porphyrio porphyrio melanotus*, *Behav. Ecol. Sociol.* **5**:331–358.

Craig, J. L., 1980a, Breeding success of a common gallinule, *Behav. Ecol. Sociobiol.* **6**:289–295.

Craig, J. L., 1980b, Pair and group breeding behaviour of a communal gallinule, the pukeko, *Porphyrio p. melanotus*, *Anim. Behav.* **28**:593–603.

Craig, J. L., and Jamieson, I. G., 1988, Incestuous mating in a communal bird: A family affair. *Am. Natural.* **131**:58–70.

Craw, R. C., 1982, Phylogenetics, areas, geology and the biogeography of Croizat: A radical view, *Syst. Zool.* **31**:304–316.

Crawley, M. C., 1982, Presidential address: Wildlife conservation in New Zealand, *N. Z. J. Ecol.* **5**:1–5.

Croizat, L., 1952, *Manual of Phytogeography*, 587 pp., W. H. Junk, The Hague.

Croizat, L., 1958, *Panbiogeography*, Volumes 1, 2a, 2b, 1721 pp., published by the author, Caracas.

Croizat, L., 1964, *Space, Time, Form: the Biological Synthesis,* 881 pp., published by the author, Caracas.

Dawson, D. G., 1984, Principles of ecological biogeography and criteria for reserve design, *J. Roy. Soc. N. Z.* **14:**11–15.

De Beer, G., 1956, The evolution of ratites, *Bull. Br. Mus. nat. Hist., Zool.* **4:**57–70.

Diamond, J., 1982, Man the exterminator, *Nature* **298:**787–789.

Diamond, J. M., and Veitch, C. R., 1981, Extinctions and introductions in the New Zealand avifauna, *Science* **211:**499–501.

East, R., and Williams, G. R., 1984, Island biogeography and the conservation of New Zealand's indigenous forest-dwelling avifauna, *N. Z. J. Ecol.* **7:**27–35.

Emlen, S. T., 1982, The evolution of helping. 1. An ecological constraints model, *Am. Natural.* **119:**29–39.

Falla, R. A., 1941, Preliminary report on excavations at the Pyramid Valley swamp, Waikari, North Canterbury: the avian remains, *Rec. Cant. Mus.* **4:**339–353.

Falla, R. A., 1953, The Australian element in the avifauna of New Zealand, *Emu* **53:**36–46.

Falla, R. A., Sibson, R. B., and Turbott, E. G., 1982, *The New Guide to the Birds of New Zealand,* Collins, Auckland.

Flack, J. A. D., 1975, The Chatham Island black robin, extinction or survival?, *XII Bull. Int. Council Bird Preservation,* pp. 146–150.

Fleming, C. A., 1939, Birds of the Chatham Islands, Part II, *Emu* **38:**492–509.

Fleming, C. A., 1950a, New Zealand flycatchers of the genus *Petroica* Swainson. Part I, *Trans. Roy. Soc. New Zealand* **78:**14–47.

Fleming, C. A., 1950b, New Zealand flycatchers of the genus *Petroica* Swainson (Aves). Part II, *Trans. Roy. Soc. New Zealand* **78:**127–160.

Fleming, C. A., 1962a, New Zealand biogeography—a palaeontologist's approach, *Tuatara* **10:**53–108.

Fleming, C. A., 1962b, History of the New Zealand land bird fauna, *Notornis* **9:**270–274.

Fleming, C. A., 1962c, The extinction of moas and other animals during the Holocene period, *Notornis* **10:**113–117.

Fleming, C. A., 1975, The geological history of New Zealand and its biota, in: *Biogeography and Ecology in New Zealand* (G. Kuschel, ed.), W. H. Junk, The Hague, pp. 1–86.

Freed, L. A., Conant, S., and Fleischer, R. C., 1987, Evolutionary ecology and radiation of Hawaiian passerine birds, *Trends in Ecology and Evolution* **2:**196–203.

Gould, S. J., 1980, Is a new and general theory of evolution emerging? *Paleobiology* **6:**119–130.

Grant, P. R., 1965, The adaptive significance of some size trends in island birds, *Evolution* **19:**355–367.

Grant, P. R., 1986, *Ecology and Evolution of Darwin's Finches,* Princeton University Press, Princeton, New Jersey.

Gregg, D. R., 1972, Holocene stratigraphy and moas at Pyramid Valley, North Canterbury, New Zealand, *Rec. Cant. Mus.* **9:**151–158.

Gurr, L., 1967, Interbreeding of *Larus novaehollandiae scopulinus* and *Larus bulleri* in the wild in New Zealand, *Ibis* **109:**552–555.

Guthrie-Smith, H., 1925, *Bird life on Island and Shore,* 195 pp., W. Blackwood & Sons, Edinburgh.

Harper, P. C., and Kinsky, F. C., 1978, *Southern Albatrosses and Petrels: An Identification Guide,* 80 pp. Victoria University Press, Wellington.

Hayes, F. N., and Williams, M., 1982, The status, aviculture and restablishment of brown teal in New Zealand, *Wildfowl,* **33:**73–80.

Hemmings, A. D., 1989, Communally breeding skuas: Breeding success of pairs, trios, and groups of *Catharacta skua lonnebergi* on the Chatham Islands, New Zealand, *J. Zool., Lond.* **218**:393–405.
Houde, P., 1986, Ostrich ancestors found in the northern hemisphere suggest new hypothesis of ratite origins, *Nature* **324**:563–565.
Jamieson, I. G., and Craig, J. L., 1985, Dominance and mating in a communal polygynandrous bird: Cooperation or indifference towards mating competitors? *Ethology* 317–327.
Jenkins, P. F., 1978, Cultural transmission of song patterns and dialect development in a free-living bird population, *Anim. Behav.* **26**:50–78.
Johnsgard, P., 1965, *Handbook of Waterfowl Behavior*, 378 pp., Cornell University Press, Ithaca, New York.
Johnson, N. K., Marten, J. A., and Ralph, C. J., 1989, Genetic evidence for the origin and relationships of the Hawaiian honeycreepers (Aves: Fringillidae), *Condor* **91**:379–396.
Karl, B. J., and Best, H. A., 1982, Feral cats on Stewart Island; their foods and their effects on kakapo, *N. Z. J. Zool.* **9**:287–294.
Kear, J., and Scarlett, R. J., 1970, The Auckland Islands merganser, *Wildfowl* **21**:78–86.
Kinsky, F. C., 1970, *Annotated Checklist of the Birds of New Zealand*, 96 pp., A. H. and A. W. Reed, Wellington.
Lack, D., 1947, *Darwin's Finches*, 204 pp., Cambridge University Press, Cambridge.
Lack, D., 1966, *Population Studies of Birds*, 341 pp., Clarendon Press, Oxford.
Latham, J., 1785, A General synopsis of Birds, Vols. 1–6, 2224 pp., White, London.
Lockley, R. M., 1980, *New Zealand Endangered Species*, 150 pp., Cassell, Auckland.
Long, J. L., 1981, *Introduced Birds of the World*, 528 pp., Reed, Sydney, Australia.
McDowell, R. M., 1978, Generalized tracks and dispersal in biogeography, *Syst. Zool.*, **27**:88–104.
Meredith, M. A. M., and Sin, F. Y. T., 1988, Genetic variation of four populations of little blue penguins, *Eudyptula minor*, *Heredity* **60**:69–76.
Merton, D. V., 1975, The saddleback: Its status and conservation, in: *Breeding Endangered Species in Captivity* (R. D. Martin, ed.), Academic Press, London.
Millener, P. R., 1981, The Quaternary avifauna of the North Island, New Zealand, Ph.D. dissertation, University of Auckland.
Millener, P. R., 1982, And then there were twelve: The taxonomic status of *Anomalopteryx oweni* (Aves: Dinornithidae), *Notornis* **29**:165–170.
Mills, J. A., 1971, Sexing red-billed gulls from standard measurements, *N. Z. Jl. mar. Freshwat. Res.* **5**:326–328.
Mills, J. A., 1972, A difference in band loss from male and female red-billed gulls *Larus novaehollandiae scopulinus*, *Ibis* **114**:252–255.
Mills, J. A., 1973, The influence of age and pair-bond on the breeding biology of the red-billed gull *Larus novaehollandiae scopulinus*, *J. Anim. Ecol.* **42**:147–162.
Mills, J. A., 1979, Factors affecting the egg size of red-billed gulls *Larus novaehollandiae scopulinus*, *Ibis* **121**:53–67.
Mills, J. A., 1990, Red-billed Gull, in: *Lifetime Reproduction in Birds* (I. Newton, ed.), Academic Press, London, pp. 387–404.
Mills, J. A., and Williams, G. R., 1979, The status of endangered New Zealand birds, in: *The Status of Endangered Australasian Wildlife*, (M. J. Tyler, ed.), Royal Zoological Society of South Australia, pp. 147–168.
Mills, J. A., Lavers, R. B., and Lee, W. G., 1984, The takahe—a relict of the Pleistocene grassland avifauna of New Zealand, *N. Z. J. Ecol.* **7**:57–70.

Newton, I., 1985, Lifetime reproductive output of female sparrowhawks, *J. Anim. Ecol.* **54**:241–253.

Nur, N., 1986, Alternative reproductive tactics in birds: Individual variation in clutch size, in: *Perspectives in Ethology,* Volume 7 (P. P. G. Bateson and P. H. Klopfer, eds.), Plenum Press, New York, pp. 49–77.

Oliver, W. R. B., 1949, The moas of New Zealand and Australia, *Dom. Mus. Bull.* **15**:1–206.

Oliver, W. R. B., 1955, *The Birds of New Zealand,* 2nd Edition, 661 pp., A. H. and A. W. Reed, Wellington.

Olson, S. L., 1984, The relationships of the extinct Chatham Island Eagle, *Notornis* **31**:273–277.

Pierce, R. J., 1984, Plumage, morphology, and hybridisation of New Zealand stilts *Himantopus* spp., *Notornis* **31**:106–130.

Pycraft, W. P., 1900, On the morphology and phylogeny of the Palaeognathae (Ratitae and Crypturi) and Neognathae (Carinatae), *Trans. zool. Soc., Lond.* **15**:149–290.

Rich, P. V., and Scarlett, R. J., 1977, Another look at *Megaegotheles,* a large owlet-nightjar from New Zealand, *Emu* **77**:1–8.

Richdale, L. E., 1957, *A Population Study of Penguins,* Clarendon Press, Oxford.

Richdale, L. E., 1965, Biology of the birds of Whero Island, New Zealand, with special reference to the diving petrel and the white-faced storm petrel, *Trans. Zool. Soc. Lond.* **27**:4–86.

Ripley, S. D., 1977, *Rails of the World: a Monograph of the Family Rallidae,* M. F. Fehely, Toronto.

Rogers, J. S., 1972, Measures of genetic similarity and genetic distance, University of Texas Publication **7213**:145–153.

Selander, R. K., 1966, Sexual dimorphism and differential niche utilization in birds, *Condor* **68**:113–151.

Sibley, C. G., and Ahlquist, J. E., 1981, The phylogeny and relationships of the ratite birds as indicated by DNA–DNA hybridization, in: *Evolution Today* (G. G. E. Scudder and J. L. Reveal, eds.), *Proc. 2nd Int. Congr. Syst. Evol. Biol.,* pp. 301–335.

Sibley, C. G., Williams, G. R., and Ahlquist, J. E., 1982, The relationships of the New Zealand wrens (Acanthisittidae) as indicated by DNA–DNA hybridization, *Notornis* **29**:113–130.

Smith, G. A., 1975, Systematics of parrots, *Ibis* **117**:18–66.

Stead, E. F., 1932, *Life Histories of New Zealand Birds,* 162 pp., Search, London.

Thomson, G. M., 1922, *The Naturalisation of Animals and Plants in New Zealand,* Cambridge University Press, Cambridge.

Triggs, S. J., Powlesland, R. G., and Daugherty, C. H., 1989, Genetic variation and conservation of kakapo (*Strigops habroptilus:* Psittaciformes), *Conserv. Biol.* **3**:92–96.

Trotter, M. M., and McCulloch, B., 1984, Moas, men, and middens, in: *Quaternary Extinctions* (P. S. Martin and R. G. Klein, eds.), University of Arizona Press, Tuscon, Arizona, pp. 708–727.

Van Wagner, C. E., and Baker, A. J., 1990, Association between mitochondrial DNA and morphological evolution in Canada geese, *J. Mol. Evol,* **31**:373–382.

Williams, G. R., 1953, The dispersal from New Zealand and Australia of some introduced European passerines, *Ibis* **95**:676–692.

Williams, G. R., 1956, The kakapo (*Strigops habroptilus* Gray): A review and reappraisal of a near extinct species, *Notornis* **10**:15–32.

Williams, G. R., 1960, The takahe, a general survey, *Trans. Roy. Soc. N. Z.* **88**:235–258.

Williams, G. R., 1962, Extinction and the land and freshwater inhabiting birds of New Zealand, *Notornis* **10**:15–32.

Williams, G. R., 1973, Birds, in: *The Natural History of New Zealand* (G. R. Williams, ed.), A. H. and A. W. Reed, Wellington, pp. 304–333.

Williams, G. R., 1976, In the flightless tradition: New Zealand, *Aust. Nat. Hist.* **18**:36–39.

Williams, G. R., 1977, Marooning—a technique for saving threatened species from extinction, *Int. Zoo. Yrbk* **17**:102–106.

Williams, G. R., and Harrison, M., 1972, The laughing owl *Sceloglaux albifacies* (Gray, 1844). A general survey of a near extinct species, *Notornis* **19**:4–19.

Worthy, T. H., 1988, A re-examination of the moa genus *Megalapteryx*, *Notornis* **35**:99–108.

Young, E. C., Jenkins, P. F., Douglas, M. E., and Lovegrove, T. G., 1988, Nocturnal foraging by Chatham Island skuas, *N. Z. J. Zool.* **11**:113–117.

CHAPTER 2

AVIAN RADIOECOLOGY

I. LEHR BRISBIN, JR.

1. INTRODUCTION

Most of the public's awareness and concern for the environmental effects of the "atomic age" can be attributed to two events separated by over 40 years of time: the detonation of the first nuclear weapons over Hiroshima and Nagasaki in August 1945, and the disastrous explosion and fire at the Chernobyl nuclear power station in the Soviet Union in April 1986. During the intervening four decades, there developed within the scientific community a multifaceted subdiscipline known as "radioecology" (Odum, 1965). As heralded by its birth in an era of nuclear weaponry and fear of the consequences of global thermonuclear war, much of the early work of radioecology was directed at studies of radiation effects and emphasized environmental and biological aspects of various post-nuclear attack scenarios. Various attempts at nuclear test-ban treaties and weapons reduction agreements have not convinced everyone that our civilization is now safe from such disasters. There has nevertheless been a growing tendency to diminish the emphasis on radiation effects studies in favor of greater attention to the long-term fate and effects of radionuclides that have been released into the environment. These releases have occurred either as a result of nuclear weapons testing or as a result of inadvertent releases from various nuclear industrial activities. Concerns for birds and the ecosystems which they inhabit have fre-

I. LEHR BRISBIN, JR. • Savannah River Ecology Laboratory, Aiken, South Carolina 29801.

quently been either implied or directly expressed by research in these areas.

Closely related to these considerations has been the use of radioactive substances as natural *in situ* "tracers" to determine the structure and function of natural food web pathways (e.g., Odum, 1971) and/or to determine the patterns and rates at which functional processes take place within natural ecosystems (Wiegert *et al.*, 1967). In such tracer studies, whether they are conducted within an individual organism's body (e.g., using radioactive iodine to study thyroid function) or at the level of population, community or ecosystem, advantage is made of the fact that radioactive elements behave both chemically and biologically like the isotopically stable counterparts of the same elements and/or their chemical analogs. Thus, for example, strontium 90 (^{90}Sr) may be used to study the cycling of calcium and its deposition in bone tissue.

Even beyond inferences concerning stable isotopes of the same element, however, radioactive tracer studies may also be used to gain better understanding of basic processes at a variety of levels of biological organization. As will be described later, for example, the gamma-emitting radioisotope cesium 137 (^{137}Cs) can be used in a variety of ways to determine the length of time that a given individual bird has resided in a naturally contaminated habitat and can thereby be used to determine its migratory status (e.g., long-term resident vs. transient visitor) (Cadwell *et al.*, 1979; Potter *et al.*, 1989). Such information can then be used, in turn, to study various aspects of the cycling of other contaminants, including those which are neither chemically nor biologically similar to the original radioactive tracer used. Here, the emphasis is directed at the basic biological process (i.e., migration) which influences the cycling of both radioactive and nonradioactive contaminants alike. Studies of radioactive isotopes have been used, for example, to better understand seasonal changes in levels of heavy metals such as mercury in waterfowl (Clay *et al.*, 1980). Studies of radioactive contaminants have also helped to provide a better understanding of both local movements as well as longer distance migrations of birds (Fig. 1; Hanson and Case, 1963; Glover, 1967) while basic studies using bird-banding techniques have, in turn, helped to provide a better understanding of the environmental fate of released radionuclides (Brisbin and Swinebroad, 1975). Thus, as predicted almost prophetically by Odum (1965), the field of radioecology has developed a two-way feedback exchange with the field of ecology itself; information gained through the study of basic ecological and biological processes helps to better understand the cycling, fate, and effects of both natural and introduced radioactive substances, and, by way of feedback, studies of the radioactive substances (e.g., tracer kinet-

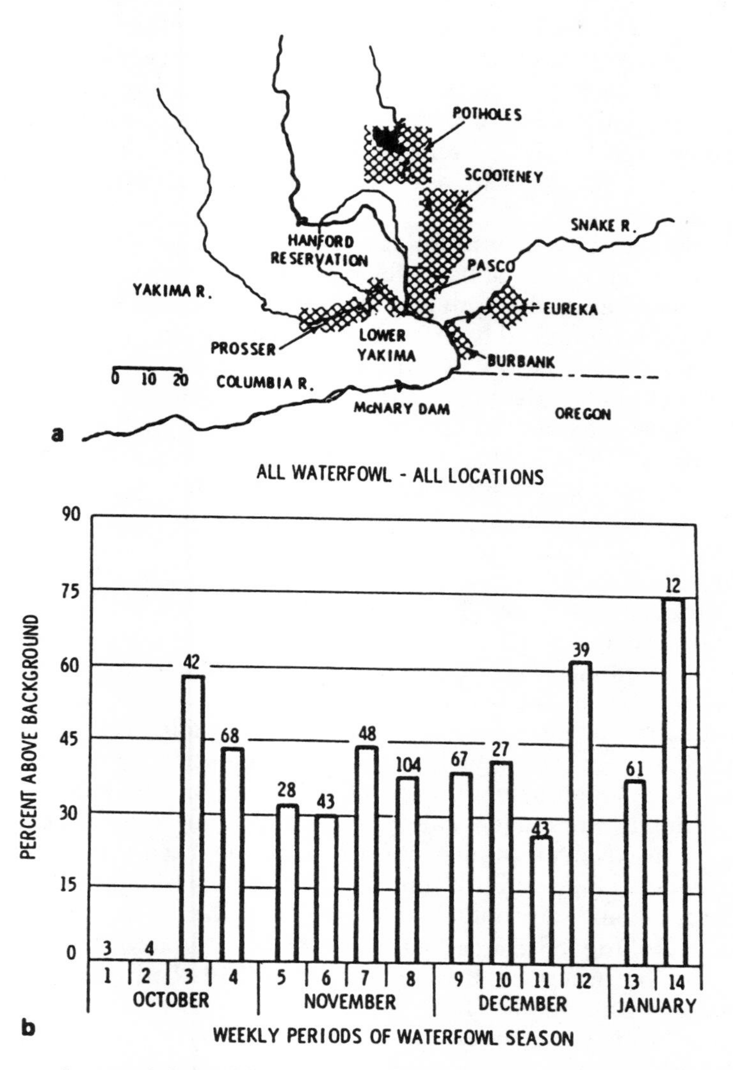

FIGURE 1. The use of elevated levels of the radioisotopes ^{32}P and ^{65}Zn to demonstrate the dispersal of waterfowl from the environs of the U.S. Department of Energy Hanford Reservation, state of Washington, to major offsite sport hunting areas, as shown by the cross-hatched areas in the upper figure (a). The lower figure (b) indicates the percent of all samples of heads of shot ducks which contained elevated (greater than background) levels of ^{32}P and ^{65}Zn. From Hanson and Case (1963).

ics) have helped to provide basic biological/ecological information that would otherwise have been either difficult or impossible to obtain by other means. This two-way exchange between basic ecology and radioecology has been evident throughout the development of each of the three major subdivisions of radioecology: (1) studies of radiation effects, (2) studies of the cycling, fate, and effects of radioactive contaminants, and (3) the use of radioisotopes as tracers to study natural biological and ecological processes.

While the field of ornithology has never developed a definitive subdivision related to radioecology *per se*, the three major areas of research listed above have all had historical impacts on various aspects of studies of birds and all have, in turn, been impacted by them (Schultz, 1974). As an example, the use of birds as "early warning" sentinels of environmental degradation, starting with the concept of the "canary in the mines" and extending through the publication of Rachel Carson's now famous *Silent Spring* (Carson, 1962), has seen birds and avian research figure prominently as long-term indicators of environmental change and/or degradation (Temple and Wiens, 1989). Currently, concerns for the potential impact of the recent Chernobyl accident upon bird populations (e.g., Desante and Geupel, 1987; Baeza *et al.*, 1988), together with concerns for the ways in which migratory waterfowl and game bird populations may come to serve as mechanisms for the global transport of radioactive contaminants released by that event (Anon., 1986), are examples of how current ornithology has now become involved with this two-way exchange between radioecology and basic ecological research.

This chapter is designed to first provide a basic overview of some of the fundamental principles of radioecology and the subdisciplines upon which it is based (e.g., nuclear chemistry, nuclear physics). Then, using these basic principles, the two-way feedback relationship between radioecology and ornithological research will be traced through the historical and conceptual development of the major areas outlined above. In linking these areas, certain "keystone" studies from the literature will be described in detail together with both published and unpublished information from the author's own research. Those studies cited should be considered as examples that demonstrate basic principles that, in some cases, may have been expanded upon (or in some cases refuted) by later work.

2. SOME BASIC PRINCIPLES OF IONIZING RADIATION

Like other forms of radiation (e.g., light or heat), ionizing radiation involves the propagation of energy through space. The ionizing radiations are those which have sufficiently high energy to cause the formation of ions by the ejection of electrons when they interact with matter. While the basic principles of ionizing radiation and its measurement may be found in general physics texts, and an excellent summary of those principles of greatest concern to radioecological research has been provided by Whicker and Schultz (1982), a brief overview and summary will be provided here.

2.1. Types of Ionizing Radiation

Three types of ionizing radiation are particularly important with respect to avian radioecology. Two of these radiations (alpha and beta) are particulate in nature while the third (gamma and X-rays) comprise electromagnetic waves. The *alpha particle* is essentially a helium nucleus without orbital electrons. It is emitted by radionuclides with high atomic numbers—particularly those in the so-called "transuranic" or "actinide" series (e.g., uranium, thorium, radium, neptunium, and plutonium). Alpha particles interact strongly with matter and therefore have the ability to cause a great deal of biological damage. However, most alpha particles travel only a few centimeters in air and a few millimeters in most biological media. Several millimeters of plastic, paper, or wood will thus suffice to shield against most alpha radiation. The other important particulate form of ionizing radiation, the *beta particle,* consists of spontaneously ejected electrons that have very little or negligible mass. Beta particles have greater penetrating power than alpha particles, and relatively high biological damage may result from exposure, but again only over a relatively short distance as compared with the more penetrating electromagnetic wave radiations.

Gamma and X-rays, the two important forms of electromagnetic wave radiations, may be considered together because they have similar properties and effects and generally behave identically in their interaction with biological materials. The two differ only in the atomic substructure from which they originate. While X-rays are commonly used as laboratory sources of radiation stress they are not generally of concern in most natural situations. It is rather the gamma radiation emitted from a variety of isotopes which are generally of primary concern with respect to environmental effects of radioactive emissions and radioactive contaminants. Gamma and X-rays have very short wavelengths but otherwise may be considered similar in many ways to other components of the electromagnetic wavelength spectrum (e.g., heat, light, or radio waves). That is, they have no mass and are propagated through space at the velocity of light. The higher energy of the gamma and X-rays however, as compared with the other wavelengths, results in their causing more extensive damage when interacting with cells and tissues.

The ability of these electromagnetic waves of ionizing radiation to penetrate great distances through biological materials and even denser substances is well known. The need to use lead-shielding against X-radiation, for example, is a familiar part of medical practice. Similarly, extensive shielding using thick layers of dense material such as concrete, soil, or metal (particularly lead) is required to substantially reduce the energy of emitted gamma rays that could otherwise travel great distances through air, attenuating their energy at a rate directly proportional to the electron density of the material through which they are passing and inversely proportional to the square of the distance traveled from their source.

In terms of risk of biological damage from radiation exposure, the hazards posed by emitted gamma radiation are generally of the most immediate concern, particularly in the case of exposure in the immediate vicinity of a weapons detonation or nuclear accident. Risk of damage from alpha and/or beta radiation, on the other hand, would be minimal from sources external to the body and would rather require actual uptake (e.g., ingestion and/or inhalation) of the alpha- and/or beta-emitting isotopes themselves.

2.2. Some Important Radionuclides and Their Characteristics

Although there are many radioactive isotopes, only a few account for the majority of concern in radioecological study and environmental assessments. Some of the most important of these and their characteristics are listed in Table I. The importance of these particular isotopes derives from any combination of three factors: (1) the predominance and/or ubiquity of the particular isotope as a component of fallout or yield from weapons explosions and nuclear accidents, (2) the tendency of the isotope to persist for long periods of time in biota and the environment, and (3) the tendency of the isotope to be rapidly and efficiently taken up and transported through food chains, and/or to become selectively concentrated within particularly important or vulnerable portions of an organism's body. The importance of ^{137}Cs is particularly noteworthy with regard to all three of the criteria presented above. This isotope is one of the most ubiquitous of all of those found in fallout from nuclear weapons testing and nuclear industrial accidents (e.g., Anspaugh *et al.*, 1988). Its physical half-life of 30 years (Table I) assures that it will persist in the environment for long periods of time, and, finally, as an analog of potassium, it is selectively concen-

TABLE I
Some General Properties of Radionuclides That Are of Potential Importance as Environmental Contaminants of Birds[a]

Radionuclide	Physical half-life	Biological retention	Principle form of emission	Mode of exposure[b]	Vertebrate target organ	Chemical analogs
^{3}H (tritium)	12 years	Low (days)	Beta	Ing, Abs, Ih	Total body	H
^{131}I	8 days	Moderate (days–weeks)	Beta	Ing, Abs, Ih	Thyroid	I
^{137}Cs	30 years	Moderate (weeks)	Gamma	Ing, Abs, Ext	Whole body, muscle	K
^{45}Ca	160 days	Moderate–high (months–years)	Beta	Ing, Abs	Bone	Ca
^{90}Sr	28 years	High (years)	Beta	Ing, Abs	Bone	Ca
^{60}Co	5.2 years	Moderate (weeks)	Gamma	Ing, Abs, Ih, Ext	GI, lung, whole body	Co
^{65}Zn	245 days	Moderate–high (months–years)	Gamma	Ing, Abs, Ih, Ext	Liver, lung, whole body	Zn
^{239}Pu	24,000 years	High (years)	Alpha	Ing, Ih, Ads	Bone, lung	None

[a]Adapted from Brisbin (1989), as adapted from Whicker and Schultz (1982).
[b]Ing = ingestion, Abs = absorption, Ads = adsorption, Ih = inhalation, Ext = external gamma exposure.

trated in the edible skeletal muscle of game species that are frequently consumed by humans as food (e.g., Narayanyan and Eapen, 1971; Brisbin and Smith, 1975; Potter *et al.*, 1989).

2.3. Radionuclide Cycling and Turnover

The *physical half-life* of an isotope (Table I) represents the amount of time required before one-half of the original quantity of the isotope disappears through the process of normal radioactive decay. Half-lives may vary from a few minutes or days to as much as thousands of years. As a general rule, the amount of time that a quantity of any isotope may persist in detectable quantities may be estimated by multiplying its half-life × 5. Thus for example, once deposited in some stable compartment of an ecosystem, a release of ^{137}Cs (half-life = 30 years) should persist at detectable levels for approximately 30 × 5 = 150 years.

In reality, however, the biological materials (e.g., soil, water, body tissues) into which radioisotopes become incorporated are usually not chemically stable in their own right. Rather, the chemical constituents of protoplasm and various ecosystem compartments are nearly always in a state of dynamic flux or turnover and these processes tend to increase the rates at which radioactive substances (or other contaminants for that matter) are moved within and eliminated from the body or ecosystem. This, then, results in a *biological half-life* which is the amount of time required for half of the quantity of isotope taken up by an organism to be eliminated from its body. In some cases, especially when deposited in tissues with low turnover rates such as bone, the biological half-life of an isotope, corrected for physical decay, may be greater than its physical half-life alone (e.g., calcium 45). In most other cases, however, the opposite is true.

Biological turnover rates, which may vary greatly even from one tissue to another within a given organism, are usually proportional to metabolic rates and are therefore often influenced by factors influencing metabolic rate. All else being equal, small organisms with high surface/volume ratios will tend to have higher rates of isotopic turnover than closely related organisms with larger body sizes (Reichle *et al.*, 1970; Table II). Similarly, all else being equal, homeotherms, with their generally higher metabolic rates, will tend to have shorter biological half-lives for various isotopes than will poikilotherms of the same body size. The biological half-life of ^{137}Cs in a 460 g Wood Duck (*Aix sponsa*) for example is only about five to eight days (Fendley *et al.*, 1977), while the biological half-life of this same isotope in a 430 g rat snake (*Elaphe obsoleta*) has been measured as 902 days (Staton *et al.*, 1974). Applying

TABLE II
Representative Biological Half-Life Values for Several Radioisotopes in a Variety of Avian Species[a]

Isotope	Species	Tissue/compartment	Biological half–life	Reference
^{137}Cs	Wood Duck (*Aix sponsa*)	Whole-body	5.6 days	Fendley *et al.* (1977)
	American Coot (*Fulica americana*)	Whole-body	7.2 days	Potter (1987)
^{65}Zn	Domestic Chicken (*Gallus domesticus*)			
	(1 day old)	Whole-body	15.0 days[b]	Suso and Edwards (1968)
	(18 weeks old)	Whole-body	43.1 days[b]	Suso and Edwards (1968)
^{137}Cs	Mallard (*Anas platyrhynchos*)	Whole-body	11 days	Halford *et al.* (1983)
^{60}Co	Mallard (*Anas platyrhynchos*)	Whole-body	67 days	Halford *et al.* (1983)
^{65}Zn	Mallard (*Anas platyrhynchos*)	Whole-body	67 days	Halford *et al.* (1983)
^{131}I	Mallard (*Anas platyrhynchos*)	Whole-body	10 days	Halford *et al.* (1983)
^{137}Cs	Northern Bobwhite (*Colinus virginianus*)	Whole-body	11 days[c]	Anderson *et al.* (1976)
^{60}Co	Northern Bobwhite (*Colinus virginianus*)	Whole-body	13 days[c]	Anderson *et al.* (1976)

[a]Values given are for the "long-component" elimination of isotopes which have actually been incorporated into the compartment designated. "Short-term" components representing the elimination of isotope from gut contents would be more rapid (see text, and Halford *et al.*, 1983).
[b]Following intramuscular injections. Half-lives varied when the isotope was given by oral dose.
[c]Following chronic feeding with the isotope for 21 days. Biological half-lives following feedings producing acute (4 hr) exposures were more rapid, due to the predominance of gut elimination (see footnote *a*).

the criterium of five half-lives to reach background levels of contamination, the snake would require over a decade to eliminate a given body burden of this isotope, while a Wood Duck, contaminated to the same level, would decline to background in less than two months. As a result of these factors, radionuclides in birds, particularly the smaller passerines, tend to have relatively shorter biological half-lives than other vertebrates, primarily as a result of their relatively small body sizes and high homeothermic metabolic rates. A summary of some representative published values for biological half-lives of several important isotopes in a variety of avian species is presented in Table II.

The rate of disappearance of a radioactive isotope from an animal's body is the result of the sum of its disappearance due to physical decay plus its loss due to biological elimination, and the sum of these two processes is the *effective half-life* of the isotope in that organism. In those cases of the many isotopes with relatively long physical half-lives, the effective half-life will essentially equal the biological half-life because the contribution of physical isotopic decay will be negligible.

Finally, radioactive isotopes released into the environment may also show an *ecological half-life* that represents the amount of time required for a given level of isotope, once established and at equilibrium within a given ecosystem component, to decrease by 50%, as a result of the isotope being either physically removed from or rendered biologically unavailable within the system (e.g., as a result of downstream movement in a watershed or downward migration of the isotope into the soil or sediment profile) (Bagshaw and Brisbin, 1984; Brisbin *et al.*, 1989a). Again, the physical half-life may or may not be greater than the realized ecological half-life of the isotope in question. Biological turnover rates however are usually not of direct relevance in this context since ecological half-lives are based on isotope levels that have reached a state of biological equilibrium. Although ecological half-lives usually tend to be much longer than biological half-lives, only a few estimates of such ecological turnover rates are available for birds (Table III; see also Brisbin and Vargo, 1982).

The values presented in Table III demonstrate the importance of environmental characteristics specific to given habitats in determining ecological half-lives, as opposed to the characteristics of the organisms themselves. For example, there is only a twofold difference between the ecological half-life of ^{137}Cs in the Wood Duck and rat snake when both occupy the same riverine swamp habitat. However, the biological (physiological) half-lives of ^{137}Cs in these organisms differ by over 150-fold! Furthermore, although the biological half-lives of ^{137}Cs in Wood Ducks and American Coots are quite similar (5.6 vs. 7.2 days, respec-

TABLE III
Comparisons of Physical, Biological and Ecological Half-Lives of ^{137}Cs in Wood Ducks, American Coots and Rat Snakes Residing in Habitats Contaminated by Releases from Nuclear Reactors

Species (habitat)	Physical half-life of ^{137}Cs	Biological half-life	Ecological half-life
Wood Ducks (*Aix sponsa*) (Bottomland River Swamp)	30 years	5.6 days[a]	1.9 years[b]
American Coots (*Fulica americana*) (Reactor Cooling Reservoir)	30 years	7.2 days[c]	>20 years[d]
Rat Snakes (*Elaphe obsoleta*) (Bottomland River Swamp)	30 years	902 days[e]	3.8 years[f]

[a]Fendley *et al.* (1977).
[b]Fendley (1978).
[c]Potter (1987).
[d]Potter (1987), based on extrapolation of a documented decline of 75% in 10 years.
[e]Staton *et al.* (1974).
[f]Bagshaw and Brisbin (1984).

tively) their ecological half-lives can differ by more than 10-fold when they occupy different habitats (Table III). These differences may be explained by the greater biogeochemical stability of the reservoir habitat in which little if any radioisotope loss occurs through downstream flow, as documented by Whicker *et al.* (1989). In the riverine swamp habitat, however, ^{137}Cs is continually lost from the system through downstream flow as well as through the covering action of uncontaminated sediments which are brought in from upstream (Brisbin *et al.*, 1989a). Thus, all else being equal, the equilibrium levels of ^{137}Cs should decrease more slowly in the avifauna and other biota in the reservoir habitat than in the swamp. The implications of these factors for the management of waterfowl and other species in such areas will be discussed in Section 4.

2.4. Radionuclide Uptake and Concentration

One of the most important aspects of avian radioecology concerns the process by which birds take up radioactive contaminants from the environment and accumulate them in their bodies. This is particularly true in the case of radionuclide contamination of skeletal muscle and eggs of species which may be eaten by man (e.g., domestic poultry or game birds). Radionuclide uptake through the food chain has been conventionally described in terms of the "classic" model of Davis and Foster (1958):

$$C_t = C_e(1 - e^{-kt}) \tag{1}$$

where C_t is the radioisotope concentration in the bird's body at time t, C_e is the contaminant concentration in the bird's body after the attainment of equilibrium, and k is the uptake rate constant. As explained by Whicker and Schultz (1982), the uptake rate constant may be calculated from the biological half-life (BHL) as:

$$k = \frac{\ln 2}{BHL} \tag{2}$$

The above model describes radionuclide uptake which follows a monotonically-increasing negative exponential curve, as depicted for example in Fendley *et al.* (1977).

Recently, a revised model for radionuclide uptake was proposed by Brisbin *et al.* (1989b) and has been applied by these investigators to waterfowl populations. This revised model is based on the Richards sigmoid growth model (Richards, 1959). Brisbin *et al.* (1989b) describe radionuclide uptake as:

$$C_t = \left[C_e^{(1-m)} - \left(C_e^{(1-m)} - C_o^{(1-m)}\right) \exp\left(\frac{-2t}{T}(m + 1)\right)\right]^{\frac{1}{1-m}} \tag{3}$$

where C_t, C_e, and t are as defined in Equation (1), C_o is the concentration of the radioisotope in the bird's body at the beginning of the uptake period (i.e., when $t = 0$), and T is the amount of time required for C_t to reach 90—95% of C_e. T is calculated as the inverse of the proportional weighted mean contaminant uptake rate, as based on a reparameterization of Richards (1959) model by Brisbin *et al.* (1986). Finally, m represents the Richards curve shape parameter (Richards, 1959). Varying the numerical value of this parameter changes the shape of the sigmoid uptake curve as explained by Brisbin *et al.* (1989b). Briefly, changes in the value of m change the time (and ratio of C_t/C_e) at which the curve attains its point of inflection (i.e., dC/dt is maximized). When $m = 0$, the Richards model becomes the classic model as defined in Equation (1). In relating Equation (1) to Equation (3) with $m = 0$, the parameter k is transformed to T as described by Brisbin *et al.* (1986), using the equation:

$$T = \frac{2(m + 1)}{k} \tag{4}$$

As will be explained later, the exact shape of the uptake curve (as defined by the value of m) may have important implications for predicting the degree to which migratory birds may accumulate and transport radioactive contaminants released by a nuclear accident (Brisbin *et al.*, 1989b).

2.5. Detection and Measurement of Radiation

The methods and equipment needed for the detection and measurement of ionizing radiation vary primarily as a function of the characteristics of the three types of radiation described in Section 2.1. Because the alpha and beta forms of radiation can penetrate biological materials for only short distances, their emissions are often effectively shielded from outside detection when they are emitted from sources deep within the body. One of the beta emitters of greatest biological importance for example is ^{90}Sr which, as an analog to Ca, is selectively taken up in bone tissue. However, with the exception of turtles which carry much of their bony tissue externally, beta emissions from ^{90}Sr within most vertebrate bone is effectively shielded from outside detection by surrounding body tissues. Thus, direct detection and quantification of such an isotope would not be possible in a living bird and would rather require the sacrifice of the individual, followed by the removal of bone tissue which would then have to be subjected to extensive (and usually costly) chemical preparations prior to beta-counting.

Gamma and X-radiations, because of their ability to more deeply penetrate biological tissues, can often be directly detected and measured, even when emitted from deep within the bodies of living organisms. It is therefore usually possible to directly quantify the amounts of such gamma-emitting isotopes (e.g., ^{137}Cs) without the sacrifice or dissection of the animal in question. Such live whole-body gamma counting procedures can in fact provide a very powerful and effective tool to conduct time-series analyses of isotope uptake, concentration and/or elimination within individual organisms, with only a minimum amount of disturbance to the subjects involved. This can in turn not only provide information concerning the cycling dynamics and environmental behavior of the isotopes themselves, but it can also provide important information concerning the basic biology of the individuals being studied, as described earlier and as will be indicated in some of the examples to follow.

It is beyond the scope of this chapter to provide detailed descriptions of the equipment and procedures used for radioisotope counting. Excellent reviews of this subject have been provided by Schultz and Whicker (1982) and Whicker and Schultz (1982). It is important to note, however, that in order for the counts obtained from isotopes in a bird's body or tissue sample to be statistically distinguishable from background, the amount of counting time must be increased in inverse proportion to the total amount of isotope in the subject. Thus, very small birds or those with low levels of contamination must be counted longer than those of larger body sizes or higher contamination levels. Because living subjects must be restrained to some degree during gamma counting procedures, live whole-body counting may not be practical if the level of isotopic concentration is too low to be statistically distinguishable from background radiation without restraining the organism for undue amounts of time. One approach to this problem has been to design more sensitive counting equipment which, by virtue of increased shielding, etc., can reduce the amount of background radiation perceived by the detector(s). Equipment for use with small birds which are to be routinely counted alive for low levels of gamma-emitting radionuclides has been described by Levy *et al.* (1976). Another approach to this problem is to increase the amount of isotopic activity in the subject. This can be done by artificial "spiking" (i.e., administering concentrated solutions of radioactive isotopes by either oral gavage or injection). Of course, artificial spiking provides no information on naturally-occurring isotope levels, and turnover rates may differ both qualitatively and quantitatively from those of naturally-acquired isotopic body burdens (Fendley *et al.*, 1977). The latter, for example, may be more extensively assimilated into various body tissue pools than would be the case with an artifically-administered spike.

The above problems point out the great scientific value and importance of moderately contaminated natural habitats that are occupied by birds and other organisms under essentially "natural" conditions, except for the fact that levels of one or more radioisotopes have been elevated by some form of accidental release. Such contaminated habitats are often viewed as environmental liabilities, but avian radioecological studies at such sites can provide important data that could not be collected in other ways since the subjects of study, as well as their predators, prey items, etc., may all show levels of radioactive contamination that would make them easily distinguishable from background within time frames feasible for live whole-body counting techniques. Because of elevated levels of contamination most such study sites are closed to direct public access; in the United States the majority are located on the lands of large nuclear industrial reservations, particularly on the National Environmental Research Parks of the U.S. Department of Energy (Table IV). Such sites, however, are accessible to qualified investigators that may wish to use them to conduct radioecological research.

2.6. Units of Radiation Measurement

The amount of radioactive material contained in a given amount of matter is measured in terms of the number of atoms undergoing radioactive disintegrations per unit time. A *curie* (Ci) represents the amount of material in which 2.2×10^{12} atoms disintegrate per min. Generally the levels of radionuclide body burdens found in biota as a result of environmental contamination processes are much lower, often at the level of *millicuries* (mCi; $= 10^{-3}$ Ci), *microcuries* (μ Ci; $= 10^{-6}$ Ci), or *picocuries* (pCi; $= 10^{-12}$ Ci). In much of the scientific literature today, use of the corresponding SI units is usually either encouraged or required. The SI unit corresponding to the Ci is the *bequerel* (Bq), with 1.0 Bq = 27 pCi.

Radiation exposure to gamma or X-rays has classically been mea-

TABLE IV
Habitats at U.S. Department of Energy National Environmental Research Parks Showing Elevated Levels of Radioactive Materials and That Have Proven to Be Useful Sites for Avian Radioecological Research

Location	Site	Habitat	References
U-Pond, Gable Mountain Pond	Hanford Reservation, Hanford, Washington	Small waste ponds, emergent aquatic vegetation	Fitzner and Rickard (1975)
White Oak Lake	Oak Ridge Reservation, Oak Ridge, Tennessee	Marshy habitat in the site of a drained lake bed	Willard (1960)
Pond-B	Savannah River Plant, Aiken, South Carolina	87-ha reactor cooling reservoir	Potter (1987); Brisbin (1989)
Steel Creek Swamp	Savannah River Plant, Aiken, South Carolina	River swamp and stream delta	Marter (1970); Brisbin *et al.* (1989a)
TRA Waste Pond	Idaho National Engineering Laboratory, Idaho Falls, Idaho	Small waste pond	Halford *et al.* (1981; 1983)

sured in terms of the *roentgen* (R), which is defined on the basis of the number of electrostatic charges that these radiations produce in dry air at 0°C and 760 mm Hg. Thus, 1 R = 2.58×10^{-4} coul/kg air. Again, more realistic environmental exposure levels are generally on the order of *milliroentgens* (mR; = 10^{-3}). Absorbed radiation dose, unlike exposure, is measured in terms of the amount of ionizing energy absorbed per unit of exposed tissue. The unit of radiation dose, the *rad*, is the absorbed dose of 100 ergs of energy per g of tissue. The corresponding SI unit is the *gray* (Gy), with 1.0 Gy = 100 rad. In considering the effects of X-rays or gamma radiation on most living tissue, 1 R of exposure will produce 1 rad or 0.01 Gy of absorbed dose (Whicker and Schultz, 1982).

3. RADIATION EFFECTS STUDIES

As discussed earlier, exposures to alpha- and beta-radiation are generally only of concern when the isotope has been taken into the bird's body, and the area of effect is then generally limited to the specific organ or tissue into which the isotope has been incorporated. Because such radiation damage is generally proportional to the number of exposed cells which are in an active state of division, the incorporation of the beta-emitter ^{90}Sr into bone is of particular concern since mitotically active cells in blood-forming bone marrow tissue could thus be exposed. On the other hand, isotopes such as the beta-emitter iodine 131 (^{131}I) would tend to become concentrated in the tissue of the thyroid gland, to which its effects would be largely confined during its brief stay in the body (physical half-life = 8 days; Table I).

Transuranic elements such as plutonium 239 (^{239}Pu) when inhaled as particles and deposited in the lung can result in intense alpha exposure to tissues in the immediate vicinity of their deposition. Although no data seem to be available on the subject, it could be hypothesized that the more efficient air flow through the avian air-sac and respiratory system would provide for a more rapid and effective removal of such particles after their inhalation. This would suggest a more limited hazard to birds in this regard as compared to the same concentration of isotopic particulates in the air being inhaled by a mammal.

The effects of exposure to alpha- and beta-emitters is thus largely an issue which is addressed at the tissue, cellular, and even molecular level. It deals with the effects of ionizing radiation and its sequelae in terms of structural damage and/or disruption of normal processes operating at these levels of organization. It is particularly at this level for example that concerns for cancer induction are most frequently addressed. Discussion of such topics, however, are beyond the scope of this report and the interested reader should consult the general treatment and references on this subject that have been provided by Whicker and Schultz (1982).

Effects studies have been most extensively developed within the area of avian radioecology with regard to gamma and X-radiation. These studies have dealt with both the effects of *chronic exposures*, which are usually delivered at low dose rates over prolonged periods of time, and *acute exposures* which involve higher exposure rates over much briefer ranges of time. Of the two, chronic exposures most closely simulate those that would be experienced by birds or other organisms living in habitats that have been contaminated with accidental releases of radioactive contaminants or atmospheric fallout from nuclear weapons testing. Acute radiation exposures at rates of $> 5-10$ R/min, on the other hand, are seldom if ever experienced by birds under natural conditions except in the rare cases of individuals residing near the site of a nuclear weapons detonation or reactor explosion. Studies of acute gamma radiation effects have thus nearly all been conducted under controlled experimental conditions. In many cases, these studies have used gamma radiation in the role of a controlled stressor that can be administered to subjects under uniform and easily quantifiable conditions in the laboratory (e.g., Latimer and Brisbin, 1987). As explained by these authors, much of this work has therefore been considered at the level of generic stress responses rather than as attempts to gather information relevant only to nuclear issues *per se*.

3.1. Mortality Responses to Radiation

Much of the early radiation effects research used domestic poultry almost exclusively. This work showed that survival following gamma and/or X-radiation exposure was related to a number of factors, the most important of which included: (1) total dose received (with survival being reduced at higher total doses), and (2) the rate at which the total dose was delivered. With regard to the latter, survival following exposure to a given total dose generally increased if the exposure was delivered at a lower rate and/or over a longer period of time (Stearner, 1951; Vogel and Stearner, 1955). The ability of normal biological repair processes to compensate for radiation-induced cellular and tissue damage seems to be enhanced when such damage occurs over a more prolonged period of time and/or at lower dose rates (Zach and Mayoh, 1986b).

The effects of age on survival following radiation exposure is less clear. Although some authors have concluded that LD_{50} values are comparable for adult chickens and young chicks (Stearner and Christian, 1968, 1969), other studies have shown some degree of radioresistance with increasing age in chickens of the New Hampshire breed (Prochazka, 1970). Moreover, Banks *et al.* (1963), studying adult White Leghorn chickens, showed a difference between the sexes in radiation sensitivity, with males showing a lower LD_{50} than females (Table V). Overall, as summarized by Prochazka and Hampl (1966), the $LD_{50/30}$ for all domestic *Gallus* seems to fall between 600–1500 R, depending on the rate of dose delivery (Table V).

In the case of nondomestic bird species, radiation sensitivities again follow similar patterns with regard to dose rate, sex, age, etc. However, few if any studies have evaluated such factors systematically. In general, large-bodied precocial species such as waterfowl have similar $LD_{50/30}$ values to domestic *Gallus,* as do the few adult passerines that have been studied (all 500–900 R; Table V). However, data of Willard (1963) suggest a much higher degree of radiation resistance on the part of nestling Eastern Bluebirds (*Sialia sialis*) with an $LD_{50/30}$ of 2,500 R (Table V). Data from other studies (e.g., Norris, 1958) also support the conclusion that nestling passerines may be much more resistant to radiation stress than other birds although no actual $LD_{50/30}$ values could be calculated in that latter study. Zach and Mayoh (1984) report that acute exposures of nestling Tree Swallows (*Tachycineta bicolor*) to levels as high as 450 R, delivered at 360 R/min, failed to produce any detectable radiation-induced mortality throughout the 20-day nestling period. Further studies by these same authors also indicate that nestling House Wrens (*Troglodytes aedon*) are apparently similarly resistant to radiation stress, again as judged on the basis of mortality through the nestling period (Zach and Mayoh, 1986a). Finally, Norris (1958) irradiated four Eastern Bluebird nestlings, at 23.5 R/min, up to total exposures as high as 1200 R, again with no sign of radiation-induced mortality during the nestling period. Zach and Mayoh (1984) note that their previous work has shown that Tree Swallow nestlings are very resistant to such stressors as chilling, lack of food, and parasitism, and they relate this to the contention of Whicker and Schultz (1982) that resistance to stress in a general sense may predispose organisms to a higher resistance to radiation stress. At this point, there are no data to indicate what radiation exposure levels would be required to produce mortality in most of these nestling passerines, but the work of Willard (1963) would suggest that the $LD_{50/30}$ probably exceeds 2,000–2,500 R.

TABLE V
Measures of Mortality of Wild and Domestic Species of Birds following Acute Exposures to X-Radiation and Gamma Radiation Stress[a]

Species (breed)	Sex/age[b]	Exposure rate	LD_{50}	Reference
Domestic Chicken (*Gallus domesticus*):				
White Leghorn	Male/adult	50 R/min	623 R	Banks *et al.* (1963)
White Leghorn	Female/adult	50 R/min	833 R	Banks *et al.* (1963)
White Leghorn	Combined/2-days	6 R/min	1460 ± 240[c]R	Latimer and Brisbin (1987)
Cobb Broiler	Combined/2-days	6 R/min	1580 ± 270[c]R	Latimer and Brisbin (1987)
Feral Bantam	Combined/2-days	6 R/min	1530 ± 200[c]R	Latimer and Brisbin (1987)
New Hampshire	Combined/adult	40 R/min	900 R	Prochazka and Hampl (1966)
Genus *Gallus* (as a Whole)	Combined/combined	40 R/min	600–900 R	Prochazka and Hampl (1966)
Other species:				
Mallard (*Anas platyrhynchos*)	Combined/4-month	51 R/min	704 R	Abraham (1972)
Mallard (*Anas platyrhynchos*)	Combined/one-year	50 R/min	630/650 R[d]	Curnow *et al.* (1970)

Blue-winged Teal (*Anas discors*)	Female/adult	51 R/min	715 R	Tester *et al.* (1968)
Green-winged Teal (*Anas crecca*)	Male/adult	51 R/min	485 R	Tester *et al.* (1968)
Northern Shoveler (*Anas clypeata*)	Combined/adult	51 R/min	894 R	Tester *et al.* (1968)
Song Sparrow (*Melospiza melodia*)	Combined/adult	90 R/min	800 R	Sturges (1968)
Dark-Eyed Junco (*Junco hyemalis*)	Combined/adult	90 R/min	900 R	Sturges (1968)
European Starling (*Sturnus vulgaris*)	Combined/adult	133 R/min	800 R[e]	Garg *et al.* (1964)
Eastern Bluebird (*Sialia sialis*)	Combined/16-day	43 R/min	2500 R	Willard (1963)
Red-billed Quelea (*Quelea quelea*)	Male/adult	500 R/min	1060 R	Lofts and Rotblat (1962)

[a]Mortality is generally measured as the $LD_{50/30}$, which is the dose required to kill 50% of the exposed birds within 30 days.
[b]Birds of the year are considered as adults.
[c]95% confidence limits.
[d]For ^{60}Co gamma and X-irradiation exposures, respectively.
[e]Produced 56% mortality when delivered as a single dose.

3.2. Growth Responses to Radiation

Although the studies reported above suggest that nestling passerines may have a higher degree of radioresistance than other birds in terms of mortality responses, the opposite seems to be true with regard to growth. Zach and Mayoh (1986b) pointed out that growth responses appear to be more sensitive indicators of radiation stress than are occurrences of gross abnormalities, fledging, or hatching success. To some degree, this is almost certainly related to the finer degree of measurement and data analysis that can be applied to changes in growth under controlled laboratory conditions. In the simplest form, such analyses involve linear regressions with changes being related to changes in radiation exposure (e.g., Brisbin, 1969; Brisbin and Thomas, 1971). More recently however, analyses based on nonlinear regression techniques applied to sigmoid growth curves have allowed even more sophisticated analyses of growth responses (e.g., Fendley and Brisbin, 1977; Zach and Mayoh, 1984; Zach and Mayoh, 1986b; Brisbin *et al.*, 1987). Analyses using the logistic growth model have been used by Zach and Mayoh (1984) to compare the growth of nestling Tree Swallows which were exposed to either acute or chronic gamma radiation stress vs. those of unirradiated controls and nestlings which died of natural causes (Fig. 2). The growth curve of the chronically exposed nestlings in their study was well below that of the control birds, and the growth curves of both chronically and acutely exposed birds were similar to that of nestlings which died of natural causes.

In general, much of the early work on poultry has shown that growth responses to sublethal levels of radiation stress have the same patterns as do mortality responses following higher level exposures. That is to say, decreases in growth rate following exposure to a given dose will be reduced if that exposure is given over a longer period of time and/or at a lower dose rate (Tyler and Stearner, 1966). Again, enhanced ability of repair mechanisms to replace damaged cells and/or tissues are likely responsible for such phenomena.

Studies of growth depression following gamma radiation exposure of poultry have shown that the magnitude of growth decrease does not always increase linearly with increasing dose. Rather, there may be ranges within which increases in radiation does cause little or no additional decrease in growth of body-weight (e.g., between 300–800 R, delivered at 7–11 R/min for the 2-day old broiler chick; Brisbin, 1969; Brisbin and Thomas, 1971). Unlike the situation with the broiler chicken however, Zach and Mayoh (1986a) reported that growth depressions of nestling House Wrens appeared to be linear with increasing ex-

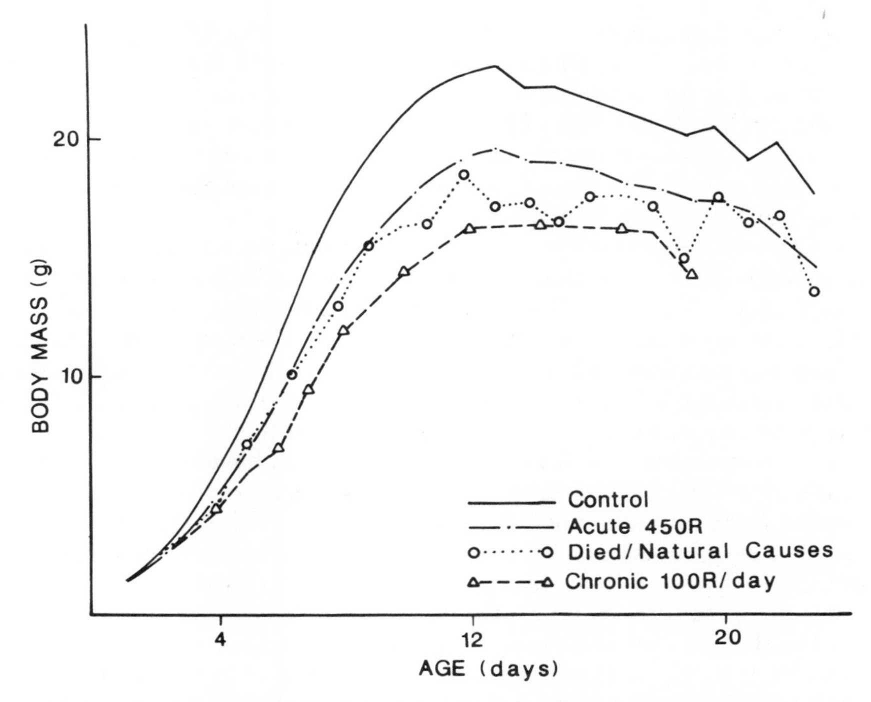

FIGURE 2. Growth curves for body weights of nestling Tree Swallows exposed to either 450 R of acute or 100 R/day of chronic gamma radiation stress, as compared to growth curves of unirradiated controls and nestlings that died after 11 days of age due to natural causes. Growth curves predicted by the logistic model are used prior to age 13 for the control and acute exposure groups. All other data plotted represent daily means. Redrawn from Zach and Mayoh (1984).

posures for 90–600 R. Analyses of the biomass of broilers which survived at least 30 days following acute exposures to 100–900 R indicated that the weight loss due to radiation stress was not significantly different in percentage composition of water, fat, lean-dry weight or caloric density, from the biomass of control birds. However, fat indices (g fat/g lean-dry weight) tended to be slightly elevated following 200 R and 700 R of exposure (Brisbin, 1972).

An additional finding of these studies was that in these broiler chickens, growth in sizes of body structures (e.g., culmen, tarsus, wing primaries) were invariably less affected by radiation exposure than was growth in body weight. This finding was later confirmed by Latimer and Brisbin (1987) and extended by them to several additional domestic and one feral breed of chicken; a similar conclusion can also be drawn from data presented by Zach and Mayoh (1984, 1986b) for nestling passerines. The latter two studies, for example, showed that exposure of Tree Swallows at either the nestling or embryonic stage of development to 320–450 R of gamma radiation decreased the asymptotic size attained by body weight by 13 to 14%, while corresponding reductions of only 3 to 4% and 3 to 5% occurred in the sizes of primary feather length and foot length, respectively.

Although reductions in sizes of body structures are generally less following radiation exposure than is the reduction in body weight, the former may often have disproportionally important consequences. Willard (1963) showed that when nestling bluebirds were exposed to as little as 300 R gamma radiation at two days of age, primary wing feather growth was stunted to such a degree that although the exposed birds fledged at the same time as their unexposed nestmates, they had impaired flying ability which almost certainly would have rendered them more vulnerable to predation after leaving the nest. Even nestlings exposed to as much as 1500 R exhibited normal behavioral development and usually left the nest box at the same time as their unirradiated siblings. However, they were completely unable to fly due to the even more severe stunting of their wing feather growth (Willard, 1963). Considering that the lethal dose for nestling Eastern Bluebirds is 2500 R (Table V), it is apparent that stunting of wing feather growth can result in *ecological lethality* at much lower levels of radiation exposure than that required to produce *physiological lethality* (i.e., that which would result in the death of the bird even when it is kept under protected laboratory conditions). Zach and Mayoh (1984, 1986a, 1986b) also found significant stunting of wing feather growth following the radiation exposure of Tree Swallow and House Wren nestlings, suggesting that this phenomenon may also occur in those species. The latter authors were unable to document postfledging mortality, however, due to their inability to relocate fledglings which had left the nest. In precocial chickens, on the other hand, Brisbin (1969) was unable to find any evidence of wing feather stunting in birds irradiated at two days of age, at exposures as high as 900 R. Moreover, any stunting of wing feather growth in the young precocial chickens would be of lesser consequence in terms of escaping predators than it would in the case of the Bluebird, since the former relies as much or more on running and hiding, as opposed to flying, to escape predators. This indicates the importance of considering basic differences in developmental patterns, such as those between altricial and precocial species, in predicting ecologically lethal consequences of the stunting of growth following radiation exposure (Brisbin, 1969).

By studying both mortality and growth responses of various breeds of chickens, Latimer and Brisbin (1987) concluded that those traits for

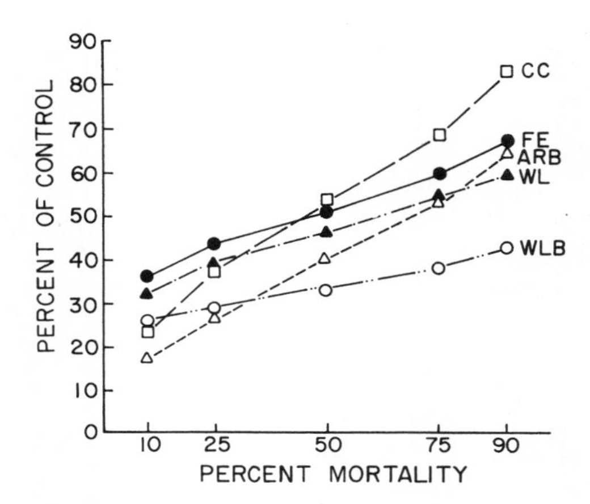

FIGURE 3. Percentage decreases in 30-day rates of growth in body weight of five breeds of chickens following exposures to varying levels of gamma radiation stress at two days of age. Higher values on the vertical axis represent progressively greater stunting of growth rate at given levels of mortality. Points represent the averages of slopes of linear regressions fitted to individual birds' body weights for 31 days following exposure. Breeds designated are: CC = Cobb commercial broilers; FE = feral bantams; ARB = Athens Randombreds; WL = White Leghorns; and WLB = White Leghorn Bantams. Cobb broilers and feral bantams are shown to exhibit greater stunting of growth rate than other breeds, at exposures producing higher levels of mortality. From Latimer and Brisbin (1987).

which a breed was most strongly selected during the course of its development seemed to be those which showed the greatest proportional decrease or impairment following radiation exposures up to levels producing equivalent levels of mortality. Cobb commercial broiler chickens which had been strongly selected for rapid growth in body weight showed proportionally greater decreases in this parameter before exhibiting given levels of mortality (Fig. 3). Feral bantam chickens, however, which like wild precocial galliforms should be strongly selected for rapid growth of structures enhancing food-gathering ability (e.g., the culmen) and escape from predators (e.g., well-developed legs and wing feathers), showed greater proportional decreases in the growth of these structures (Fig. 4). This again emphasizes the importance of understanding avian responses to radiation stress in the context of the full suite of selection pressures, both natural or artificial, under which the particular species or breed population evolved or was developed.

3.3. Responses of Reproduction to Radiation

Like mortality, reproductive characteristics have proved to be relatively insensitive indicators of radiation response in birds, as compared to growth (Zach and Mayoh, 1986b). Many of the studies in this area

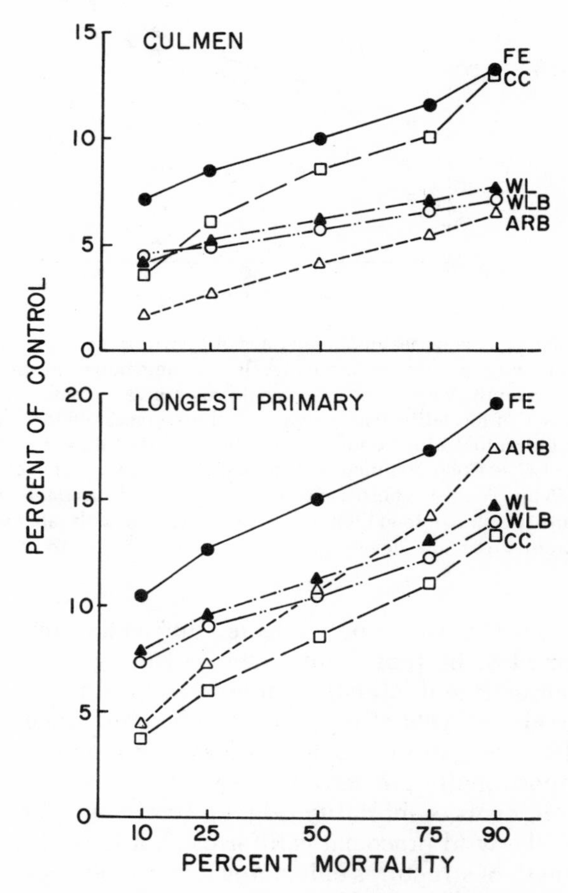

FIGURE 4. Percentage decreases in the sizes of culmens and longest primary wing feathers of five breeds of chickens following exposures to varying levels of gamma radiation stress. Higher values on the vertical axes represent progressively greater stunting of the culmen or longest primary at given levels of mortality. Points plotted represent the averages of sizes of these structures at 31 days of age following exposures at two days of age. Breeds are as designated in Fig. 3. Feral bantams are shown to exhibit greater stunting in the size of these structures than other breeds, at exposures producing equivalent levels of mortality. From Latimer and Brisbin (1987).

have dealt with wild passerines in addition to those using domestic poultry (Table VI). Studies of wild species have often involved birds reproducing under natural conditions in areas of elevated gamma radiation exposure which have been created by experimental longterm continual chronic irradiations of various habitats (e.g., Woodwell, 1963).

Perhaps the most intriguing findings of this work have been those of Zach and Mayoh (1982), which had suggested that both Tree Swallows and House Wrens were somehow able to sense and then avoid selecting nest boxes in areas of elevated gamma radiation exposure. Although these birds were apparently able to respond to radiation levels as low as 100 times background, it was not clear whether they were actually responding to the continuous gamma radiation itself or to one of its secondary effects. As breeding density of birds increased between years, more of the birds in this study were forced to nest in boxes located in "suboptimal" habitat (i.e., that with higher radiation exposure levels). However, even under the most unfavorable conditions (mean exposure rate of nests = 0.03 mR/min), there were no effects on the breeding performances of either the wrens or swallows, as judged by clutch size, hatching success, young fledged per nest, incubation time, or nestling period (Zach and Mayoh, 1982).

In terms of sample sizes and experimental design, the study by Zach and Mayoh (1982) is the most thorough treatment available of the effects of radiation exposure on the breeding performance of wild birds. However, a number of other studies (Table VI) have provided information on the fate of nesting efforts under conditions of even higher radiation exposures. Much of this information has been somewhat anecdotal and has involved observations of only single pairs of birds. In addition, in many cases breeding failures could not be attributed to radiation effects *per se* vs. natural causes. However, taken together, this information has suggested that wild passerine species can successfully nest and fledge young under conditions of gamma and/or X-radiation exposure at levels which, as will be shown later, greatly exceed those which would be produced by the levels of radionuclide fallout resulting from environmental contamination by nuclear weapons testing or nuclear industrial accidents. However, as pointed out by Zach and Mayoh (1982), citing Mraz and Woody (1973), the successful fledging of young *per se*, is no guarantee that such young will later survive to reproduce normally since some female chickens irradiated as embryos, even at low exposure rates, later showed reduced egg production as adults. However, in the case of wild populations such effects would only be demonstrable by multi-generation studies with exceptionally large sample sizes, none of which have yet been conducted.

TABLE VI
Studies of Reproductive Performance of Wild and Domestic Species of Birds Exposed to Gamma or X-Radiation Stress

Species (subjects)	Exposure level	Radiation Effects	Reference
Domestic Chicken (*Gallus domesticus*) (breeding females)	5–45 R/min (total = 400–800 R)	Lay fewer eggs	Maloney and Mrax (1969)
Domestic Chicken (*Gallus domesticus*)(incubating eggs)	15 mR/min (total = 432 R)	Reduced egg hatchability	Mraz (1971)
Red-billed Quelea (*Quelea quelea*) (breeding males)	500 R/min (total = 420 R)	Incomplete spermatogenesis	Lofts and Rotblat (1962)
Eastern Bluebird (*Sialia sialis*) (breeding females)	24 R/min[a] (total = 200/600 R)	None	Norris (1958)
Eastern Bluebird (*Sialia sialis*) (incubated eggs)	24 R/min[a] (total = 420 R)	6/9 eggs failed to produce hatchlings which fledged	Norris (1958)
Tufted Titmouse (*Parus bicolor*) (incubating female)	24 R/min[a] (total = 600 R)	None	Norris (1958)
Tree Swallow (*Tachycineta bicolor*) (adults and nest site)	13 mR/min[b] (total= 558 R)	None	Wagner and Marples (1966)

Rufous-sided Towhee (*Pipilio erythropthalmus*) (adults and nest site)	15 mR/min[b] (total = 497 R)	Nest deserted	Wagner and Marples (1966)
	27 mR/min[b] (total not given)	Nest deserted	Wagner and Marples (1966)
Brown Thrasher (*Toxostoma rufum*) (adults and nest site)	10 mR/min[b] (total not given)	Nestling disappeared from nest	Wagner and Marples (1966)
Eastern Bluebird (*Sialia sialis*) (adult female and nest)	90 mR/min[b] (total > 780 R)	Female found dead	Wagner and Marples (1966)
House Wren (*Troglodytes aedon*) and Tree Swallow (*Tachycineta bicolor*) (adults and nest site)	0.03 mR/min[c] (continuous)	None	Zach and Mayoh (1982)
Tree Swallow (*Tachycineta bicolor*) (adults and nest site)	69 mR/min[d] (continuous)	Reduced hatching but not fledging success	Zach and Mayoh (1982)
House Wren (*Troglodytes aedon*) and Tree Swallow (*Tachycineta bicolor*) (incubted eggs)	360 R/min[e] (total = 400/3200 R)	None	Zach and Mayoh (1986b)

[a]Subjects removed to laboratory where irradiations were administered in 2–3 fractionated acute exposures.
[b]Exposed under natural conditions 20 hr/day.
[c]Mean exposure rate for 94 nests.
[d]One nest only.
[e]Eggs removed to laboratory where irradiations were administered in a single acute dose.

3.4. Radiation Effects and Hormesis

No discussion of the effects of radiation exposure on birds would be complete without a consideration of the concept of *hormesis*. As described and extensively documented by Luckey (1980), hormesis refers to the phenomenon by which low doses of harmful agents produce responses that are exactly opposite to the effects produced by these same agents at higher doses. While hormesis has been studied for exposures to a variety of environmental contaminants and other stressors, it has probably been best documented with regard to radiation stress (Luckey, 1980). In the case of growth for example, it is clear that exposure to high doses of gamma or X-radiation can interfere with normal development and cause stunting (e.g., Willard, 1963; Brisbin, 1969; Zach and Mayoh, 1986b). However, low-level radiation exposures have long been known to stimulate growth as has been documented for a variety of plant and animal species (e.g., Lorenz *et al.*, 1954). Indeed, one of the most common popular caricatures of the effects of radiation exposure has been gigantism and the production of oversized "monsters" from exposed individuals.

In the case of birds, Fendley and Brisbin (1977) present data which demonstrate the enhancement of certain aspects of growth in captive-reared Wood Ducks hatched from eggs collected from nest boxes located in a South Carolina swamp delta contaminated with elevated levels of ^{137}Cs as a result of releases from nuclear ractor effluents. As compared to birds hatched from eggs laid in uncontaminated habitats elsewhere in South Carolina or Georgia, ducklings from the population inhabiting the contaminated area grew significantly faster and had earlier points of inflection for growth curves defined by the logistic model, although they grew to lower (but not significantly lower) levels of asymptotic body weight (Table VII). In reviewing the application of the concept of hormesis to the growth of waterfowl and other biota exposed to a variety of stressors, Brisbin *et al.*, (1987) urge caution and a clear delineation of which aspects of the growth process are being considered. For example, in a number of cases cited by the latter authors various aspects of the growth process were shown to respond differently to varying levels of stressor exposure.

Besides growth, responses indicating hormesis have also been documented in factors such as reproduction, longevity, and disease resistance (Luckey, 1980). In the case of birds Lofts and Rotblat (1962) have shown that low levels of X-radiation exposure (50 R) caused precocious testis metamorphosis in male Red-billed Queleas (*Quelea quelea*), although exposure to higher dose levels (420 R and above) resulted

TABLE VII

Growth Parameters of Captive-Reared Wood Ducks Hatched from Eggs Collected from Radionuclide Contaminated and Uncontaminated Habitats[a]

Growth parameter	Sex	Uncontaminated habitats: South Carolina[c]		Uncontaminated habitats: Georgia[d]		Contaminated swamp delta[e]	
Asymptotic weight (g)[b]	Males	508.1**	(26)	525.4**	(11)	503.2**	(23)
	Females	459.1	(24)	479.1	(10)	468.9	(29)
Growth-rate constant (k)[b]	Males	0.0934**	(26)	0.1034**	(11)	0.1143**	(23)
	Females	0.1073	(24)	0.1127	(10)	0.1158	(29)
Age at point of inflection (days)[b]	Males	20.3**	(26)	28.1**	(11)	24.6**	(23)
	Females	26.6	(24)	25.6	(10)	23.9	(29)

[a]From Fendley and Brisbin (1977).
[b]Mean values calculated for parameters of the logistic model.
[c]16 km east of Columbia, Richland County, South Carolina.
[d]Fort Gordon Military Reservation, Richland and Columbia Counties, South Carolina.
[e]Steel Creek swamp delta of the Savannah River Plant site, Barnwell County, South Carolina (Table IV; Fig. 7).
**Asterisks indicate significant differences between the sexes ($p \leq 0.01$), and values in parentheses indicate sample sizes. Mean values underscored by the same line do not differ significantly.

in testicular degradation similar to that which normally occurs at the end of the breeding season (Table VI). These authors have discussed the possibility that these radiation-induced changes, both hormetic and degradative, are mediated through pituitary stimulation.

4. CYCLING OF RADIOACTIVE CONTAMINANTS

Probably no aspect of avian radioecology is as relevant to current-day environmental concerns as is the uptake and cycling of radioactive contaminants. Not only do studies in this area relate directly to the fate and effects of radionuclides produced by nuclear accidents, but, as explained earlier, these studies also have important implications for understanding the cycling of nonnuclear contaminants. The general principles relating to this area have been dealt with in Sections 2.3 and 2.4 and in particular involve the models for contaminant uptake and concentration as described by Equations (1) through (4).

Rather than presenting a specific review of the literature (which is actually rather scanty) on the subject, as was done for studies of radiation effects, the principles of avian radionuclide cycling will be presented primarily through a description of the author's own work in this area—specifically that dealing with the radionuclide contamination processes operating with the avifauna of the United States Department of Energy's Savannah River Plant (SRP), a nuclear weapons production and research facility located near Aiken, South Carolina (Fig. 5). Within these studies, a particular emphasis will be given to work with the site's waterfowl populations. This segment of the avifauna is particularly important because of the ubiquity of these birds in aquatic habitats of the SRP where a number of inadvertent radionuclide releases have occurred, and the possibility that such birds could quickly move off of the site to be shot and used by hunters as food only a brief time after leaving radioactively contaminated habitats—thus serving as potential vectors of SRP-released radionuclide contaminants to the food chain of man.

Later, in Section 5, the principles developed through these and related studies that will be cited from the literature will be applied to an assessment of the potential of migratory waterfowl to act as global vectors of radionuclide contaminants to the food chain of man, following their release by the Chernobyl nuclear accident in the Soviet Union in 1986. In both this case and the studies of waterfowl on the SRP, the focus will be on the long-lived gamma-emitter ^{137}Cs because of its ubiquity in the environment following most nuclear accidents and be-

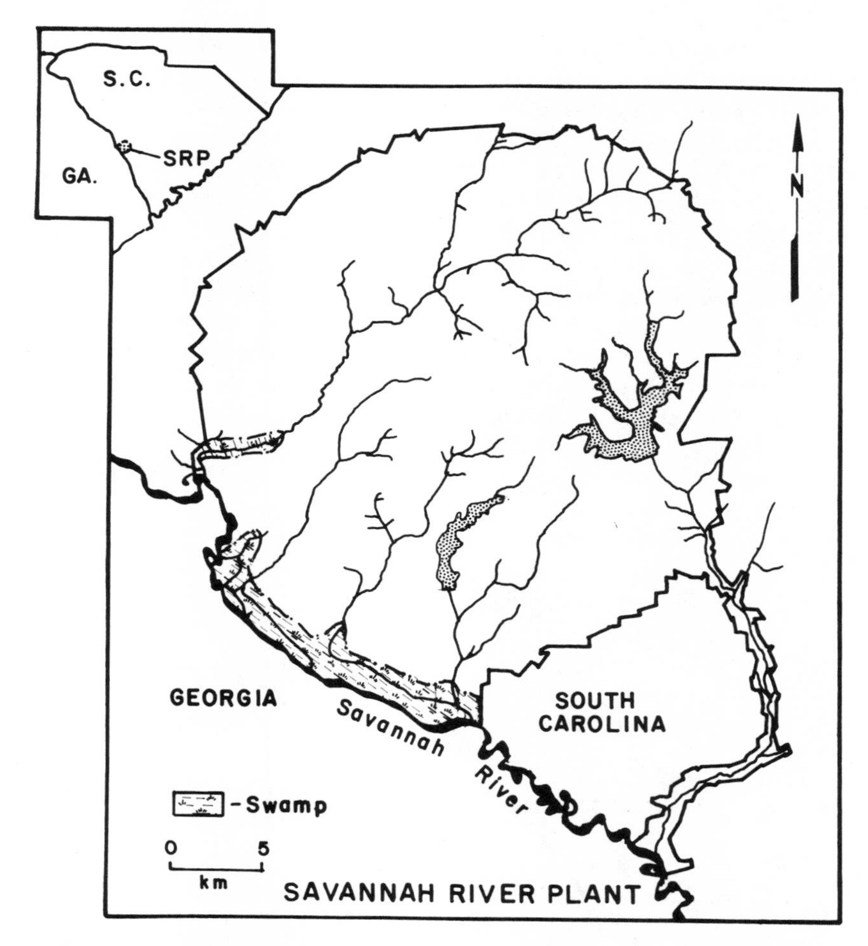

FIGURE 5. Map showing the location and wetland habitats of the U.S. Department of Energy's Savannah River Plant. Stippled areas represent reactor cooling reservoirs.

cause of its importance as a contaminant of human food chains based on wild game, as explained in Section 2.2.

4.1. The Savannah River Plant and Its Avifauna: A Case Study

The Savannah River Plant is a 750 km^2 nuclear production and research facility of the United States Department of Energy. Located in portions of Aiken, Allendale, and Barnwell Counties, South Carolina (Fig. 5), the lands of the SRP were closed to public access in 1952, when the site was established as a center for the production of tritium and plutonium for nuclear weapons. Five nuclear production reactors have operated at the SRP over varying periods of time since its establishment. Unlike power reactors, these production reactors discharge all of their waste heat into their cooling water effluents which flow into the several natural stream water courses that traverse the site and eventually empty into an extensive riverine swamp fringing the northern shore of the Savannah River. One of these streams was dammed in 1958 to produce the 1130 ha Par Pond reservoir system (Fig. 6). Until recently, heated water from one of the site's reactors was sent to one of this reservoir's three major arms, (the so-called "Hot Arm") and after cooling to ambient temperatures, water was then pumped-back from the reservoir's "West Arm" through the reactor to provide closed-loop cooling.

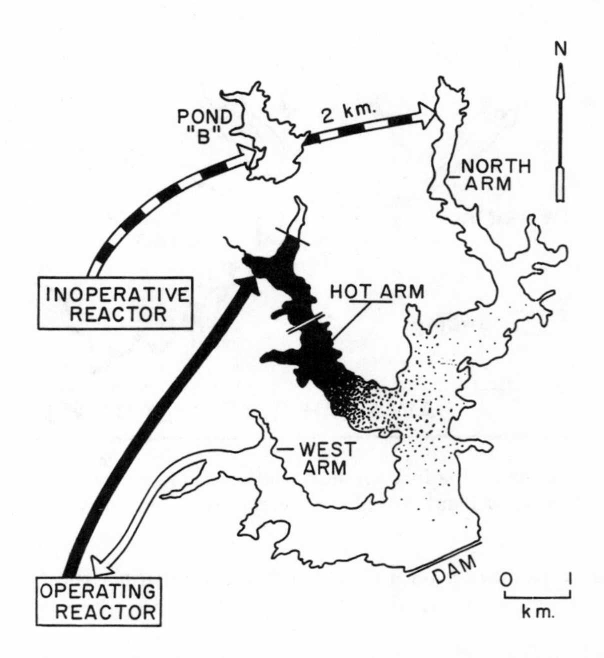

FIGURE 6. Diagrammatic representation of the Par Pond reactor cooling reservoir system of the U.S. Department of Energy's Savannah River Plant (Fig. 5). The intensity of stippling is roughly proportional to the degree of thermal influence of reactor cooling effluents which until 1988 were introduced into the Hot Arm to provide closed-loop cooling. Pond B and the North Arm of Par Pond received similar effluents between 1961–1964, during which time some contamination of these areas with ^{137}Cs and other radioisotopes occurred.

Another now-inoperative production reactor also passed heated cooling water effluents to the "North Arm" of the Par Pond reservoir until operation of that reactor ceased in 1964 following the inadvertent release of radioactively contaminated cooling water between 1954–64. Although small amounts of this release occurred over an extended period of time, the principle source was the rupture of an experimental fuel element in 1957 (Alberts *et al.*, 1979). Effluents containing this contamination entered the North Arm extension of the Par Pond reservoir system until the reactor ceased operation. Prior to this time, however, the canal/stream system connecting the failing reactor to the North Arm was also dammed, creating the 87 ha "Pond B" reservoir in 1961 (Fig. 6). For the next three years, the majority of the radioactive contaminant release was received by this impoundment, and since 1964, Pond B and the canal system connecting it to the now-inoperative reactor have shown the highest levels of radionuclide contamination, with lower contaminant levels occurring in the North Arm of Par Pond itself. Since the shutdown of the now-abandoned reactor in 1964, there has been no evidence of any additional releases of radioactive contaminants to any other part of the Par Pond system, and, as will be shown later, contaminant levels in the Hot Arm and West Arm have not been elevated above background.

An inventory of radionuclides released to Pond B and the present distribution of these contaminants in the sediments, water, and biota of this reservoir have been provided by Whicker *et al.* (1989). Although several transuranic elements, ^{90}Sr, and a few other minor radionuclides are still present, the principle concern with respect to the avifauna has been the release to this system of approximately 150 Ci of the 134 and 137 gamma-emitting isotopes of cesium, collectively known as "radiocesium." Because of the shorter (2.1 year) half-life of cesium 134 (^{134}Cs), however, the amount of this isotope now remaining in the system is relatively small, and even in the late 1960's the ratio of ^{134}Cs to ^{137}Cs was only 1:20 in Par Pond waterfowl (Marter, 1970). Thus, although radiocesium body burdens were determined by counting the combined emissions of both the 134 and 137 isotopes, in the studies to be described here the longer-lived ^{137}Cs was essentially the only isotope remaining of any biological significance (Table I).

In addition to the releases of ^{137}Cs to the SRP reservoirs, additional releases of this isotope have also contaminated certain areas of the site's bottomland river swamp habitat (Fig. 5; Table IV). Between 1961 and 1970, effluents containing both cooling water and purge water from the fuel storage basins of one of the site's reactors were contaminated with about 261 Ci of radiocesium and smaller amounts of other gamma-emitting radionuclides (Marter, 1970). These effluents were introduced into Steel Creek, a 20 km long stream which flows into the SRP's river swamp, where its outflow, along with those of two other SRP streams, are partially dammed by levee deposits of the Savannah River creating a swamp delta approximately 16 km long and 2.5 km wide. The water and sediments of the lower reaches of this swamp delta, together with those of considerable portions of Steel Creek's lateral floodplain, all received contamination from the reactor effluents. The shutdown of the offending reactor and the consequential decrease in stream flow exposed areas of contaminated mudflats and islands which were rapidly colonized by vegetation. Succession over the intervening years has since created a mosaic of both aquatic and terrestrial communities whose contamination with ^{137}Cs have been the subject of a number of studies (e.g., Marter, 1970; Brisbin *et al.*, 1989a), including several which have dealt specifically with birds, particularly waterfowl (Straney *et al.*, 1975; Fendley, 1978; Fendley and Brisbin, 1977; Fendley *et al.*, 1977).

The waterfowl of the SRP have been described in detail by Mayer *et al.* (1986) and may be generally divided, for purposes of radioecological study, into two groups: those inhabiting the riverine swamp system, and those frequenting the more open waters of the reactor cooling reservoirs. The waterfowl community of the swamp system consists almost exclusively of dabbling duck species (Anatidae: Anatinae), including the site's only resident breeding waterfowl, the Wood Duck. Extensive long-term studies of the population biology of this species have been undertaken on the SRP (Hepp *et al.*, 1989) and are continuing.

In addition to a permanent resident and breeding population of Wood Ducks, the SRP riverine swamp is also utilized by large numbers of winter migrant waterfowl, principally Mallards, which use the swamp system for both protected nighttime roosts and foraging. Aerial surveys and night roost-flight counts show peak waterfowl use of the SRP swamp system to occur in January and February (Mayer *et al.*, 1986). Over the four-year period from 1981–1985, these authors found that midwinter Mallard numbers increased 72.6% in the SRP swamp system while corresponding Mallard numbers decreased by 33.0 and 70.4%, respectively, in the Atlantic Flyway and state of South Carolina as a whole, thus indicating the importance of this protected SRP swamp habitat as a refuge for these birds. The combined numbers of Mallards and Wood Ducks censused often exceeded 1200–1500 individuals for single aerial surveys and, when combined with roost counts, indicated total resident winter populations on the SRP that might well exceed 2000–2500 individuals at any one time (Mayer *et al.*, 1986).

In contrast to the SRP swamp system, the reactor cooling reservoirs of the site are inhabited principally by diving ducks (Anatidae: Aythyinae) and American Coots. Almost all of these waterfowl are winter visitors to the SRP, arriving between September–November and departing between March–April, with peak numbers occurring in December–February (Brisbin, 1974; Mayer *et al.*, 1986). Brisbin (1974) estimated coots to comprise 71% of the Par Pond waterfowl community, with a total winter resident population of between 5000–10000 birds (Brisbin *et al.*, 1973). Within the diving duck group, Mayer *et al.* (1986) found Lesser Scaup (*Aythya affinis*) to be the most abundant species, followed by Ring-necked Ducks (*Aythya collaris*), Ruddy Ducks (*Oxyura jamaicensis*), and Bufflehead (*Bucephala albeola*). As in the case of the swamp habitat, the importance of the "sanctuary effect" resulting from the isolation of the SRP lands from human disturbance and hunting is also evident in the numbers of waterfowl resident on the SRP reactor cooling reservoirs, as compared to other areas of the southeast. Aerial counts of the Par Pond reservoir system from 1981–1985 have shown a 74.7% increase in the number of wintering waterfowl (not including coots), while comparable midwinter numbers declined 19.0% and 24.5% in the Atlantic Flyway and state of South Carolina as a whole, respectively (Mayer *et al.*, 1986). During this period, the four most abundant diving duck species found on the SRP reservoirs by these authors averaged over 1900 more individuals than were censused for these species on all inland bodies of water for South Carolina combined, as based on the Zone 1 midwinter waterfowl counts of the United States Fish and Wildlife Service. Thus, between both the riverine swamp and reactor cooling reservoir systems, the protected wetland habitats of the SRP represent an immensely important resource to the declining waterfowl populations of the region, and this fact underlines even more clearly the importance of carefully assessing both the fate and effects of site-released radionuclide contaminants which may be encountered by these birds during their stay on the site.

4.2. Radionuclide Distribution in the SRP Avifauna

Because the accidental release of radionuclides has not been uniform across all habitats of the SRP site, an opportunity exists to sample birds from areas of both high and low contamination and thereby test the value of the site's avifauna as "indicator species" of both general and localized environmental contamination levels (Straney *et al.*, 1975; Fig. 7). Although the high flight mobility of birds might argue against their use as indicators of specific sites of contaminant release, an appar-

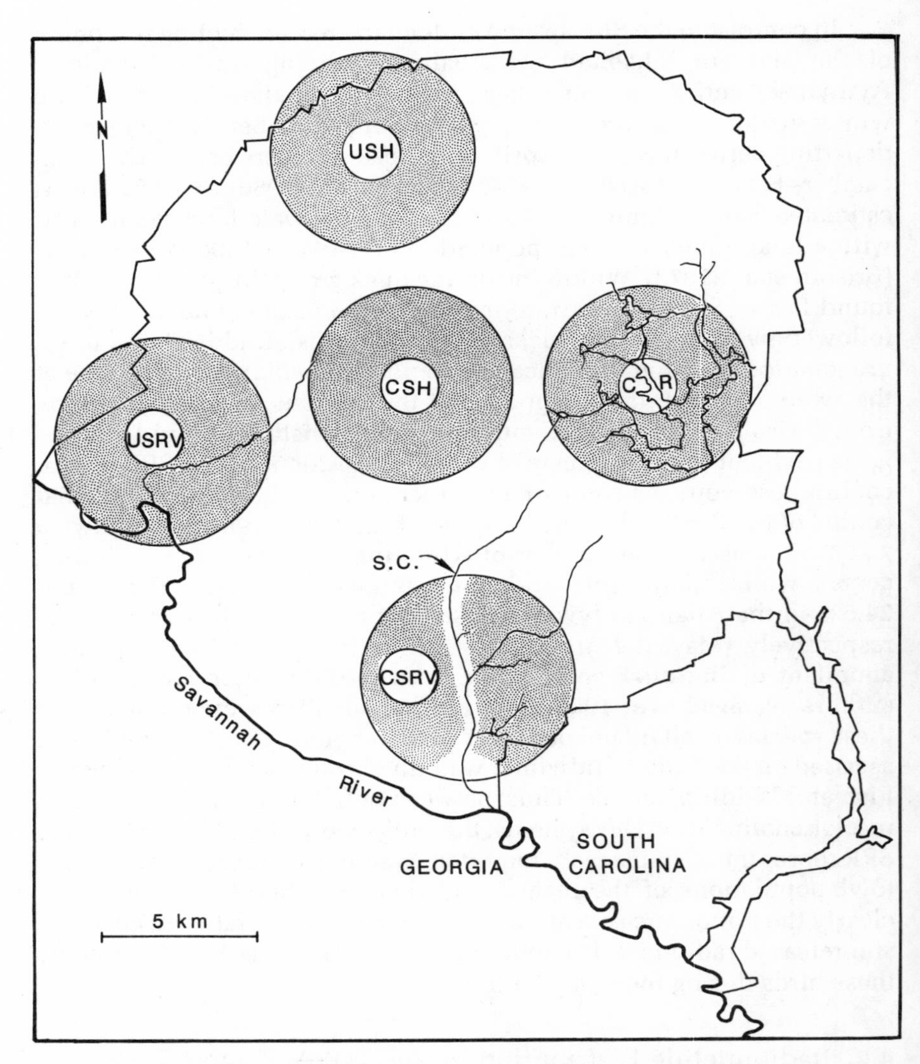

FIGURE 7. Sampling areas used by Straney *et al.* (1975) to study radiocesium levels in birds from the U.S. Department of Energy's Savannah River Plant. Major geographic features of the Savannah River Plant are shown in Fig. 5. SC = contaminated Steel Creek floodplain and delta; CSRV = contaminated Savannah River Valley (lowland) area; USRV = uncontaminated Savannah River Valley (lowland) area; CSH = contaminated sand hills (upland) area; USH = uncontaminated sand hills (upland) area; CR = reactor cooling reservoir (Par Pond). Redrawn from Straney *et al.* (1975).

ently high degree of territoriality and site fidelity of the birds sampled by these authors resulted in significant differences between the radiocesium levels of birds from contaminated vs. uncontaminated habitats, even when separated by distances as small as several hundred m (Straney *et al.*, 1975). This study, which determined radiocesium levels in 255 summer birds of 53 species, showed that birds collected as close as 200 m to a contaminated reactor effluence stream (Steel Creek; Table IV; Fig. 7) had consistently lower body burdens than those taken on the stream's delta and floodplain where radiocesium levels were elevated in the soils and vegetation.

This same study also indicated clearly the importance of having adequate ornithological knowledge concerning the birds to be examined for radionuclide uptake, since the differences in body burdens due to contaminated vs. noncontaminated locations were demonstrable only in adults and not in juvenile birds-of-the-year. Greater variability in body burdens of the juveniles within sampling locations prevented the detection of significant differences between contaminated vs. uncontaminated habitats. Such a need to identify and eliminate juvenile birds from any sampling program designed to use birds as indicators for radiocesium argues strongly for the importance of including in such a program investigators with appropriate expertise in field ornithology (e.g., those able to distinguish adults from juveniles for the resident bird species). Frequently, however, studies of environmental contamination processes are undertaken solely by individuals with expertise in such fields as chemistry or engineering and the input of such ornithological expertise is often lacking. This is an important theme for those concerned about the future and importance of ornithology in today's era of increased emphasis on environmental monitoring activities and will be addressed later.

Even in the case of nonterritorial winter visitors to the SRP, high degrees of foraging site fidelity have also resulted in certain bird species proving useful as indicators of spatial patterns of environmental contamination on a microgeographic scale. For example, on the SRP's reactor cooling reservoirs color-marking studies have demonstrated that winter resident American Coots show a high degree of site fidelity to specific foraging locations on the reservoir's shoreline (Potter, 1987). Although individual coots may move nearly one km to nightly roosts in emergent stands of vegetation, daytime sightings of the same birds while foraging may be as little as 10 m apart for periods as long as 22 days (Potter, 1987). Moreover, this high degree of site fidelity can extend across years, with up to 90% of all returning marked birds in the latter study being sighted for as many as three successive winters on exactly the same section of reservoir shoreline. One of the consequences of this high degree of long-term site fidelity is the exposure of certain individuals to long-term contaminant uptake if they happen to be returning to a location where elevated contaminant levels occur. This would be particularly important in the case of those radionuclides such as ^{90}Sr which accumulate in bone or other body compartments with slow turnover rates and which therefore require long periods of exposure to reach equilibrium levels (Table I). While most migrant waterfowl wintering on the SRP would not be likely to remain at any specific contaminated location long enough to build up equilibrium levels of such long-lived radionuclides, the constant reexposure of certain site-faithful individuals of species such as the coot would place them at a higher risk for showing elevated levels of such contaminants.

Even within a single winter's stay at the SRP, high winter site fidelity of the coot can produce a spatial mosaic in radiocesium body burdens of this species, with individuals from the more contaminated portion's of Par Pond's North Arm (Fig. 6) showing consistently higher body burdens than those from either the Hot or West Arms (Brisbin *et al.*, 1973; Fig. 8). This ability of the coot to serve as an indicator of microgeographic patterns of radionuclide contamination would not be shared by other waterfowl species of the SRP such as Mallards or many of the diving ducks, which are generally less site-faithful in their foraging habits (Mayer *et al.*, 1986). Again, the importance of basic ornithological understanding and expertise is emphasized with respect to either the selection of an indicator species or the interpretation of contaminant data from avifauna from an area such as the SRP.

Studies of the distribution of radiocesium in the coot vs. other species of winter resident waterfowl from the SRP also illustrate the importance of understanding the distribution of such contaminants between tropic levels within a given biotic community. On its wintering grounds on the SRP, stomach content analyses indicate that the coot is almost completely herbivorous, feeding on algae and various forms of aquatic macrophytes (Potter, 1987). However, the radiocesium body burdens of coots are consistently higher than those of other species of winter-resident waterfowl which show more omnivorous or carnivorous (piscivorous) dietary habits (Brisbin *et al.*, 1973, Fig. 9). Thus, at least with respect to this particular waterfowl community and radiocesium, the phenomenon of trophic level concentration or biomagnification does not seem necessarily always to occur in upper trophic levels (Brisbin *et al.*, 1973), even though typical biomagnification of heavy metal contamination with mercury does indeed occur in these same birds (Clay *et al.*, 1980). The concept of trophic level concentra-

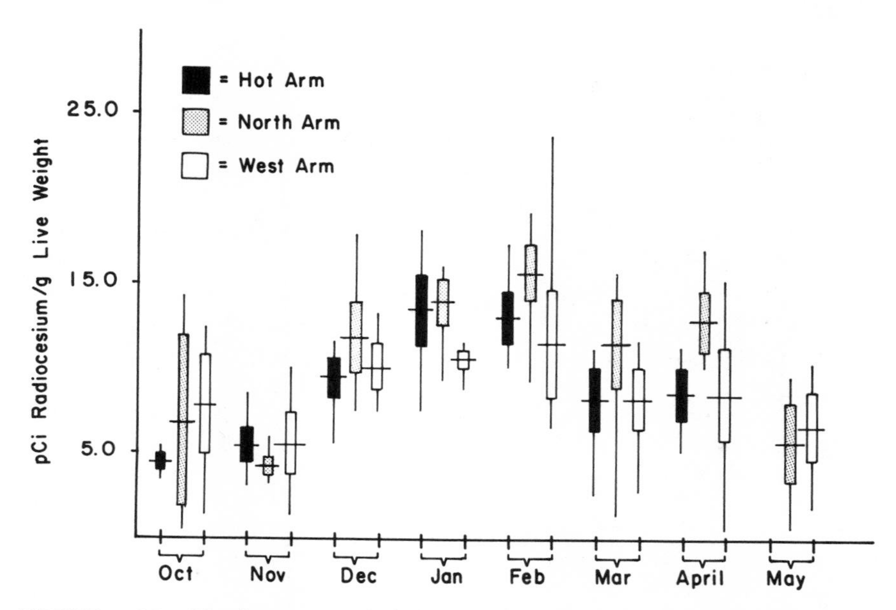

FIGURE 8. Monthly changes in whole-body radiocesium burdens of American Coots wintering in the three major arms of the Par Pond reservoir of the U.S. Department of Energy's Savannah River Plant, as shown in Fig. 6. Means of samples of ten coots collected per month per reservoir arm are indicated by horizontal lines; rectangles represent ± two standard errors and vertical lines indicate ranges. From Brisbin *et al.* (1973).

tion is perhaps one of the most basic tenets of much of the "popular wisdom" concerning the distribution of contaminants in the environment, particularly pesticides (e.g., Carson, 1962), and even in the case of radiocesium some empirical evidence does exist for the phenomenon (Jenkins and Fendley, 1968). However, data for the distribution of radiocesium in the wintering waterfowl community of Par Pond (Brisbin *et al.*, 1973; Fig. 9), as well as that of Straney *et al.* (1975) for the birds of the SRP as a whole, indicate that a blanket assumption cannot be made concerning the occurrence of this phenomenon in all avian communities for all classes of environmental contaminants. In the latter study herbivorous birds again showed higher levels of radiocesium than did more predatory species (in this case mostly insectivores), with omnivores generally showing intermediate levels of radiocesium contamination (Fig. 10). The exceptionally high levels of ^{137}Cs in the omnivores collected from the periphery of the SRP's principle radioactive waste burial ground (area CSH of Fig. 7) were mostly due to elevated levels in both Common Crows (*Corvus brachyrhynchus*) and Fish

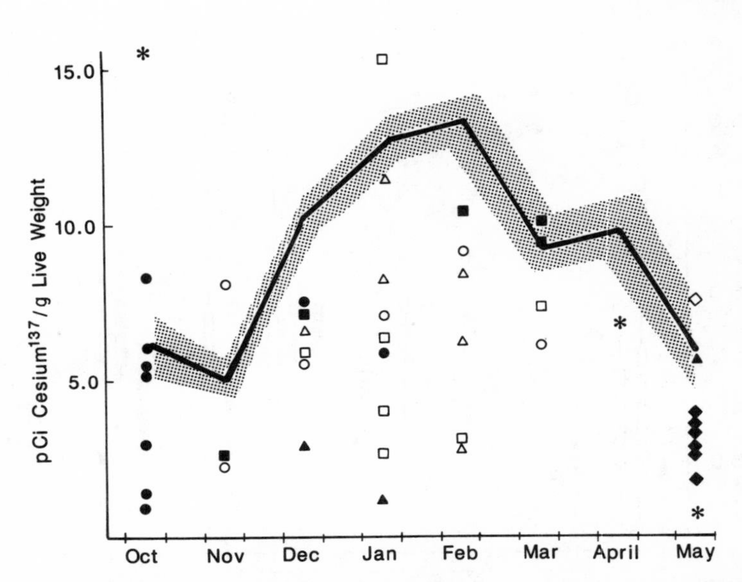

FIGURE 9. Radiocesium whole-body burdens of American Coots vs. those of other species of waterfowl and aquatic birds wintering on the Par Pond reservoir of the U.S. Department of Energy's Savannah River Plant. The solid line connects monthly means, and the shaded band represents the 95% confidence interval about the collective mean for all coots sampled from all reservoir arms, as depicted in Fig. 8. *, Common Gallinules (*Gallinula chloropus*); ●, Pied-bills Grebes (*Podilymbus podiceps*); ○, Horned Grebes (*Podiceps auritus*); ▲, Ring-necked Ducks (*Aythya collaris*); △, Lesser Scaup (*Aythya affinis*); ■, Bufflehead (*Bucephala albeola*); □, Ruddy Ducks (*Oxyura jamaicensis*); ◆, Common Terns (*Sterna hirundo*); and ◇, a Black Tern (*Chlidonias niger*). From Brisbin *et al.* (1973).

Crows (*Corvus ossifragus*) which tend to forage over much larger areas than most of the other smaller passerine species sampled, and which were likely to have actually foraged within the interior of the waste burial ground itself (Straney *et al.*, 1975). Reichle *et al.* (1970) and Anderson *et al.* (1973) report a progressive decrease in the concentration of radiocesium in food chains involving terrestrial arthropods, which could help explain the failure of some of the insectivorous species examined by Straney *et al.* (1975) to show radiocesium biomagnification.

Another phenomenon which is often not considered in assessments of the environmental distribution of radionuclides and other contaminants is what might be termed the "Jenkins Effect" (Jenkins and Fendley, 1968). This refers to the tendency of contaminants which are distributed atmospherically across broad geographic regions (such as has occurred in the distribution of global fallout from nuclear weapons testing) to concentrate, all else being equal, at higher levels in wildlife species (e.g., deer, waterfowl, or upland game birds) than in domestic species of livestock and poultry in the same region. This phenomenon is almost certainly a result of the dependence of wildlife food webs upon natural vegetation that is more dependent on nutrients deposited by rainfall and dry atmospheric deposition to balance nutrient budgets. In the case of cultivated crop plants or pasturage that would form the food base for domestic species however, supplementation with commercial fertilizer would tend to "dilute" the effects of atmospherically-deposited radioactive elements by increasing the availability of their stable element counterparts or chemical analogs (Whicker and Schultz,

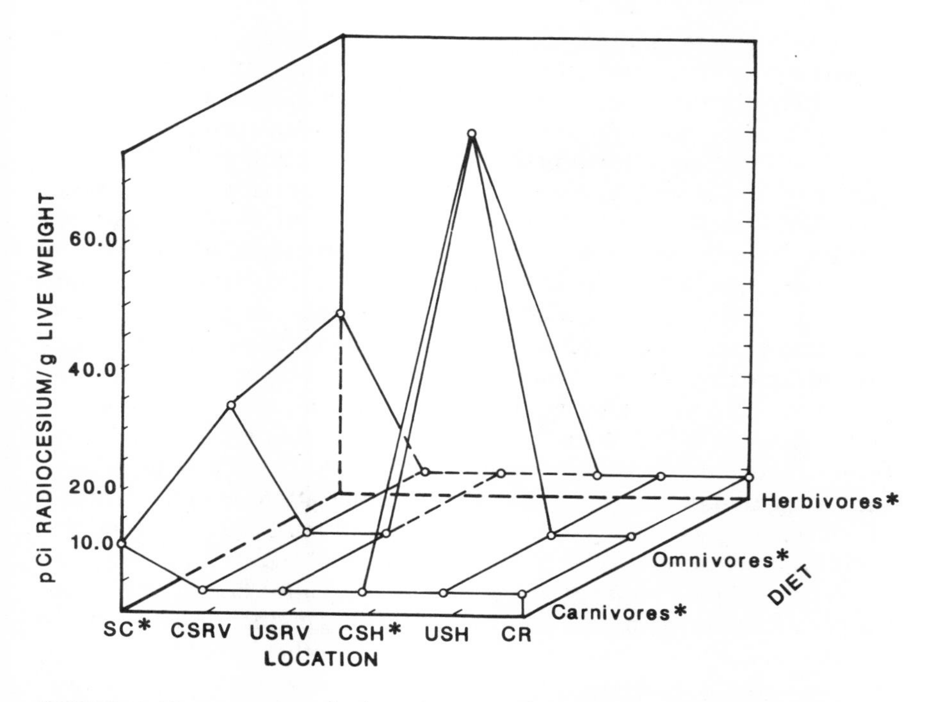

FIGURE 10. Representation of a three-dimensional "response surface" depicting whole-body radiocesium body burdens of adult summer birds collected on the U.S. Department of Energy's Savannah River Plant. (Total = 255 individuals, representing 53 species.) Locations of collection are as indicated in Fig. 7. Lines are joined to facilitate interpretation and do not imply functional relationships. Asterisks indicate categories of diet or locality within which significant differences were detected ($p \leqq 0.05$). From Straney *et al.* (1975).

1982). This mechanism would produce the greatest contamination differential between wildlife species and domestic livestock in those regions where soil nutrient reserves are either low or unavailable to plants. As summarized by Whicker and Schultz (1982), such areas include the nutrient poor soils of the United States' southeastern coastal plain and the regions of permafrost and/or shallow soils underlying alpine or arctic tundra. The importance of this phenomenon as a factor determining the environmental distribution of radionuclide contaminants that may enter the human food chain through wild game birds will be discussed later with regard to radioactive contaminants released by the Chernobyl nuclear accident.

4.3. Radionuclide Uptake and Dispersal by SRP Waterfowl

The temporal pattern of changes in the levels of radiocesium shown by coots during their winter stay on the Par Pond reservoir system clearly suggests the uptake of this radionuclide by these birds during the first three to four months of their residence on the site (Figs. 8 and 9). Previously, this uptake would have been described with the classic uptake model of Davis and Foster (1958), as described by Equation (1). Indeed, earlier attempts by Fendley *et al.* (1977) to produce empirical data in support of this model for the uptake of radiocesium by free-franging Wood Ducks in the SRP's Steel Creek watershed indicated that this model provided a close approximation of the amount of time that Equation (1) would predict would be required for 90% of equilibrium radiocesium body burdens to be reached (20.9 days predicted vs. 23.4 days observed). In addition, data reported by Brisbin and Swinebroad (1975) for individually banded and recovered coots resident on the Par Pond reservoir confirmed that individuals were increasing in radiocesium whole-body concentrations at rates comparable to those suggested by the body burdens of monthly cross-sectional samples obtained from the population by shooting.

With the development of the Richards sigmoid model as a means for describing the contaminant uptake of free-ranging birds (Brisbin *et al.*, 1989b) a study was designed to test the hypothesis that Equation (1) was indeed the most appropriate for describing the uptake of radionuclides by free-ranging birds foraging in naturally contaminated habitats. This study (Potter, 1987; Brisbin *et al.*, 1989b) used coots which were captured in the uncontaminated West Arm of Par Pond and therefore had low or negligible radiocesium body burdens. These birds were then transplanted to the more contaminated Pond B reservoir (Fig. 6) after first marking them with neck collars to allow later identification and wing-scissoring to prevent return flights. Least-squares fits to data for the radiocesium body burdens of these birds as they were periodically recaptured from the Pond B habitat showed that the Richards sigmoid model, as described by Equation (3), gave a statistically superior fit to the uptake data as compared to the classic model of Equation (1) (Brisbin *et al.*, 1989b; Fig. 11). In this particular case, the most appropriate curve shape parameter had a value of $m = 1.2$, rather than $m = 0$ as would have been predicted if Equation (1) had been more

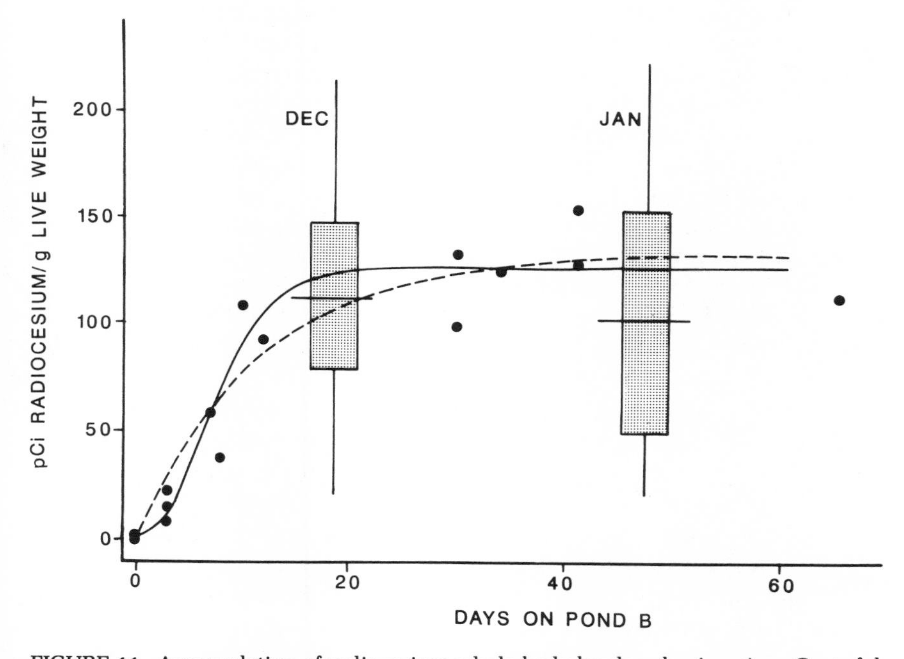

FIGURE 11. Accumulation of radiocesium whole-body burdens by American Coots following their release in December 1987 on the Pond B reservoir of the U.S. Department of Energy's Savannah River Plant, after being wing-scissored and transplanted from an uncontaminated portion of the nearby Par Pond reservoir (Fig. 6). Each point represents the recapture of a different individual, except at day = 0 which represents the average initial body burden of all individuals at the time of release. Curves represent least-squares fits to the data points using either Equation (1) (dotted line) or Equation (3) with the curve shape parameter, $m = 1.2$ (solid line). The goodness of fit of the solid line was significantly greater than that of the dotted line ($p \leqq 0.05$). Horizontal lines, shaded rectangles, and vertical lines represent, respectively, the means, ± two standard errors, and range of corresponding whole body burdens of two monthly samples of 10 wild free-ranging coots per month that were also collected from Pond B in December 1987 and January 1988. Redrawn with additional data from Brisbin *et al.* (1989b), as based on studies described by Potter (1987).

appropriate. An m value of 1.2 suggests a curve shape closer to the Gompertz sigmoid model as would occur when m approaches a value of 1.0 (Richards, 1959). As indicated by Fig. 11, the classic uptake model of Equation (1) tended to overestimate early radiocesium uptake and then underestimate it later as asymptote was approached. The greatest disparity between the two models occurred after the birds had been in the contaminated habitat for only three to four days. It is noteworthy that in suggesting that Equation (1) gave an adequate fit to data for the uptake of radiocesium by Wood Ducks, Fendley *et al.* (1977) relied on a data set that had no values for this early part of the uptake process when the two models would have diverged to the greatest extent. Most of the latter study's data were from birds that were either approaching or had passed the time predicted to attain asymptote—a period when the two models agreed much more closely in the coot study (Fig. 11). In the latter case, asymptotic contamination levels of 4.98 vs. 4.73 Bq/g live weight were predicted by Equations (1) and (3), respectively, with corresponding times to attain approximately 90% of asymptote (T) being 23 and 14 days (Brisbin *et al.*, 1989b). After only a three-day stay on the contaminated reservoir, however, the body burdens predicted by the two models differed by over 280%. It is precisely just such short stays that would be most common in the case of waterfowl which encounter contaminated wetlands in the course of their migratory journeys, and the consequence of the differences between the predictions of the two uptake models will be demonstrated later in Section 5 in discussing the implications of the Chernobyl nuclear accident.

While the exact cause of the initial lag in radiocesium uptake by the coots is not known exactly, it likely involves both behavioral factors which are expressed only under free-ranging conditions, as well as mathematical consequences related to modeling multicompartment contaminant uptake processes, as discussed in detail by Brisbin *et al.* (1989b). There is little likelihood, however, that the sigmoidal radiocesium uptake demonstrated by the coots (Fig. 11) is either an artifact or unique to this particular species, contaminant, or habitat, since Brisbin *et al.* (1989b) have demonstrated similar sigmoidal characteristics in the uptake of radiocesium by free-living yellow-bellied turtles (*Pseudemys scripta*) in the Pond B reservoir and in the uptake of mercury from contaminated aquarium water by captive mosquitofish (*Gambusia affinis*).

Finally, previously unpublished data from the author's own studies indicate that sigmoidal uptake models may also be appropriate for describing the uptake of radiocesium by birds foraging in contaminated terrestrial habitats. In this case, feral bantam chickens representing a strain whose development has been described in detail by Latimer (1976) were released to forage freely in radiocesium-contaminated floodplain habitat fringing the same SRP Steel Creek study area used by Fendley *et al.* (1977). These birds were periodically recaptured and subjected to live whole body counting techniques using equipment and procedures similar to those described by Fendley *et al.* (1977). Predation on these birds prevented the collection of uptake data long enough for the attainment of an asymptote sufficient for analysis with the Richards model. However, a qualitative assessment of this uptake data (Fig. 12) clearly shows a concave-upward curve with an initial delay in contaminant uptake relative to the prediction of Equation (1)—again the same result as indicated earlier for coots from the Pond B reservoir (Fig. 11).

A frequency distribution of the radiocesium body burdens attained by these bantam chickens after 70 days of foraging in the contaminated floodplain habitat (Fig. 13) shows a positive skewness with only relatively few individuals having higher levels of radiocesium and the majority showing much lower levels of contaminant uptake. The same positively skewed frequency distribution has been found by Pinder and Smith (1975) to typify the distribution of radiocesium levels in a variety of forms of biota including, for example, those of coots collected and studied by Brisbin *et al.* (1973). Thus, even though the "surrogate" bantam chickens were not native to the site where they were introduced and although they were continually provided with access to a source of uncontaminated commercial game bird feed, they still displayed the same basic patterns of contaminant uptake and distribution as native bird species, suggesting that they might serve as satisfactory analogs in which these processes may be studied under controlled experimental conditions.

Because of the flight mobility and migratory status of birds such as Wood Ducks and American Coots, a frequency distribution of body burdens such as that shown by Fig. 13 is frequently explained in these species by presuming that the most contaminated individuals had resided in the contaminated habitat for longer periods of time than those with lower body burdens. Comparing Fig. 12 with Fig. 13, however, indicates that this need not necessarily be the case, since the same distribution pattern was produced in the bantam chickens even though all individuals had been in the contaminated habitat for exactly the same amount of time, and during that time they had foraged together as a closely knit and socially organized flock. This suggests that differences between individuals such as those shown in Fig. 12 and Fig. 13 must either be due to physiological differences in the way that the

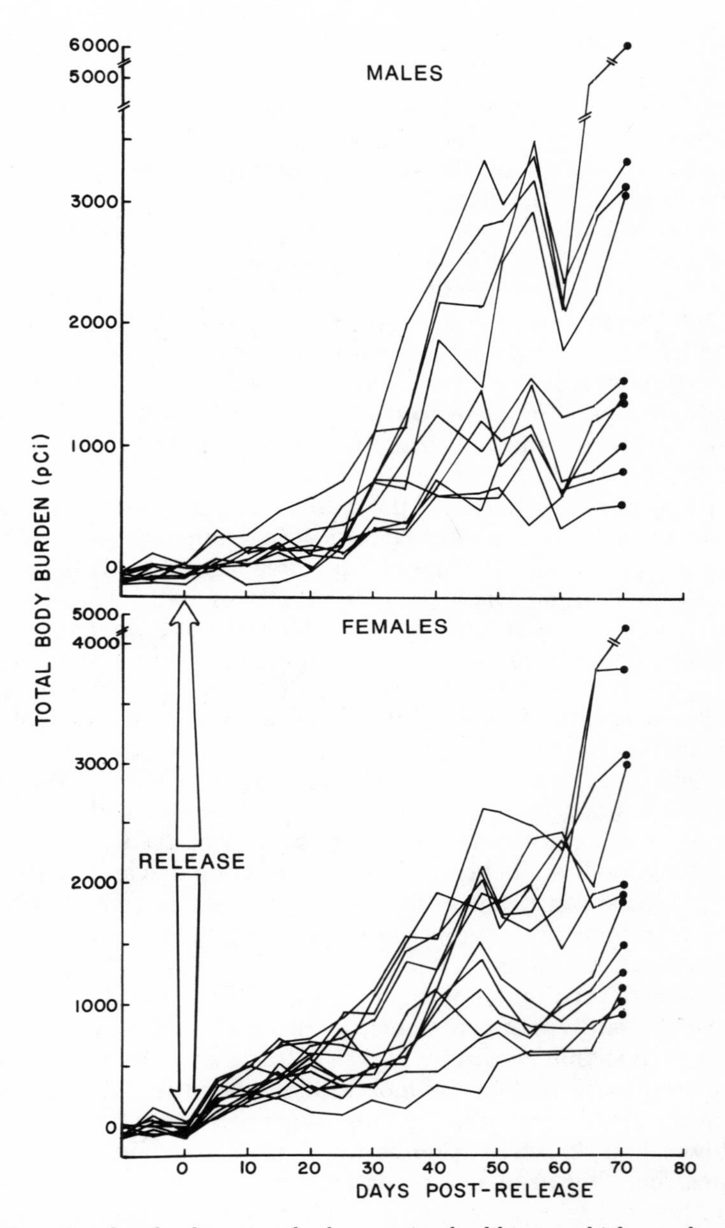

FIGURE 12. Uptake of radiocesium by free-ranging feral bantam chickens released into the contaminated floodplain forest bordering Steel Creek on the U.S. Department of Energy's Savannah River Plant (area SC of Fig. 7). Birds were allowed free access to uncontaminated game-bird feed during the uptake study but, as indicated by their increasing whole-body burdens, gradually increased their use of contaminated food resources in the surrounding habitat.

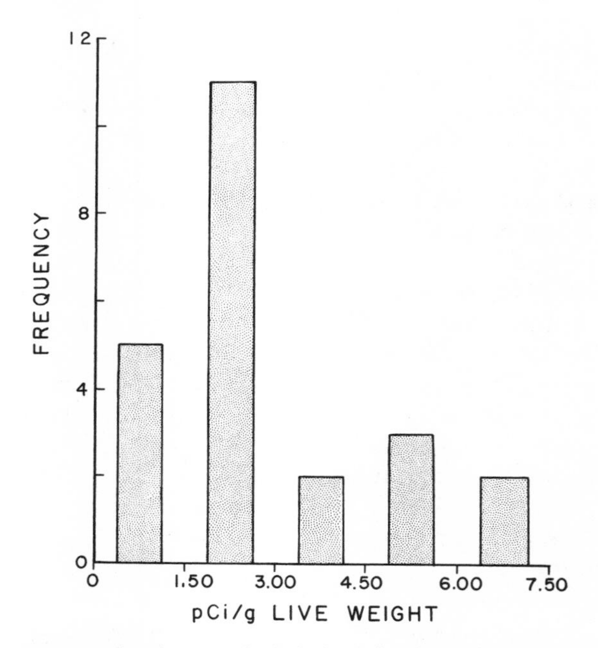

FIGURE 13. Frequency distribution of whole-body burdens of radiocesium in feral bantam chickens after foraging for 70 days in the contaminated floodplain forest bordering Steel Creek on the U.S. Department of Energy's Savannah River Plant (area SC of Fig. 7). This frequency distribution was generated from the final body burdens of all birds whose radiocesium uptake curves are presented in Fig. 12.

isotope behaves within the individual birds' bodies or to more subtle differences in behavioral or ecological aspects of the ways that individual birds exploit the habitat's contaminated food resources. Tamed and easily-observed birds such as bantam chickens offer an opportunity to study such factors to a much more thorough degree than would be possible in wild native species such as ducks or coots.

Following the initial increase in radiocesium in coots wintering on the SRP reservoirs, there is a period of brief stabilization of body burdens in December and January, followed by a decline from February through May. As a result, the last coots to leave the reservoir in spring show radiocesium levels that are not significantly higher than those of the first incoming birds to arrive in fall (Brisbin *et al.*, 1973; Fig. 8). (It should be noted here that an inadequately designed monitoring program that only compared body burdens of the first birds to arrive vs. those of the last to leave would have erroneously concluded that no bioaccumulation of this isotope was occurring in the coots from this habitat.) While the later winter through spring decline may suggest some form of excretion of the isotope—due perhaps to seasonal changes in physiology or dietary habits—a closer examination of the basic natural history of the birds involved provides a simpler explanation.

As demonstrated by Brisbin *et al.* (1973) the sex ratios of monthly samples of coots collected from the Par Pond reservoir varies significantly from month to month throughout their winter stay. Sex-ratios of early-arriving birds in October and those departing later in April and May did not differ significantly from 1 : 1, while those collected between November–March contained significantly more males—sometimes comprising as much as 87% of the sample. This suggests that, like many other species of waterfowl, coots show differential migration patterns of the sexes, with females tending to move on farther south in fall, leaving male-dominated population cohorts in the more northern parts of the wintering range (Welty, 1962; Johnsgard, 1968). As a result these changing sex ratios can be used as "markers" of different population cohorts, and the declining average radiocesium body burdens of coots sampled in late winter–spring can now be seen as a consequence of population "dilution," with the resident male-dominated cohorts of December and January achieving asymptotic levels of contamination and then departing for more northern breeding grounds while being replaced by more female-dominated cohorts from the south. Since these latter birds had not wintered on the reservoir, their radiocesium body burdens would be lower and the resulting population average would decline.

While the preceding scenario would seem relatively simple and straightforward to most ornithologists, it must be remembered that the American Coot is not obviously sexually dimorphic, and, once again, were it not that a monitoring program using these birds included the input of individuals with the proper basic ornithological expertise, possibly erroneous conclusions could have been reached concerning the causes and implications of the monthly changes observed in radiocesium contents (Fig. 8). Furthermore, as indicated by Clay *et al.* (1980), an understanding of the basic natural history and migratory behavior of Par Pond coots can also be used to better interpret monthly changes in nonnuclear contaminants such as mercury, despite the differences in both the spatial and temporal cycling dynamics of these two contaminants in these same birds on Par Pond. This provides still another example of the benefits of "closing the loop" between radioecological study and both basic and applied ecological concerns outside of the nuclear field.

The seasonal uptake, concentration, and dispersal of radiocesium by coots and other migratory waterfowl wintering on the SRP's Par Pond reservoir system has resulted in the annual removal of an estimated 3.8×10^{-5} Ci of this isotope from the site, as calculated by Brisbin *et al.* (1973) on the basis of contamination levels found in the winter of 1971–1972. Although this isotope has presumably been redistributed by these birds along their migratory pathways, the radiocesium levels of all of the birds sampled in this study (Figs. 8 and 9) were well within the range of radiocesium levels reported for other species of wild game from elsewhere throughout the southeastern coastal plain of the United States (Jenkins and Fendley, 1968). The highest individual radiocesium value ever recorded for Par Pond waterfowl was 171 pCi/g live weight for a Green-winged Teal (*Anas crecca*), as reported by Marter (1970). Even at this level, however, the radiocesium contents of such a bird would be below the levels that would, according to the calculations of Jenkins and Fendley (1968), suggest a health hazard to persons who might eat them as food, considering the amount of wild waterfowl meat that an average person would be likely to consume in a year. Moreover, cooking such meat would even further reduce its ^{137}Cs concentration (Halford, 1987).

4.4. Long-Term Changes in Radionuclide Levels of SRP Waterfowl

Once an accidental environmental release of a radionuclide has occurred, the first question of immediate practical importance concerns the rate, pattern, and extent to which the contaminant may be taken up and accumulated in biota which may either be consumed by man or be directly harmed themselves by the contaminants. Once equilibrium levels have been achieved, the next important question concerns the length of time that such contamination levels can be expected to persist either above background or at levels which might pose a hazard. Answers to such questions require knowledge of the loss-rate functions by which levels of radioactive contaminants may be expected to decline under natural (and/or perhaps remediated) environmental conditions. The basic principles related to such declines have been presented in Section 2.3, in which the importance of environmental characteristics in determining these rates and patterns of contaminant decline have been demonstrated (e.g., Table III). Thus, in order to be able to estimate the times that will be required for given declines in radionuclide contamination levels to occur following various accident scenarios, these processes must be studied in the biota of as wide a variety of naturally-contaminated habitats as possible (Section 2.5; Table IV). The SRP has two habitats where such studies can and indeed have taken place and are still continuing with regard to resident waterfowl and other species; some of the findings of these studies have been presented earlier (Table III; Section 2.3).

In addition to the results reported above, Brisbin and Vargo (1982) have studied both the spatial and temporal aspects of the long-term decline in radiocesium levels of coots wintering on the Par Pond reservoir. Their results have shown that the spatial pattern of radiocesium distribution between coots collected from the three major arms of the reservoir (Fig. 6) has not changed as radiocesium levels of coots declined uniformly in all parts of the reservoir habitat (Fig. 14). Thus, within the Par Pond reservoir itself, there has been no tendency for the contaminant levels in the coots to reflect any "downstream" movement of radiocesium toward the Par Pond dam or the point of cooling water pump-out in the West Arm (Fig. 6). This is in keeping with the findings of Potter (1987) which showed that the reservoir habitat of Pond B reduced the tendency of radiocesium to be lost downstream, as compared to the Steel Creek swamp delta where the long-term decline of radiocesium in Wood Ducks has been more rapid as was described earlier (Fendley, 1978; Table III).

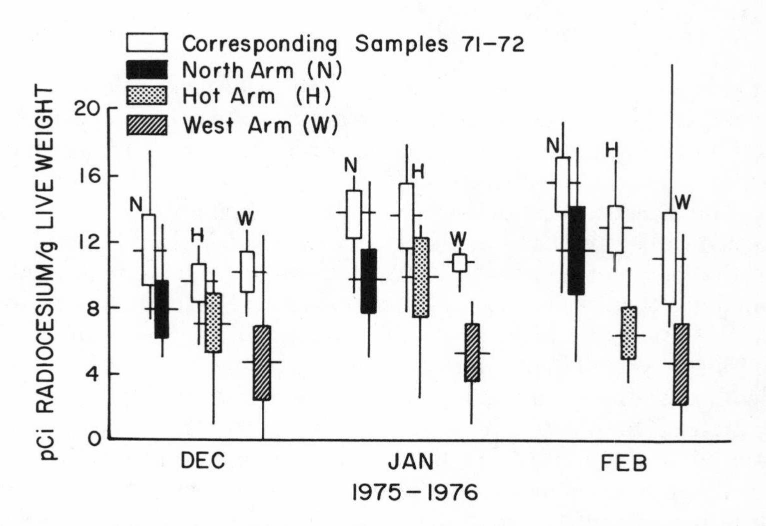

FIGURE 14. Four-year declines in whole-body burdens of radiocesium in American Coots sampled from the three major arms of the Par Pond reservoir of the U.S. Department of Energy's Savannah River Plant (Fig. 6). Horizontal lines represent means, rectangles represent ± two standard errors, and vertical lines indicate ranges of samples of ten birds each. From Brisbin and Vargo (1982).

These findings have important implications for choosing between alternate strategies for managing wetland habitats which may have been contaminated by radioactive releases, and which might prove attractive to migratory waterfowl. Several such strategies have been outlined by Fendley (1978) for the Steel Creek swamp delta, where his data indicate that three factors combine between December and February to increase the potential for food chain transfer of radiocesium to man through the consumption of waterfowl, which (1) use this area in peak numbers during this winter period, (2) show maximal levels of radiocesium body burdens during these months, and (3) are subject to legal shooting by hunters during open game seasons which occur during this same time period. While building a dam to convert the lower reaches of the Steel Creek watershed to reservoir habitat would almost certainly slow the loss of radiocesium downstream and offsite to the Savannah River, the data reported above suggest that such an action would substantially decrease the rate of long-term radiocesium decline in the waterfowl of the area. The waterfowl community would then also change in species composition as diving duck species, which frequent such open waters, would come to replace the dabbling duck species as water levels in the swamp rise, decreasing the foraging value of the habitat for the latter (Fendley, 1978). However, proper engineering of the basin configuration of such an impoundment could, by increasing overall water depth and the steepness of shallow water profiles to the maximum extent possible, also decrease the foraging value of the area to most diving ducks. In the case of the species of diving ducks found on Par Pond, for example, nearly all foraging usually occurs in water of less than ten meters in depth (Smith *et al.*, 1986).

As in the case of waterfowl leaving Par Pond, Wood Ducks collected in the present Steel Creek swamp delta would not be a hazard to human health if consumed as food, as again suggested by dose estimates calculated by Jenkins and Fendley (1968). However, the data and strategies suggested by this work can provide a model for the management of wetlands contaminated by higher levels of radiocesium release such as almost certainly has occurred in the case of the Chernobyl nuclear accident (Anspaugh *et al.*, 1988).

5. APPLIED AVIAN RADIOECOLOGY: THE CHERNOBYL ACCIDENT

The April 1986 Chernobyl nuclear accident in the Soviet Union has provided a classic and now well studied case of global contaminant cycling resulting from processes based on physical (principally meteorological) transport mechanisms. What has so far remained largely unstudied, however, is the potential for the biotic transport of radioactive materials away from the heavily contaminated habitats in the area immediately surrounding the reactor site. A number of studies have dealt with the uptake of Chernobyl contaminants by biota in regions to which the contaminants were carried as a result of physical/ meteorological transport processes (e.g., Ruiz *et al.*, 1987; Anspaugh *et al.*, 1988; Baeza *et al.*, 1988). However, this is a quite different situation from biotic transport *per se,* in which mobile biota first become contaminated in the immediate vicinity of the reactor site itself and then later move the contaminants contained within their bodies elsewhere after they leave the area. The contrast between the physical and biotic transport of such contaminants is particularly important if, as will be indicated in the case of the Chernobyl accident, the two (physical vs. biotic) processes result in transport in different directions and/or at different rates.

The biotic transport of radioactive contaminants also becomes particularly important if it results in these substances being moved over long geographic distances while being simultaneously introduced into human food chains in ways that may normally be difficult to assess or predict. All of these conditions prevail in the case of migratory waterfowl. On their fall and spring migratory journeys such birds may fly hundreds or even thousands of km, sometimes in only a few days, and this would result in their eliminating only a small proportion of the radionuclide body burdens which they may have accumulated during a stopover in the wetlands surrounding the Chernobyl site. Furthermore, while wild waterfowl are occasionally shot and eaten as food by hunters throughout the world, in many of the underdeveloped countries through which Chernobyl migrants may move (e.g., those in the Middle East and/or North Africa), the consumption of wild waterfowl and other birds may be more than casual. In many such countries, the use of wild birds and other game may be important from a subsistence point of view and many species which may seldom be eaten in other parts of the world (e.g., coots and other rails) may actually serve as staples of the diet (Ripley, 1976). While both waterfowl and upland game birds may thus become potential contaminant vectors, the tendency of reactor-released radionuclides to quickly enter and become concentrated in wetland areas make waterfowl and other species of aquatic birds particularly likely to be exposed, as has been documented by Fendley (1978) and Straney *et al.* (1975). Considering these factors, it is unfortunate that while considerable effort has been focused on documenting the uptake and concentration of Chernobyl-released contaminants by domestic livestock and their products (e.g., WHO, 1986; Anspaugh *et al.*, 1988), relatively little attention has been paid to commonly consumed waterfowl and other wild game in the same regions. This is even more ironic when it is considered that, as explained earlier, the contamination of areas inhabited by both domestic livestock and wild game will, all else being equal, result in higher levels of radionuclide contamination being shown in the wild game (Jenkins and Fendley, 1968).

The importance of the extensive wetlands found in the immediate vicinity of the Chernobyl reactor site as habitat for migratory waterfowl was recognized and documented as much as two decades ago by Isakov (1966). Generally known as the "Pripyat Marshes," or on some maps the "Pinsk Marshes," these wetlands are shown by satellite imagery to extend to and immediately surround both the Chernobyl reactor site itself and its cooling reservoir (Edwards and Raymer, 1987; Sadowski and Covington, 1987). They also extend for over 300 km to the northwest, principally along the fringes of the Pripyat River to near the Polish border. The total area of wetlands actually involved may be as much as 15,000 km^2 as estimated from standard geographic maps (Hall, 1981). The central 300 km^2 of this area surrounding the reactor site has already been designated as an "ecological reserve" for studies of radiation impact, and evacuation of residents occurred over an area of more than 2000 km^2 (Anon., 1988; WHO, 1986).

While no published data seem to exist for radionuclide contamination levels in soil, sediments, water, or biota of these wetlands as such, more general studies of radionuclide deposition rates and exposure levels suggest that, particularly in the wetlands adjoining the reactor site itself, it would not be unreasonable to expect extremely high levels of ^{137}Cs, one of the most abundant and long-lived of the radioisotopes released by the accident (Hohenemser *et al.*, 1986; Medvedev, 1986; Anspaugh *et al.*, 1988). The importance of ^{137}Cs as a contaminant of waterfowl and other wild game is further emphasized by its tendency to accumulate in highest concentrations in the animal's skeletal muscles (Narayanyan and Eapen, 1971; Brisbin and Smith, 1975; Potter *et al.*, 1989) which are precisely the tissues that would later be cooked and eaten by humans as food (Halford, 1987; Table I).

Since no published information is available to document radionuclide levels in waterfowl or other wetlands biota from the Pripyat Marshes either before or after the Chernobyl accident, indirect methods must be used to estimate the contamination levels which might conceivably be expected under different exposure and/or uptake scenarios in these areas. This may be done using those principles and calculations developed through studies of radiocesium cycling in waterfowl in other contaminated habitats such as those of the Steel Creek Swamp and Par Pond reservoir system of the Savannah River plant, as described in Section 4. Both of these major wetland habitat types at the SRP are likely to resemble their respective counterparts at the Chernobyl site in many ways, corresponding to the wetlands of the Pripyat Marshes and the Kiev reservoir south of the Chernobyl site, respectively.

5.1. Estimation of Radionuclide Transport by Waterfowl

The purpose of the calculations presented here is to determine the possibility of waterfowl residing in or visiting the Pripyat Marshes and/or Kiev reservoir to serve as vectors of ^{137}Cs contamination to human consumers along their distant migratory pathways, at levels that might equal or exceed those currently permitted in these areas for this isotope in human diets. Specifically, using information published for ^{137}Cs uptake, concentration, and elimination by SRP waterfowl, indirect estimates have been made (Table VIII) of the proportion of the 3.7×10^{16} Bq of this isotope estimated to have been deposited in European Russia following the Chernobyl accident (Anspaugh *et al.*, 1988) that would have to have been released directly into the 1,000 km^2 of wetlands immediately adjacent to and/or surrounding the reactor site in order to result in a reasonable probability of contaminating migratory waterfowl to a level exceeding 600 Bq ^{137}Cs/kg fresh weight of muscle, the maximum level generally permitted in meat for human consumption in Europe (EEC, 1986).

The effects on the birds' ^{137}Cs levels were also estimated for a migratory flight of up to 3000 km. At this distance, and considering the general flyway patterns described by Isakov (1966), contaminated birds would have had an opportunity to reach the vicinity of most of the major population centers of southern and southwestern Europe as well as their wintering grounds in much of the Middle East and North and Northwest Africa (Fig. 15). As indicated by Isakov (1966), waterfowl found in such areas as Scandinavia and the British Isles would not be expected to include the Chernobyl region within their migratory journeys. They would therefore normally only be exposed to Chernobyl contamination which had first been diluted by atmospheric processes while being transported to these areas. The reporting of ^{137}Cs waterfowl body burdens which are either very low or nondetectable in such regions (Masconzoni, 1987; Hancock and Woollam, 1987) is therefore not surprising. However, such findings do not preclude the possibility of much higher ^{137}Cs levels occurring in wintering waterfowl from

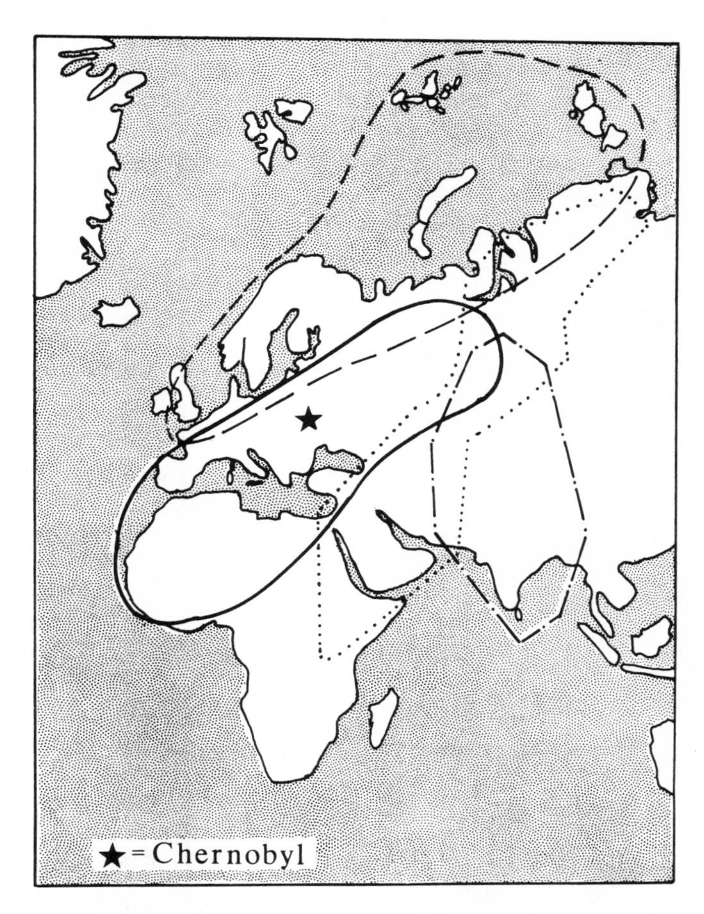

FIGURE 15. Major waterfowl migratory flyways of Europe and European Russia in relation to the site of the Chernobyl nuclear accident. The solid line outlines the only flyway which includes the Chernobyl site within its boundaries, while broken lines outline other flyways which do not include the Chernobyl site. Redrawn with modifications from Isakov (1966).

southern Europe and/or northern Africa where such birds have a greater likelihood of having had direct contact with the contaminated wetlands of the Pripyat Marshes around the reactor site itself (Fig. 15). It is important to note here that while the principal direction of the physical transport of the Chernobyl radionuclide release by atmospheric factors was initially to the northwest, the principal direction of transport by waterfowl of radionuclides away from the site would be to the southwest.

TABLE VIII
Estimation of Radiocesium Releases to the Wetlands Surrounding the Chernobyl Nuclear Power Station That Would Result in Significant Levels of Waterfowl Contamination (Annotations Indicated by Superscripts Are Given in Text)

	Waterfowl established as residents that attain a steady-state equilibrium body burden	Waterfowl making a three-day stopover on migration	
		"Classic" negative-exponential uptake model[a]	Richards sigmoidal uptake model[b]
Maximum level of radiocesium permitted by the EEC in meat for human consumption (Bq/kg cooked breast muscle)	600[c]	600[c]	600[c]
Corresponding level in uncooked waterfowl skeletal muscle (Bq/kg)	822[d]	822[d]	822[d]
Corresponding whole-body burden of bird at time of harvest by a hunter on the wintering grounds, after an assumed migratory journey of 3000 km in 3 days[f] (Bq/kg live weight)	457[e]	457[e]	457[e]
Corresponding whole-body burden of bird at the time of departure from Chernobyl wetlands (Bq/kg live weight)	607[g]	607[h]	607[h]
Corresponding equilibrium whole-body burden (Bq/kg live weight)	607[i]	2437[j]	6064[k]
Corresponding equilibrium inventory of ^{137}Cs in all waterfowl resident in the Pripyat Marsh wetlands (Bq/km^2)	76.5×10^2[l]	30.7×10^4[l]	76.4×10^4[l]
Corresponding total radiocesium deposition in the Pripyat Marsh wetlands (Bq/km^2)	17.7×10^{10}[m]	71.2×10^{10}[m]	17.7×10^{11}[m]
Radiocesium estimated to actually have been deposited in European U.S.S.R. as a result of the Chernobyl accident (Bq/km^2)	6.98×10^9[n]	6.98×10^9[n]	6.98×10^9[n]

The calculations made in Table VIII are often complex and require further explanations which are indicated by superscripts that will be described in detail below:

(a) As described by Equation (1).
(b) As described by Equation (3), using a curve shape parameter value of $m = 1.2$, as determined by Brisbin *et al.* (1989b).
(c) EEC (1986).
(d) Calculated on the basis of cooking causing a 27% reduction in ^{137}Cs concentration (Halford, 1987).
(e) Calculated from the relationship between ^{137}Cs levels in whole body burdens vs. skeletal muscles of American Coots as provided by Potter *et al.* (1989):

$\bar{x}$ (Bq/g fresh muscle)/(Bq/g whole body burden) = 1.8.

(f) Based on information provided by Welty (1962), indicating that migratory passages of ducks and geese are capable of covering distances as great as 2400–2700 km at overall rates of between 40–80 km/hr. The calculations made here are based on the minimum overall rate of travel of 40 km/hr.
(g) Calculated on the basis of a depuration half-life of 11 days for ^{137}Cs, as determined by Halford *et al.* (1983) for Mallards; calculations assume that due to the more rapid turnover of body tissue resulting from increased energetic demands for sustained migratory flight, ^{137}Cs will be lost at a rate 1.5 times as great as that given above, during the three days of migratory movement from the Pripyat Marshes.
(h) In the case of birds which make only a three-day stopover in the contaminated area, these contamination levels would be considerably less than the projected equilibrium levels given below, which would never actually be attained (see superscripts *j* and *k*).
(i) In the case of established residents, the body burden at the time of departure would be the same as the final equilibrium level of contamination which was attained.
(j) Calculated from the equation given in superscript *b*, by setting C_t = 607 Bq/kg, as explained in superscript *h*, and $t = 3$ days, and solving for C_e. A value of $m = 0$ was used to produce the classic negative exponential model, and T was set = 23 days, as determined by Brisbin *et al.* (1989b) for the uptake of radiocesium by waterfowl when $m = 0$. C_o, the "background" level of ^{137}Cs whole body burdens shown by mi-

grants which had just arrived at the Pipyat Marshes, would reflect levels of global fallout of this isotope, and was set = 62 Bq/kg on the basis of the average body burden of 14 coots collected from uncontaminated habitats on the Savannah River Plant (Brisbin *et al.*, 1989b).

(k) Calculated as explained in superscript *j*, but using a value of $m = 1.2$, as indicated in superscript *b*, and setting $T = 14$ days, as determined by Brisbin *et al.* (1989b) for the uptake of radiocesium by waterfowl when $m = 1.2$.

(l) Assuming that the carrying capacity of the wetlands of the Pripyat Marshes for waterfowl is similar to that of the Pond B reservoir of the United States Department of Energy Savannah River Plant which supports an average of 1.26 kg of live waterfowl biomass/ha (Whicker *et al.*, 1989). This value was estimated by these authors, using the method of Brisbin (1989) and waterfowl census data from Brisbin (1974) and Mayer *et al.* (1986).

(m) Calculated on the basis of $4.3 \times 10^{-5}\%$ of the total equilibrium ecosystem inventory of ^{137}Cs residing within the resident waterfowl population as estimated by Whicker *et al.* (1989) for the Pond B reservoir.

(n) From Anspaugh *et al.* (1988).

It should be noted that events in Table VIII are listed in the reverse order from which they would normally occur. That is to say, the table begins with the accepted level for ^{137}Cs in cooked meat from a bird collected by a hunter on its migratory wintering grounds after having traveled 3000 km from contaminated wetlands in the vicinity of the Chernobyl reactor site, where it either made a migratory stopover or resided long enough to attain equilibrium levels of ^{137}Cs concentration.

5.2. Conclusions: The Chernobyl Accident and Risks to Man and Bird

The calculations presented in Table VIII indicate that there is less than two orders of magnitude difference between the amount of radiocesium deposition actually estimated to have occurred and that amount which would produce levels that would exceed the maximum permitted for human consumption (EEC, 1986), after harvesting the birds on their southwestern European/north African wintering grounds. The 1000 km² of wetlands most closely surrounding the Chernobyl site represent less than 0.1% of all of the area estimated for the European

U.S.S.R. as a whole (Anspaugh *et al.*, 1988). This is less than half of the area over which evacuation of all residents was undertaken following the Chernobyl accident (WHO, 1986), and it is probably not unreasonable to suggest that radiocesium deposition, on an areal basis, in those 1000 km^2 of Pripyat Marsh wetlands closest to the Chernobyl site may have actually been more than two orders of magnitude higher than that which has been estimated for all of the European U.S.S.R. as a whole.

Such deposition rates would actually have occurred in this area if the amount of this isotope released to these 1000 km^2 of marsh wetlands exceeded between 0.18–1.8% of the total ^{137}Cs inventory which has been estimated to have been released by this accident (Table VIII). To put this in perspective, Anspaugh *et al.* (1988) have estimated that ^{137}Cs deposition in the European U.S.S.R. alone has accounted for 38% of the total release of this isotope from the accident. Considering these facts and the degree of variation possible in many of the assumptions underlying the calculations in Table VIII, a discrepancy of only two orders of magnitude or less would seem to be scant assurance that no threat would be posed to public health by the events postulated here. In other words, these calculations, if later supported by ^{137}Cs data collected from the Pripyat Marshes themselves, suggest that large numbers of migratory waterfowl, whose ^{137}Cs levels exceed those permitted for human consumption, might well be moving from the region of the Chernobyl accident to wintering grounds throughout southwestern Europe, north Africa, and the Mediterranean region (Fig. 15).

Whether or not these calculations indicate a credible expectation will ultimately depend on the development of more detailed models for the micro- and mesogeographic deposition of Chernobyl fallout in these wetlands and above all, upon the evaluation of data for ^{137}Cs concentrations in water, soils, sediment, and particularly biota from these portions of the Pripyat Marshes. At this point however, there appears to be sufficient justification to at least warrant the call for an international effort to collect and evaluate such data in as expeditious a fashion as possible. When combined with more detailed information obtained from banding and other basic biological studies of the waterfowl of the Pripyat Marshes, such information would help to more fully determine the extent to which a possible pathway of radionuclide contamination of the human food web may have been overlooked.

Although no published data seem to be available for levels of radiocesium or other radionuclides in birds from the wetlands in the immediate vicinity of the Chernobyl site, several studies have quantified Chernobyl-caused isotopic body burdens in avifauna from more distant areas—particularly in Europe (e.g., Ruiz *et al.*, 1987; Baeza *et al.*, 1988). As mentioned earlier, however, the distribution of radioactive materials in such avifauna away from the Chernobyl site has been, in many if not most cases, the result of the uptake of contaminants which were first transported away from the reactor site and deposited in European habitats by the physical processes of global atmospheric circulation, rainfall deposition, etc. In such situations it will generally not be clear whether or not the radioisotopic body burdens observed in individual birds were obtained *in situ* or elsewhere on their migratory journeys unless specific basic ornithological data, especially that based on banding data, exist to document the situation for the given population under study. In the case of the Mallards studied by Baeza *et al.* (1988) there is a strong likelihood that many of the birds sampled were from a nonmigratory population resident in southern Europe (Cramp, 1977). The failure to find substantial levels of contamination in such birds cannot therefore be necessarily construed as evidence that the waterfowl–radionuclide transport scenario outlined in Table VIII is not realistic, even though this region lies squarely within the flyway which includes the Chernobyl site and Pripyat Marsh wetlands (Fig. 15). Once again the importance of basic ornithological expertise to such a study is clear.

Regardless of whether or not observed avian radionuclide body burdens are the result of physically vs. biotically-transported Chernobyl contamination, a natural concern exists for the possible deleterious effects that such contaminants may have upon the avifauna in which they have been concentrated. Even in North America, where the deposition of Chernobyl-generated radionuclides has been much lower than in the case of Europe (Anspaugh *et al.*, 1988), it has been suggested that phenomena such as breeding failures of certain bird populations may be related to exposure to Chernobyl contaminant deposition (DeSante and Geupel, 1987). In the latter case, a breeding failure was documented in a land bird population under study at the Palomarin Field Station in northern coastal California, with the breeding failure occurring at the same time (May 1986) that globally-distributed Chernobyl fallout was detected in this portion of North America. It should be noted from the outset that the ability to detect such a breeding failure was the result of a long-term history of thorough studies of the breeding bird populations at that site. While breeding failures had occurred in previous years in this area, the failure in the spring of 1986 could not be attributed to any of the factors which had previously been associated with the appearance of abnormally low numbers of juveniles being recorded in netting operations after the nesting season. Under these conditions, although there was no direct evidence of excessive levels of radionuclide contamination of the area, the authors noted that although it may have merely been a coincidence, ". . . the time of the passage of the Chernobyl cloud coincided remarkably well with the timing of the onset of the reproductive failure at Palomarin", and that "the geographical area over which substantial rainfall was coincident with the passage of the cloud appears, at first glance, to coincide with the geographical areas that experienced some reproductive failure . . ." (Desante and Geupel, 1987, p. 651).

Unfortunately, even in the absence of any conclusive evidence of radionuclide contamination of the area in question, the mere suggestion of such a coincidence resulted in widespread public attention and concern. As a result, a sample of land birds collected at the Palomarin Field Station between June and October, 1986 was made available for determination of whole body radiocesium burdens with counting equipment and procedures used routinely at the Savannah River Ecology Laboratory as described in detail by Potter (1987). Of the total sample of 28 birds representing 16 species, only three showed levels of radiocesium which were significantly elevated above background. The maximum whole body burden of radiocesium was 38 pCi/g live weight, shown by a Wilson's Warbler (*Wilsonia pusilla*). Even at this level, however, the internal dose which would be received from such a level of gamma contamination would be well below that which would be expected to be of significant health concern to that bird. Halford *et al.* (1982) for example, present calculations leading to the conclusion that waterfowl using a radioactively contaminated waste pond in Idaho were not exposed to radiation doses from either external or internally-deposited radionuclides that would pose a hazard to either them or hunters who may eat them as food, even though these waterfowl averaged over 900 pCi radiocesium per g live weight and also contained other radioisotopes—levels of contamination over 20 times as great as the most contaminated bird from the Palomarin site. Halford *et al.* (1982) also estimated that the waterfowl studied by them could have received an average dose of 700 mrad over a period of about one year after leaving the waste pond. This would be the equivalent of about 0.0013 mrad/min or over four orders of magnitude less than the levels of exposure that have been shown to cause any detectable change in avian reproduction (Table VI).

Since radiation counts were made on the Palomarin birds in the spring of 1988, it would not have been possible to detect any ^{131}I that might have been in these individuals in 1986, due to the much shorter half life of this isotope as compared to radiocesium (Table I). However, considering the radiocesium levels observed, it is extremely unlikely that any significant degree of ^{131}I contamination could have occurred without concurrently elevating radiocesium levels above the detection limits used in these counting procedures.

Demonstrating the low likelihood of Chernobyl-generated radionuclide contamination contributing to the land bird reproductive failure described above does not diminish the importance of carefully investigating and documenting levels of radionuclide contamination each time that such a scenario might be suggested in the future. Rather, it argues strongly for the importance of clearly understanding the basic principles of both radioecology and basic ornithology, and then using them both to provide a context within which the risks associated with events such as the Chernobyl nuclear accident may be clearly evaluated and understood for both man and birds alike.

6. SUMMARY

The field of radioecology, which combines the basic principles and approaches of both basic ecology and radiation biology, was initially established in response to the need to understand the environmental fate and effects of radioactive materials released by nuclear weapons testing and nuclear industrial accidents. The Chernobyl nuclear accident of 1986 focused renewed attention on this field and has demonstrated the importance of including studies of all environmental components, including avifauna, in radioecological research.

A review of the basic principles of radiation physics and chemistry indicates that there are three basic forms of ionizing radiation. Alpha and beta emissions are effective over only short distances of exposure. Gamma radiation and the closely-related X-radiation however, exert impacts over much greater exposure distances and require more extensive shielding to prevent harm to biological materials. Radionuclides of greatest ecological importance are those which are most ubiquitous in environmental releases as well as on those which have long physical half-lives and therefore remain as environmental contaminants for extended periods of time. Also of concern are those isotopes which concentrate in particularly vulnerable portions of an organism's body and/or portions such as skeletal muscle or eggs which may be eaten by humans. Some of the most important radionuclides with regard to these factors are the transuranic elements, the isotopes of iodine and, particularly, cesium 137. The latter isotope acts as an analog of potassium and is concentrated in skeletal muscle. It is also one of the most abundant of the radionuclides released by most nuclear accidents and has a long physical half-life (30 years).

Although the majority of studies of radiation effects on birds have dealt with domestic poultry, the few studies which have used other species suggest that nestling passerines may be considerably more resistant to radiation stress than are the adults of larger-bodied precocial species. Despite a higher resistance to radiation-induced mortality however, nestling passerines frequently show proportionally greater disturbances of growth as the result of sublethal radiation exposures. Radiation-induced disturbances of growth can often have important consequences. In cases where wing feather growth is stunted for example, mortality can result from the inability to escape predators due to impaired flight. This would result in ecological lethality at levels of radiation stress which are much lower than those required to produce physiological lethality under laboratory conditions.

Studies of the uptake, concentration and transport of radionuclide contaminants by free-living birds have shown that some of the popular assumptions that have been made on the basis of laboratory studies do not apply when examined under natural conditions. The phenomenon of trophic level concentration, or biomagnification, for example, does not seem to apply in certain waterfowl communities contaminated with radiocesium. Similarly, the negative exponential (monomolecular) model has classically been assumed to describe radionuclide uptake pattern and rate. However, field studies of migratory waterfowl inhabiting a contaminated nuclear reactor cooling reservoir have shown that the curve shape assumed by this model is not shown under free-living conditions. A modified Richards sigmoid model demonstrates a lag in the occurrence of maximal rates of contaminant uptake, which is likely due to behavioral factors which are not expressed under conditions of continuous contaminant intake under controlled laboratory conditions.

The implications of these findings are discussed with respect to the potential for the contamination of waterfowl inhabiting wetlands surrounding the site of the Chernobyl nuclear accident in the Soviet Union. When combined with the information concerning radiation effects in birds as discussed above, this information suggests that there is little cause for concern for direct mortality or even significant reproductive impacts on waterfowl or other bird populations as a whole, although localized effects may be severe in limited areas. Of greater concern, however, is the potential for waterfowl (which may serve as common food items in many underdeveloped parts of the world) to be contaminated with levels of released radionuclides that would result in their posing a potential hazard to the diets of persons in such areas.

Evaluating the published literature and recent studies being conducted in the area of avian radioecology indicates the importance of a two-way feedback between applied concerns for the fate and effects of radioactive contaminants on one hand, and considerations of basic principles of ornithological science on the other. Like the field of radioecology as a whole, avian radioecology therefore requires the combination of studies and understanding in both basic and applied research areas, and a failure to understand the importance of this interaction can substantially impair the ability to use such research to address issues of immediate environmental concern to both man and birds alike.

ACKNOWLEDGMENTS. From the outset, my thinking in the area of radioecology has been guided by the teaching, research, and friendship of Drs. Eugene P. Odum, James H. Jenkins and F. Ward Whicker. For more than two decades, this work has been almost wholly supported by ongoing contracts (currently DE-AC09-76SROO-819) between the former United States Atomic Energy Commission, now the United States Department of Energy, and the University of Georgia. Over these years, Drs. William S. Osburn, Robert L. Watters, and D. Heyward Hamilton have provided particularly valuable guidance and advice in the development of my research program. None of this work would have been possible, however, without the hard work and contributions of the dedicated technicians and graduate students who have been more than colleagues to me over the years. Deserving of special mention have been the contributions made by Brenda E. Latimer, Peyton R. Williams, Dr. Tim Fendley, Deborah Clay Harris, Eric L. Peters, and Catherine M. Potter. Finally, the support of Drs. Michael H. Smith, Domy C. Adriano, Carl Strojan, John E. Pinder III, and Anthony L. Towns is gratefully acknowledged along with that of the other members of the research and support staffs of the Savannah River Ecology Laboratory. The use of facilities provided by the Reese Library and Department of biology at Augusta College, Augusta, Georgia is also acknowledged, as is the artwork for this and many of the other papers cited here which has been provided by Jean B. Coleman. Invaluable editorial assistance in manuscript review and preparation was provided by Edith Towns.

REFERENCES

Abraham, R. L., 1972, Mortality of Mallards exposed to gamma radiation, *Rad. Res.* **49**:322–327.

Alberts, J. J., Tilly, L. J., and Vigerstad, T. J., 1979, Seasonal cycling of cesium-137 in a reservoir, *Science* **203:**649–651.

Anderson, G. E., Gentry, J. B., and Smith, M. H., 1973, Relationships between levels of radiocesium in dominant plants and arthropods in a contaminated streambed community, *Oikos* **24:**165–170.

Anderson, S. H., Dodson, G. J., and Van Hook, R. I., 1976, Comparative retention of ^{60}Co, ^{109}Cd and ^{137}Cs following acute and chronic feeding in bobwhite quail, in: *Radioecology and Energy Resources* (Cushing, C. E., Jr., ed.), Special Publ. No. 1, Ecol. Soc. Amer., Dowden, Hutchinson, and Ross, Inc., Stroudsburg, Pennsylvania, pp. 321–324.

Anspaugh, L. R., Catlin, R. J., and Goldman, M., 1988, The global impact of the Chernobyl reactor accident, *Science* **242:**1513–1519.

Anon, 1986, Fears of more fallout from Chernobyl, *Discover* **7:**14, 16.

Anon, 1988, Chernobyl area to be ecological reserve, *Science* **240:**877.

Baeza, A., del Rio, M., Miró, C., Paniagua, J. M., Moreno, A., and Navarro, E., 1988, Radiocesium concentration in migratory birds wintering in Spain after the Chernobyl accident, *Health Phys.* **55:**863–867.

Bagshaw, C., and Brisbin, I. L., Jr, 1984, Long-term declines in radiocesium of two sympatric snake populations, *J. Appl. Ecol.* **21:**407–413.

Banks, W. C., Hensley, J. C., II, Krueger, W. F., and Runnels, S., 1963, The lethal effects of cobalt^{60} on mature White Leghorn chickens (an $LD_{50/30}$ study), *Southwestern Vet.* **17:**15–19.

Brisbin, I. L., Jr., 1969, Responses of broiler chicks to gamma-radiation exposures: changes in early growth parameters, *Rad. Res.* **39:**36–44.

Brisbin, I. L., Jr., 1972, Changes in composition and caloric density of whole body homogenates of broilers exposed to acute gamma radiation stress, *Poul. Sci.* **51:**915–920.

Brisbin, I. L., Jr., 1974, Abundance and diversity of waterfowl inhabiting heated and unheated portions of a reactor cooling reservoir, in: *Thermal Ecology* (J. W. Gibbons and R. R. Sharitz, eds.), AEC Symp. Ser. (CONF-73505). Springfield, Virginia, pp. 579–593.

Brisbin, I. L., Jr., 1989, Radiocesium levels in a population of American alligators: A model for the study of environmental contaminants in free-living crocodilians, in: *Proceedings of the Eighth Working Meeting of the Crocodile Specialist Group, SSC/IUCN*, Gland, Switzerland, pp. 60–73.

Brisbin, I. L., Jr., and Smith, M. H., 1975, Radiocesium concentrations in whole-body homogenates and several body compartments of naturally contaminated white-tailed deer, in: *Mineral Cycling in Southeastern Ecosystems* (F. G. Howell, J. B. Gentry, and M. H. Smith, eds.), ERDA Symp. Ser. (CONF-740513), Springfield, Virginia, pp. 542–556.

Brisbin, I. L., Jr., and Swinebroad, J., 1975, The role of banding studies in evaluating the accumulation and cycling of radionuclides and other environmental contaminants in free-living birds, *EBBA News* **38:**186–192.

Brisbin, I. L., Jr., and Thomas, M. G., 1971, Responses of commercial broiler chicks to acute gamma radiation radiation stress in the range of 900–1600 R, *Poul. Sci.* **50:**397–402.

Brisbin, I. L., Jr. and Vargo, M. J., 1982, Four-year declines in radiocesium concentrations of American Coots inhabiting a nuclear reactor cooling reservoir, *Health Phys.* **43:**266–269.

Brisbin, I. L., Jr., Geiger, R. A., and Smith, M. H., 1973, Accumulation and redistribution of radiocaesium by migratory waterfowl inhabiting a reactor cooling reservoir, in:

Proceedings of the International Symposium on the Environmental Behavior of Radionuclides Released in the Nuclear Industry, IAEA Symp., IAEA-SM-17272, Vienna, pp. 373–384.

Brisbin, I. L., Jr., White, G. C., and Bush, P. B., 1986, PCB intake and the growth of waterfowl: Multivariate analyses based on a reparameterized Richards sigmoid model, *Growth* **50:**1–11.

Brisbin, I. L., Jr., McLeod, K. W., and White, G. C., 1987, Sigmoid growth and the assessment of hormesis: A case for caution, *Health Phys.* **52:**553–559.

Brisbin, I. L., Jr., Breshears, D. D., Brown, K. L., Ladd, M., Smith, M. H., Smith, M. W., and Towns, A. L., 1989a, Relationships between levels of radiocesium in components of terrestrial and aquatic food webs of a contaminated streambed and floodplain community, *J. Appl. Ecol.* **26:**173–182.

Brisbin, I. L., Jr., Newman, M. C., McDowell, S. G., and Peters, E. L., 1989b, The prediction of contaminant accumulation by free-living organisms: Applications of a sigmoidal model, *Environ. Chem. & Toxicol.* **9:**141–149.

Cadwell, L. L., Schreckhise, R. G., and Fitzner, R. E., 1979, Caesium^{137} in coots (*Fulica americana*) on Hanford waste ponds: Contribution to population dose and offsite transport estimates, in: *Proceedings Health Physics Society Twelfth Midyear Topical Symposium on Low-Level Radiation Waste Management*, Williamsburg, Virginia.

Carson, R., 1962, *Silent Spring*, Houghton Mifflin Co., Boston.

Clay, D. L., Brisbin, I. L., Jr., Bush, P. B., and Provost, E. E., 1980, Patterns of mercury contamination in a wintering waterfowl community, *Proc. Ann. Conf. S.E. Assoc. Fish & Wildl. Agencies* **32:**309–317.

Cramp, S. (ed.), 1977, *Handbook of the Birds of Europe, the Middle East and North Africa*, Volume 1, Oxford University Press, Oxford.

Curnow, R. D., Whicker, F. W., and Glover, F. A., 1970, Radiosensitivity of the Mallard duck [*Anas platyrhynchos*], Tech. Paper 13, 4th Intern. Congr. Radiol., Evian, France.

Davis, J. J., and Foster, R. F., 1958, Bioaccumulation of radioisotopes through aquatic food chains, *Ecology* **39:**530–535.

Desante, D. F., and Geupel, G. R., 1987, Landbird productivity in central coastal California: The relationship to annual rainfall and a reproductive failure in 1986, *Condor* **89:**636–653.

Edwards, M., and Raymer, S., 1987, Chernobyl—one year after, *Nat. Geog. Mag.* (May 1987): 632–653.

EEC, 1986, Derived Reference Levels as a Basis for the Control of Foodstuffs Following a Nuclear Accident: a Recommendation from the Group of Experts Set Up under Article 31 of the Euratom Treaty, Regulation 1707/86, Commission of Economic European Community Printing Off., Brussels.

Fendley, T. T., 1978, The ecology of Wood Ducks [*Aix sponsa*] utilizing a nuclear production reactor effluent system, Ph.D. Thesis, Utah State University, Logan, Utah.

Fendley, T. T., and Brisbin, I. L., Jr., 1977, Growth curve analyses: A potential measure of the effects of environmental stress upon wildlife populations, *Proc. XIII Intern. Congr. Game Biol.*, pp. 337–350.

Fendley, T. T., Manlove, M. N., and Brisbin, I. L., Jr., 1977, The accumulation and elimination of radiocesium by naturally contaminated Wood Ducks, *Health Phys.* **32:**415–422.

Fitzner, R. E., and Rickard, W. H., 1975, *Avifauna of Waste Ponds, ERDA Hanford Reservation Benton County, Washington*, Publ. BNWL-1885-UC-70, of the Battelle Pacific Northwest Lab., Richland, Washington.

Garg, S. P., Zajanc, A., and Bankowski, R. A., 1964, The effect of cobalt-60 on Starlings [*Sturnus vulgaris vulgaris*], *Avian Dis.* **8**:555–561.

Glover, F. A., 1967, Distribution of Mallards from the Columbia Basin region as indicated by the presence of zinc-65 in birds shot by hunters in the Pacific and Central Flyways, Final Prog. Rept. to the U.S. Atomic Energy Comm., Contract AT(1101)-1514, Colorado State Univ., Fort Collins, Colorado.

Halford, D. K., 1987, Effect of cooking on radionuclide concentrations in waterfowl tissues, *J. Env. Radioactivity* **5**:229–233.

Halford, D. K., Millard, J. B., and Markham, O. D., 1981, Radionuclide concentrations in waterfowl using a liquid radioactive waste disposal area and the potential radiation dose to man, *Health Phys.* **40**:173–181.

Halford, D. K., Markham, O. D., and Dickson, R. L., 1982, Radiation doses to waterfowl using a liquid radioactive waste disposal area, *J. Wildl. Mgt.* **46**:905–914.

Halford, D. K., Markham, O. D., and White, G. C., 1983, Biological elimination rates of radioisotopes by Mallards contaminated at a liquid radioactive waste disposal area, *Health Phys.* **45**:745–756.

Hall, A. J. (ed.), 1981, *National Geographic Atlas of the World*, 5th Ed., Nat. Geog. Soc., Washington D.C.

Hancock, R., and Woollam, P. B., 1987, Radioactivity Measurements on Live Bewick's Swans, Report TPRD/B/0897/R87 of the Berkely Nuclear Laboratories, Central Electricity Generating Board, Gloucestershire, U.K.

Hanson, W. C., and Case, A. C., 1963, A method of measuring waterfowl dispersion utilizing phosphorus-32 and zinc-65., in: *Radioecology*, Proc. First Nat. Symp. on Radioecology (V. Schultz and A. W. Klement, Jr., eds.), Reinhold, New York, pp. 451–453.

Hepp, G. R., Kennamer, R. A., and Harvey, W. F. IV, 1989, Recruitment and natal philopatry of Wood Ducks, *Ecology* **70**:897–903.

Hohenemser, C., Deicher, M., Ernst, A., Hofsäss, H., Lindner, G., and Recknagel, E., 1986, Chernobyl: An early report, *Environ.* **28**:6–43.

Isakov, Y. A., 1966, MAR project and conservation of waterfowl breeding in the USSR, in: *Proceedings of the Second European Meeting on Wildfowl Conservation*, Publ. Intern. Waterfowl Res. Bur., Slimbridge, England, pp. 125–138.

Jenkins, J. H., and Fendley, T. T., 1968, The extent of contamination, detection, and health significance of high accumulations of radioactivity in southeastern game populations, *Proc. Ann. Conf. S.E. Assoc. Game & Fish Comm.* **22**:89–95.

Johnsgard, P. A., 1968, *Waterfowl, Their Biology and Natural History*, Univ. Nebraska Press, Lincoln, Nebraska.

Latimer, B. E., 1976, Growth and mortality responses of five breeds of chickens to gamma radiation stress, M.S. Thesis, Univ. of Georgia, Athens, Georgia.

Latimer, B. E., and Brisbin, I. L., Jr., 1987, Early growth rates and their relationships to mortalities of five breeds of chickens following exposure to acute gamma radiation stress, *Growth* **51**:411–424.

Levy, C. K., Youngstrom, K. A., and Maletskos, C. J., 1976, Whole-body gamma-spectroscopic assessment of environmental radionuclides in recapturable wild birds, in: *Radioecology and Energy Resources* [C. E. Cushing, Jr., ed.], Dowden, Hutchinson and Ross, Stroudsburg, Pennsylvania, pp. 113–122.

Lofts, B., and J. Rotblat, 1962, The effects of whole-body irradiation on the reproductive rhythm of the avian testis, *Intern. J. Rad. Biol.* **4**:217–230.

Lorenz, E., Jacobsen, L. O., Heston, W. E., Shimkin, M., Eschenbrenner, A. B., Deringer, M.

K., Doniger, J., and Schwersthal, R., 1954, Effects of long-continued total body gamma irradiation on mice, guinea pigs and rabbits. III. Effect on life span, weight, blood picture and carcinogenesis, and the role of the intensity of radiation, in: *Biological Effects of External X and Gamma Radiation*, Part 1 [Zirkle, R. E., ed.], McGraw-Hill, New York, p. 24.

Luckey, T. D., 1980, *Hormesis with Ionizing Radiation*, CRC Press Inc., Boca Raton, Florida.

Maloney, M. A., and Mraz, F. R., 1969, The effect of whole body gamma irradiation on survivors' egg production in the White Leghorn, *Coturnix coturnix japonica*, and Bobwhite Quail, *Poul. Sci.* **48**:1939–1944.

Marter, W. L., 1970, *Radioactivity in the Environs of Steel Creek*, Publ. DPST-70-435, of the Savannah River Lab., Aiken, South Carolina.

Mascanzoni, D., 1987, Chernobyl's challenge to the environment: A report from Sweden, *Sci. Total Env.* **67**:133–148.

Mayer, J. J., Kennamer, R. A., and Hoppe, R. T., 1986, Waterfowl of the Savannah River Plant, Publ. SREL-22UC-66e, of the Savannah River Ecology Lab., Aiken, South Carolina.

Medvedev, Z. A., 1986, Ecological aspects of the Chernobyl nuclear plant disaster, *Trends Ecol. and Evol.* **1**:23–25.

Mraz, F. R., 1971, Effect of continuous gamma irradiation of chick embryos upon hatchability and growth, *Rad. Res.* **48**:164–168.

Mraz, F. R., and Woody, M. C., 1973, The effect of dose rate upon reproductive performance of White Leghorn hens and their progeny, *Rad. Res.* **54**:549–555.

Narayanyan, N., and Eapen, J., 1971, Gross and subcellular distribution of cesium-137 in pigeon (*Columba livia*) tissues with special reference to muscles, *J. Rad. Res.* **12**:51–55.

Norris, R. A., 1958, Some effects of x-irradiation on the breeding biology of Eastern Bluebirds, *Auk* **75**:444–455.

Odum, E. P., 1965, Feedback between radiation ecology and general ecology, *Health Phys.* **11**:1257–1262.

Odum, E. P., 1971, *Fundamentals of Ecology*, 3rd ed., W. B. Saunders Co., Philadelphia.

Pinder, J. E., III, and Smith, M. H., 1975, Frequency distributions of radiocesium concentrations in soil and biota, in: *Mineral Cycling in Southeastern Ecosystems*, (F. G. Howell, J. B. Gentry, and M. H. Smith, eds.), ERDA Symp. Ser. (CONF-740513). Springfield, Virginia, pp. 536–542.

Potter, C. M., 1987, Use of reactor cooling reservoirs and cesium-137 uptake in the American Coot, M.S. Thesis, Colorado State Univ., Fort Collins, Colorado.

Potter, C. M., Brisbin, I. L., Jr., McDowell, S. G., and Whicker, F. W., 1989, Distribution of ^{137}Cs in the American Coot (*Fulica americana*), *J. Env. Radioactivity* **9**:105–115.

Prochazka, Z., 1970, The dependence of radiosensitivity of domestic fowl on its age. *Radiobiolog. Radiotherap.* **11**:51–55.

Prochazka, Z., and Hampl, J., 1966, A contribution to the estimation of radiosensitivity in the domestic fowl [*Gallus domesticus*], *Radiobiolog. Radiotherap.* **7**:231–235.

Reichle, D. E., Dunaway, P. B., and Nelson, D. J., 1970, Turnover and concentrations of radionuclides in food chains, *Nuclear Safety* **11**:43–55.

Richards, F. J., 1959, A flexible growth function for empirical use, *J. Exp. Bot.* **10**:290–300.

Ripley, S. D., 1976, Rails of the world, *Am. Scientist* **64**:628–635.

Ruiz, X., Jover, L., Llorente, G. A., Sanchez-Reyes, A. F., and Fabrian, M. I., 1987, Song Thrushes (*Turdus philomelos*) wintering in Spain as biological indicators of the Chernobyl accident, *Ornis Scand.* **19**:63–67.

Sadowski, F. G., and Covington, S. J., 1987, Processing and Analysis of Commercial Satellite Image Data of the Nuclear Accident near Chernobyl, USSR, US Geol. Surv. Bull. 1785, US Govt. Print. Off., Washington, D.C.

Schultz, V., 1974, *Ionizing Radiation and Wild Birds: a Selected Bibliography,* Publ. TID-3919, U.S.ERDA, Nat. Tech. Info. Serv., Springfield, Virginia.

Schultz, V., and Whicker, F. W., 1982, *Radioecological Techniques,* Plenum Press, New York.

Smith, L. M., Vangilder, L. G., Hoppe, R. T., Morreale, S. J., and Brisbin, I. L., Jr., 1986, Effect of diving ducks on benthic food resources during winter in South Carolina, U.S.A., *Wildfowl* **37**:136–141.

Staton, M. A., Brisbin, I. L., Jr., and Geiger, R. A., 1974, Some aspects of radiocesium retention in naturally contaminated captive snakes, *Herpetol.* **30**:204–211.

Stearner, S. P., 1951, The effect of variation in dosage rate of roentgen rays on survival in young birds, *Am. J. Roentgenol. Rad. Therap.* **65**:265–271.

Stearner, S. P., and Christian, E. J. B., 1968, Radiation effect on the microcirculation: Relation to early mortality in the chick embryo, *Rad. Res.* **234**:138–152.

Stearner, S. P., and Christian, E. J. B., 1969, Early vascular injury in the irradiated chick embryo: Effect of exposure time, *Rad. Res.* **38**:153–160.

Straney, D. O., Beaman, B., Brisbin, I. L., Jr., and Smith, M. H., 1975, Radiocesium in birds of the Savannah River Plant, *Health Phys.* **28**:341–345.

Sturges, F. W., 1968, Radiosensitivity of Song Sparrows and Slate-colored Juncos, *Wilson Bull.* **80**:108–109.

Suso, F. A., and Edwards, H. M., Jr., 1968, A study of techniques for measuring ^{65}Zn absorbtion and biological half-time in the chicken, *Poul. Sci.* **47**:991–999.

Temple, S. A., and Wiens, J. A., 1989, Bird populations and environmental changes: Can birds be bioindicators? *Am. Birds,* **43**:260–270.

Tester, J. R., McKinney, F., and Siniff, D. B., 1968, Mortality of three species of ducks—*Anas discors, A. crecca* and *A. clypeata*—exposed to ionizing radiation, *Rad. Res.* **33**:364–370.

Tyler, S. A., and Stearner, S. P., 1966, Growth in the chicken after high-rate and low-rate cobalt-60 gamma irradiation, *Rad. Res.* **29**:257–266.

Vogel, H. H., Jr., and Stearner, S. P., 1955, The effects of dose rate variation of fission neutrons and of Co^{60} gamma rays on survival in young chicks, *Rad. Res.* **2**:513–522.

Wagner, R. H., and Marples, T. G., 1966, The breeding success of various passerine birds under chronic gamma irradiation stress, *Auk* **83**:437–440.

Welty, J. C., 1962, *The Life of Birds,* W. B. Saunders Co., Philadelphia.

Whicker, F. W., and Schultz, V., 1982, *Radioecology: Nuclear Energy and the Environment* (2 vols.), CRC Press. Boca Raton, Florida.

Whicker, F. W., Pinder, J. E. III, Bowling, J. W., Alberts, J. J., and Brisbin, I. L., Jr., 1989, Distribution of ^{137}Cs, ^{90}Sr, ^{238}Pu, ^{239}Pu, ^{241}Am and ^{244}Cm in Pond B, Savannah River Site, Publ. SREL-35UC-66e, of the Savannah River Ecology Lab., Aiken, South Carolina.

WHO, 1986, Chernobyl Reactor Accident, Information Received on Public Health Measures as of 18 May 1986, Rept. World Health Org., Copenhagen.

Wiegert, R. G., Odum, E. P., and Schnell, J. H., 1967, Forb-arthropod food chains in a one-year experimental field, *Ecol.* **48**:75–83.

Willard, W. K., 1960, Avian uptake of fission products from an area contaminated by low-level atomic wastes, *Science* **132**:148–150.

Willard, W. K., 1963, Relative sensitivity of nestlings of wild passerine birds to gamma radiation, in: *Radioecology* (Schultz, V., and Klement, A. W., Jr., eds.), Reinhold, New York, pp. 345–349.

Woodwell, G. M., 1963, Design of the Brookhaven experiment on the effects of ionizing radiation on a terrestrial ecosystem, *Rad. Bot.* **3**:125–133.

Zach, R., and Mayoh, K. R., 1982, Breeding biology of Tree Swallows and House Wrens in a gradient of gamma radiation, *Ecology* **63**:1720–1728.

Zach, R., and Mayoh, K. R., 1984, Gamma radiation effects on nestling Tree Swallows, *Ecology* **65**:1641–1647.

Zach, R., and Mayoh, K. R., 1986a, Gamma-radiation effects on nestling House Wrens: A field study, *Rad. Res.* **105**:49–57.

Zach, R., and Mayoh, K. R., 1986b, Gamma irradiation of Tree Swallow embryos and subsequent growth and survival, *Condor* **88**:1–10.

CHAPTER 3

FACULTATIVE MANIPULATION OF SEX RATIOS IN BIRDS

Rare or Rarely Observed?

PATRICIA ADAIR GOWATY

1. INTRODUCTION

It is intuitively obvious and compelling that the sex ratio of offspring affects parental fitness (Darwin, 1871). The essentials for understanding sex ratio evolution were enumerated by Fisher (1930). Sex ratio is frequency dependent. The fact that each offspring in diploid species has one mother and one father insures that the reproductive value of daughters and sons is equal to the parents. For as, say, son-producers increase in populations, the reproductive value of daughters increases, providing positive selection on daughter-producers. As these daughter-producers increase, selection for son-producers increases, and so on. In cases where such forces alone are acting, the average sex ratio will be 1 : 1. However, if some other selective pressure results in greater costs to the parents through the production of one sex or the other, selection will act to equalize parental expenditures on sons and daughters. Such selection should lead to differing numbers of sons and daughters. Theorists since Fisher (see Charnov, 1982) have confirmed his ideas mathematically and empiricists working with a wide variety of organisms

PATRICIA ADAIR GOWATY • Department of Biological Sciences, Clemson University, Clemson, South Carolina 29634-1903.

have demonstrated empirical agreement with Fisherian theory. The Fisherian rule says simply that the sex ratio (M/F) will vary inversely with the cost of individual sons relative to individual daughters. These essential sex ratio arguments apply to ideas about total daughter and son investments collectively within populations, but they are silent about what individual parental strategy within populations should be (Williams, 1979). In contrast, theories of facultative sex ratio manipulation posit a number of circumstances in which it would be profitable for parents to skew the sex ratio of their offspring (e.g., Trivers and Williard, 1973). Fisher's ideas deal with selection for the sex ratio within a population; ideas about facultative manipulation have to do with what individuals should do. There are few explicit discussions of how individuals should respond to Fisherian pressures for sex ratio evolution and other sources of selection on the sex ratio. In fact, the scarcity of ideas about facultative manipulation of progeny sex ratios in birds may reflect a common difficulty in understanding the relationship between Fisherian rules (Fisher, 1930) about selection on the equilibrium (population-wide) sex ratio and ideas about what individuals might do. [This difficulty is not unique to sex ratio theory, but is a general problem also with evolutionarily stable strategy, or ESS, theory (Parker, 1984), in that a prediction about what populations ought to do seldom provides prediction about individual behavioral options.] Or, researchers may have been loath to speculate about a phenomenon that has been difficult to observe.

This chapter is about the possibility of facultative sex ratio manipulation in birds; it is about what individual birds can or might do to increase the numbers of grandchildren they produce through their relative allocation of resources to son-function and daughter-function. The paper contains a definition of facultative manipulation of sex ratios; a review of some influential opinions about that possibility in birds; a catalog of the difficulties in observing facultative manipulation; a critical examination of the use of the binomial test in the search for facultative variation of sex ratios; and ideas and data from the literature that might profitably be pursued in a search for facultative manipulation of progeny sex ratios in birds.

I have written this work primarily as a response to the notion that nothing of much interest occurs with regard to progeny sex ratios in birds. The following quotation sums up the prevailing opinion.

> The data on bird sex ratios suggest that the sex ratio is typically very near equality at the egg or nestling stage. Even though some data says (sic) that one sex may sometimes be more costly to the parents, there is really no evidence for the overproduction of the cheaper sex (but even less evidence

suggesting that sons and daughters are of unequal cost). In addition, there is virtually no evidence supporting the notion of short-term adaptive alterations in the sex ratio. A major problem is that even when deviations from equality are observed, they are typically very small (say 10% or less). Changes of this order of magnitude are almost impossible to study. (Charnov, 1982, p. 114)

The years since publication of Charnov's important, instructive, and influential book have provided evidence at issue with each of the main points made in the paragraph quoted above. What seems clear to me is that researchers with different orientations may look on the magnitude of variations differently. For example, if my primary interest were in testing the generality of sex allocation theory, I would not choose any bird species as my study species. Alternatively, I think it is appropriate to use sex allocation theory to aid in the understanding of life history variation, demography, and behavior of birds. So, though I agree that small changes in sex ratios are difficult, sometimes discouraging, to study—especially when our colleagues who study wasps, for example, deal with enormous skews—I do not find it impossible to study the small variations in sex ratios that seem to characterize birds. Indeed, the last several years have shown that the lack of evidence for unequal son and daughter cost may be more apparent than real; the difficulty may be that we have been examining the wrong "cost" variables (Stamps, 1990; P. A. Gowaty, personal observations). Since Charnov's book, several avian examples of sex ratio variation consistent with sex allocation theory have accumulated (Gowaty and Lennartz, 1985; Emlen *et al.*, 1986; Lessells and Avery, 1987; Bortolotti, 1986; Bednarz and Hayden, 1990) and many of these can only be understood as "short-term adaptive alterations" (facultative adjustments) of sex ratios. I hope that this chapter convinces readers that facultative sex ratios occur in birds (or, at least that the issue is not yet closed), and that it will help re-ignite interest in studies of avian sex ratio variation.

2. WHAT IS FACULTATIVE MANIPULATION OF SEX RATIOS?

2.1. Definition

"Facultative manipulation occurs when individuals change sex ratios of their offspring to capitalize on ecological circumstances" (Burley, 1982). Facultative manipulation is also meant by the term "short-term adaptive alterations" (Charnov, 1982). "Ecological circumstances" is a phrase that to my mind includes social circumstances (e.g., the sex ratios other individuals in the population are producing), variation in nutritional resources (Trivers and Williard, 1973), variation in resources essential to reproduction, etcetera. The challenge for students of birds will be to identify what ecological variations may be affecting sex ratios in particular species.

2.2. Mechanisms of Facultative Manipulation

Debates about the possibility of facultative manipulation in birds often focus on the possible mechanisms of manipulation. These debates sometimes have been unproductive in that they focus attention on unknown proximate mechanisms rather than on the possibility of facultative adjustment. When facultative control is observed or suspected, mechanisms will immediately suggest themselves, providing testable hypotheses for their confirmation or rejection.

In birds, sex of the offspring is determined maternally; if meiosis is normal and sex of gametes a randomly determined event, 50% of a mother's gametes will be female, 50% male. Facultative manipulation of sex ratios could occur as a result of nonrandomly determined meiosis or as sex-specific selection against gametes, fertilized eggs, nestlings, or fledglings. In birds, there is the additional possibility that chromosomal sex and phenotypic sex need not match, so that when facultative variation is called for, developmentally controlled variation may provide proximate mechanisms.

Differential destruction or differential investment in eggs or chicks by sex are usually posited mechanisms of sex ratio skew in birds. In fact, sex differences in juvenile mortality when food is scarce suggests that parental manipulation is at least sometimes involved (Clutton-Brock, 1986). Yet, despite attempts to document differential allocations to son or daughter function (either in terms of numbers of sons and daughters or provisioning decrements to one sex of offspring) in sexually dimorphic birds, few have been discovered (see Stamps, 1990, for a recent review). On the other hand, in size monomorphic birds differential allocations (in terms of provisioning rates) to daughters and sons seems to have appeared in every case in which such possibilities have been investigated. For example, sexually size monomorphic Zebra Finches (*Poephila guttata*) vary provisioning in response to social circumstances, apparently in association with their own perceived attractiveness relative to that of their partner (Burley, 1988). Among size monomorphic Budgerigars (*Melopsittacus undulatus*) male caregivers fed their daughter-biased broods more than their son-biased broods

(Stamps *et al.*, 1987). In size monomorphic Eastern Bluebirds (*Sialia sialis*), fathers, but not mothers, fed their daughters more often than their sons (D. L. Droge and P. A. Gowaty, personal observations).

At any rate, the mechanisms of facultative manipulation of sex ratios may be subtle. In this chapter I assume that mechanisms exist for short-term adaptive alterations of sex ratios, and I will not review mechanisms any further; however, readers should keep in mind that nonrandom sex allocation can occur at any stage of development from gamete to offspring independence from parental care.

2.3. Ecological Correlates of Facultative Manipulation

Many ecological circumstances could affect facultative manipulation of sex ratios. Variation in food supplies, perceived attractiveness of mates (Burley, 1986), resources for reproduction, or availability of mates, are only a few examples. A more general statement is that any ecological variation that is "predictably" associated with the greater production of grandchildren through production of one sex or the other should affect the sex ratio of mothers who find themselves in these particular circumstances.

If food availability varies and has an effect on the condition of mothers, and mother-offspring condition is positively correlated, mothers in polygynous species (i.e., those assumed to have higher variance in male than in female reproductive success) should facultatively adjust the sex ratios of their offspring to favor sons when they are in good condition and daughters when they are in poor condition (Trivers and Williard, 1973). This hypothesis, the maternal condition hypothesis, is the most quoted hypothesis for facultative modification of the sex ratio. This hypothesis states that when one sex of offspring predictably experiences greater variance in reproductive success, mothers in good condition should preferentially invest in the sex with the greatest variance in reproductive success. When mothers are in relatively poor condition, they should preferentially invest in offspring with the lower variance in reproductive success. Some avian data have been interpreted as consistent with the maternal condition hypothesis (e.g., Howe, 1977; Blank and Nolan, 1983), but apparently there have been few successful attempts at testing this hypothesis in wild-living birds (though see Cronmiller and Thompson, 1981) and mammals. One successful field test demonstrated increased son production among nutritionally augmented field-living opossums (Austad and Sonquist, 1988). Opossums carry their undeveloped young in pouches, making it relatively easy for field workers to documents sex ratios of the young. Furthermore, female opossums have small home ranges, making it easy for field workers to predict their foraging routes. In contrast, it is almost impossible for students of birds routinely to assign newly hatched chicks to sex, and even territorial birds will forage over relatively large, often unpredictable areas. Nevertheless, a study designed to evaluate the effects of nutrition on maternal condition and the sex ratio could be carried out in captive populations. It is surprising that apparently no one has attempted such a study.

2.4. How Large Should Facultative Skews in Sex Ratios Be?

In populations with an excess of one sex, the optimum progeny would be 100% of the rarer sex (Williams, 1979). This hopeful prediction leads researchers to expect dramatic skews in some progeny sex ratios. The prediction is no doubt so, but only when parents themselves can predict (in an evolutionary sense, not in a conscious sense) with a high degree of precision the sex ratio of the population from which their offspring will find mates. This assumption of Williams' important ideas is probably rarely met in the cases of highly vagile animals like birds. Therefore, we do not observe population-wide progenies of 100% sons or 100% daughters among birds.

Haplodiploidy is typical of species with dramatic skews in sex ratios (Hamilton, 1967). In contrast, heterochromatic sex determination is a conservative force in sex ratio evolution that opposes large skews in sex ratios (see Charnov, 1982). However, much remains to be discovered about chromosome evolution, so this conservative selective force should be considered a hypothesis only.

Facultatively manipulated sex ratios should be the ones likely to result in the maximum number of grandchildren for parents (Williams, 1979). This restatement of Fisher's idea appears simple, yet I know of no empirical demonstration in birds that some sex ratio distributions actually result in more grandchildren than others. Yet, this strikes me as a not too difficult question to answer in given cases. This focus on the adaptive advantage of some mixes of sons and daughters versus other mixes, suggests that sex ratio variation need not be highly dramatic in order to be very interesting. For example, in species where same-sex sibling coalitions are advantageous (e.g., Acorn Woodpeckers, *Melanerpes formicivorous*: Koenig and Mumme, 1987), overproduction of sex combinations favoring the production of coalitions may be adaptive and more common than we have suspected (see the discussion below about lions, Packer and Pusey, 1987).

The above considerations suggest to me that the magnitude of facultative sex ratio skews will be settled only by empirical investigation; nevertheless, it seems certain that these skews will not be of the magnitude observed in haplodiploid species.

2.5. Tests for Facultative Manipulation of Sex Ratios

Williams (1979) argued that the expected binomial distribution of sex ratios among progenies of certain sizes was especially useful as a null hypothesis to the alternative of adaptive or facultative sex ratios. Using this statistical yardstick, he found no support for any theory of adaptive sex ratio in outcrossed vertebrates, i.e., he found no differences in observed sex ratio distributions when compared to binomially distributed sex ratio distributions. William's (1979) paper has been influential in that it has affected the choices of statistical methods used by some researchers interested in facultative manipulation (Gowaty, 1980; Lombardo, 1982). I labored under its constraints for several years before I realized that failure to reject the hypothesis of binomially distributed sex ratios was not adequate for concluding that facultative adjustment did not occur. This should have been immediately obvious in that it is a trivial statistical point: failure to reject a hypothesis is rarely adequate grounds for acceptance of the null hypothesis. But, the point that now interests me the most in regard to tests against binomially expected sex ratios is not statistical, but biological. If one asks what the distribution of sex compositions within broods of a given size should be in a large population of outcrossing vertebrates in which *there has been selection for the ability to modify sex ratios adaptively* I argue that the answer should be binomially distributed sex combinations, not nonbinomial. There are two reasons for this. (1) In a large outcrossed population one would fundamentally expect to find frequency-dependent sex ratios, because it is within large outcrossed populations that individuals' grandchildren probably will find mates. (2) When one examines a large *population* without reference to the variation in ecological circumstances of *individuals relative to other individuals* in that population, one effectively overlooks any variation in (frequency-dependent) sex ratios that might be facultative. (For example, consider a theoretical population made up of polygynously mated males and females that lay clutches of four eggs. Imagine further that food availability on the breeding grounds varies, such that some breeding females are relatively well-fed and thus relatively healthier than other breeding females. It seems reasonable to me to predict in this circumstance that the population sex ratio would be binomially distributed, but that relatively better-fed breeders would produce significantly more sons than poorly-fed breeders in the population.) These two insights, one trivial and statistical, the other subtle and biological, prompt me to advocate that tests against binomial expectations might most profitably be used to explore, rather than confirm, facultative manipulation of sex ratios.

Below I discuss data that have lead me to the idea that tests against the binomial will only rarely be germane to questions about facultative sex ratio manipulation in birds. In this attempt I present data on lions (*Panthera leo*) (Packer and Pusey, 1987), data on Red-cockaded Woodpeckers (*Picoides borealis*) (Gowaty and Lennartz, unpublished); Snow Geese (*Chen caerulescens*) (Harmsen and Cooke, 1983) and Eastern Bluebirds. I also demonstrate my points using published data from a variety of other bird species. These examples show in what situations the test against the binomial can be useful (e.g., it has utility as an exploratory analysis when it may indicate that something interesting may have occurred, and it has utility as a confirmatory analysis when data are collected specifically to test hypotheses about broods of particular sizes or about particular combinations of sons and daughters). More important are the examples suggesting that when there has been selection for short-term adaptive modifications of the sex ratio, the expected variation in large populations is binomial, rather than nonbinomial.

3. THE BINOMIAL TEST

The distribution of broods of different sex compositions can be compared with the binomial distribution to test a variety of hypotheses (Sokal and Rohlf, 1981). In the most common case, the expected binomial distribution of sexes within a given brood size is computed based on the standard expectation of 50% sons. The hypothesis being tested in this case is said to be "extrinsic" (Sokal and Rohlf, 1981) to the data, because the expected frequencies are calculated based on a hypothetical (often, but not always, 50% sons), rather than empirically observed, sex ratio. Should one observe significant deviation from the binomial in this case (Fig. 1), two conclusions are possible: (1) the sex ratio differs from 50% sons (the usually taken conclusion); or (2) the distribution of the sexes within broods of this size is nonrandom. In another, less common case, the expected distribution of sexes within a given brood size is computed from the observed sex ratio for each brood size (e.g., Harmsen and Cooke, 1983). Say that one has observations of sex ratios from 65 broods of four offspring and that the sex ratio within

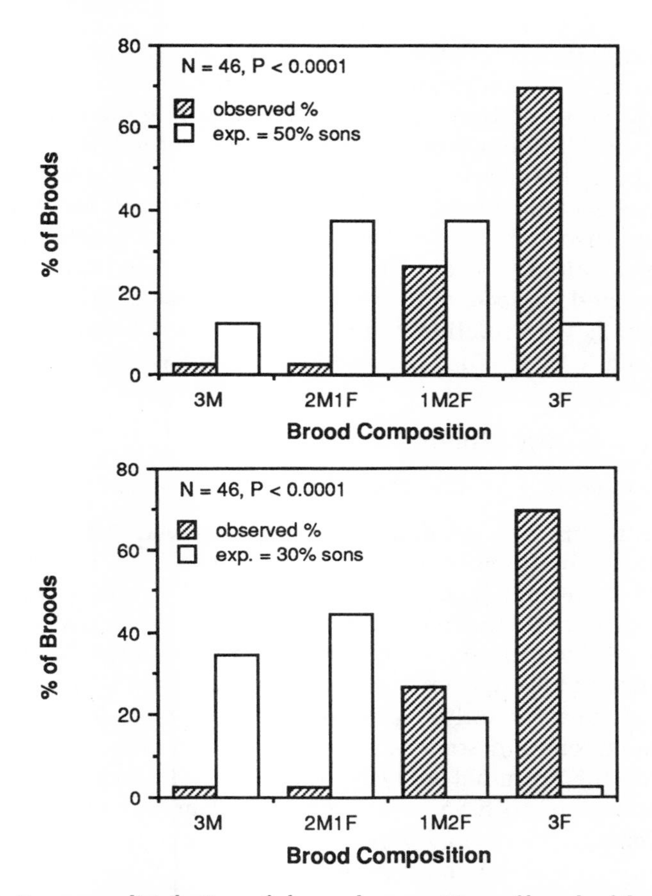

FIGURE 1. Frequency distributions of observed compositions of broods of three offspring tested against a hypothesis "extrinsic" to the data (top histogram): the distribution of sex compositions is distributed binomially with a 50% probability of sons. The bottom histogram tests the same observed distribution against a hypothesis "intrinsic" to the data: the distribution of sex compositions is distributed binomially with a 30% probability of sons. See text for discussion of the conclusions that are possible from each of these tests.

these broods is 30% sons. The expected binomial distribution is calculated based on 30% sons and 70% daughters. Here the hypothesis being tested is "intrinsic" to the data. Figure 1 graphs such a comparison. The significant difference between observed and expected values in this case indicates *only* nonrandom combinations of sons and daughters ("adaptive sex ratios"). Because variation in the sex ratio was controlled for in this test, a significant difference would not indicate that the sex ratio was different from 30% sons, only that the distribution of brood compositions was nonrandom. The assumption of such a test from the perspective of facultative manipulation is that there is something ecologically special about females who produce broods of size four or about offspring that come in groups of size four. Should these assumptions be warranted, the test for binomial variation is an ideal test for nonrandom distribution of the sexes. However, as a general, exploratory test for facultative sex ratios the test against the binomial may be misleading and unnecessarily restrictive to those of us who work on vagile animals with relatively long lives in which skews, if they exist, are expected to be small. I illustrate these points below.

Sex Ratios within Broods

Is There Something Special about Broods of a Given Size?

Lions. Figure 2 shows distributions of observed and expected litter compositions for lions from Packer and Pusey (1987). I have re-analyzed data from their Figure 2. In their original figure, they computed expected distributions from a 50 : 50 sex ratio. I calculated expected distributions from the sex ratios observed in each brood size. Because the expected values in each of these tests was taken from the observed sex ratio of cubs, the only hypothesis being tested is that the distribution of daughters and sons in litters of a given size is random. Consistent with their original analysis, litters of size four occur in a nonrandom distribution, with an overproduction of broods with three males and one female. Is this nonbinomial variation associated with anything biologically significant about lions? Conveniently, yes.

> Lions in the Serengeti National Park and Ngorongoro Crater, Tanzania, live in stable social groups and the reproductive success of males depends on the number of like-sexed companions they have. Males from coalitions of up to seven individuals act as a unit in competitions against other coalitions. A successful coalition gains temporary exclusive access to a group of females ("pride") for up to several years before being ousted by another coalition. Larger coalitions are more likely to gain residence in a pride, remain in residence longer, and gain access to more females than do small coalitions. The success of larger coalitions is sufficiently high that the per capita lifetime reproductive success of males increases strikingly with increasing coalition size. (Packer and Pusey, 1987, p. 636).

Further, the probability of living in a large group as an adult depends primarily on the size of the cohort in which an individual was reared. These male cohorts disperse as groups. Packer and Pusey go on to report that males in litters with three males have a significantly

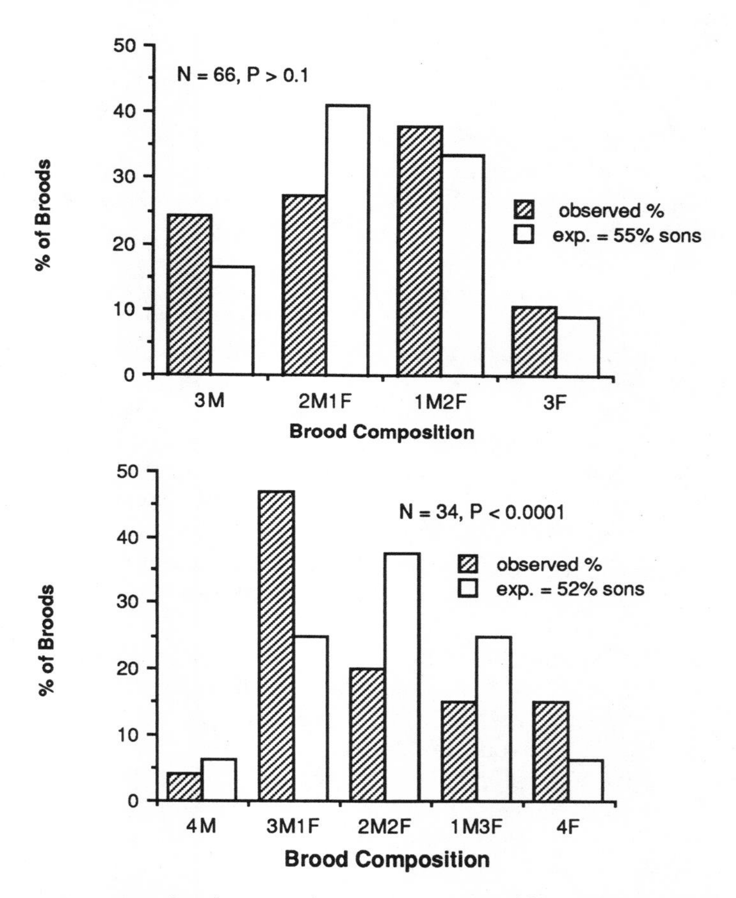

FIGURE 2. Frequency distributions of sex compositions of lions in the Serengeti from litters of three (top histogram) and four (bottom histogram). Each of the observed distributions was tested against the binomially expected distributions computed from observed sex ratios for litters of three and four respectively. The expected values were computed from the observed sex ratios reported in Figure 2 of Packer and Pusey (1987).

higher chance of becoming members of cohorts of three or more than do those with only two males or only one male. Therefore, the most productive composition of a litter of four is three sons and one daughter, since this yields the highest expected number of grandchildren. A similar analysis indicates that the most productive composition of a litter of three is three sons.

So here the test against the binomial distribution of the sexes within litters of three and four has biological relevance because of special characteristics of lions. These special characteristics include strong daughter philopatry, son dispersal, and the formation of coalitions among sons, with reproductive advantages for cohorts of three sons. These significant tests against the binomial distribution of sexes in litters of three and litters of four seem to confirm and extend biologically relevant notions about the evolutionary significance of social behavior in lions.

Exploratory Analyses

Red-cockaded Woodpeckers. Red-cockaded Woodpeckers are cooperative breeders (Lennartz *et al.*, 1987; Walters *et al.*, 1988; Walters, 1990) that produce nestling and fledgling sex ratios favoring sons (Gowaty and Lennartz, 1985). In Figure 3 are the observed and expected distributions of the sexes in broods of size two and three of Red-cockaded Woodpeckers. There data are previously unpublished results from the Francis Marion Wildlife Refuge collected by M. Lennartz and analyzed by myself (details of methods are in Gowaty and Lennartz, 1985). In broods of size three the distribution of the sexes is binomial. In broods of size two the distribution is significantly different from the binomial. The largest contribution to the test statistic came from the overproduction of broods of one son and one daughter. What might this mean? The expected distribution of the sexes in broods of two and three were computed from the observed sex ratios of these brood sizes of 70% and 55% sons, respectively, so the only hypothesis tested is that the distribution of sexes within these broods is nonrandom.

Is there anything special about broods of size two in Red-cockaded Woodpeckers? Broods of size two are more common than other brood sizes. Yet most broods of size two begin as clutches of size three. This nonbinomial variation in Red-cockaded Woodpeckers may indicate a parental strategy to maximize the number of grandchildren individuals have; however, at this point, M. R. Lennartz and I lack the behavioral and demographic data necessary to examine this hypothesis. I suspect that broods of one son yield greater numbers of grandchildren than broods with more sons, despite Fisherian selection favoring sons (the less costly sex in Red-cockaded Woodpeckers: see Gowaty and Lennartz, 1985). This may be due to competition for breeding status among philopatric sons or to competition between mothers and their adult sons for opportunities to breed. Because most Red-cockaded Woodpecker broods start out as clutches of size three, a possible mechanism of facultative sex ratio skew may be brood reduction by sex of offspring.

I will return to ideas about short-term adaptive modifications of

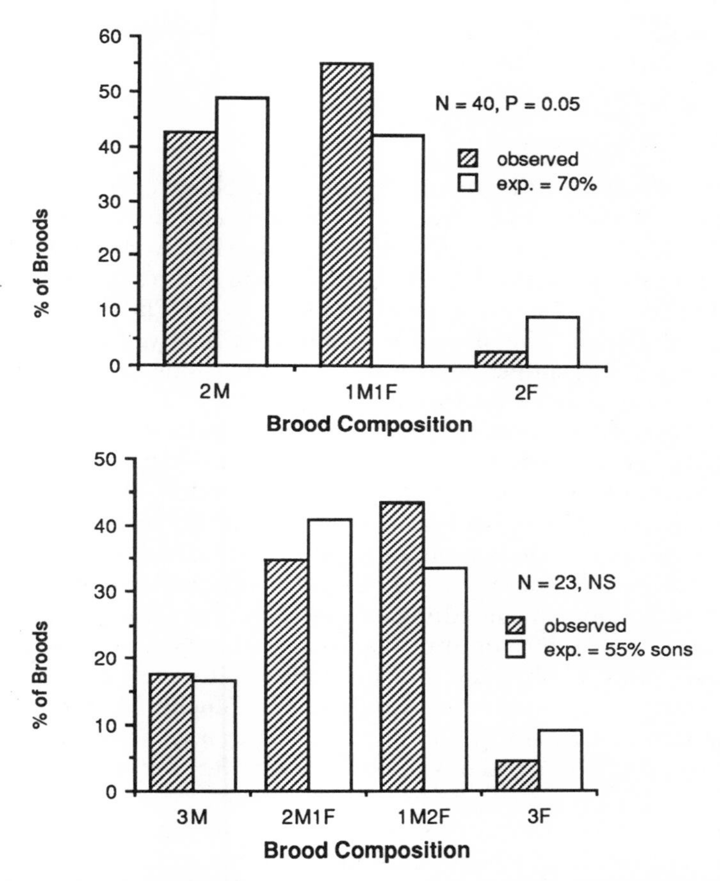

FIGURE 3. Frequency distributions of sex compositions of Red-cockaded Woodpeckers from broods of two (top) and three (bottom) nestlings. The expected distributions were computed using the observed sex ratios for broods of two and three.

the sex ratio in Red-cockaded Woodpeckers later in the chapter. Here I want to stress what we have learned from the test against the binomial. It does appear that groups producing two offspring may facultatively adjust their sex ratios. However, we are not much closer to knowing why they do this. Can we conclude from the binomial variation that we observe in other brood sizes that the groups producing these brood sizes do not facultatively adjust their sex ratios? I think not, for reasons I describe below in relation to Eastern Bluebirds. However, in the current case the test against the binomial has been a useful exploratory tool. It has raised more interesting questions than we had at the start, and the result fostered the formation of other testable hypotheses about life history variation in these birds.

In Search of Hypotheses

Eastern Bluebirds. Figure 4 is a graph of the observed and expected distributions of the sexes in broods of size four and five produced by Eastern Bluebirds on my South Carolina study sites from 1977–1981. Here again I calculated the expected distribution of the sexes from that observed in broods of size four and five respectively. The distributions are clearly binomial. What does this mean? Specifically, does this indicate that facultative adjustment of sex ratios by bluebirds producing broods of size four and five does not occur?

Most bluebird females in South Carolina produce broods of size four in both the summer and spring breeding periods; broods of size five are most common in the spring breeding period. Both sons and daughters disperse, but dispersal is over greater distances for daughters, and sons are relatively more philopatric than daughters. We know of no advantage associated with like-sexed coalitions in bluebirds and it is hard to distinguish something special about breeding females that produce broods of size four and five because these are such common brood sizes. Thus, I belive that this test tells me little or nothing about the possibility of facultative manipulation by bluebird parents. I believe this is so, because this test was not cast against anything that is particularly socially or ecologically relevant to sex allocation in bluebirds.

More important, I think that this result is one also predicted by a hypothesis of facultative manipulation of sex ratios. Consider a large, widely distributed population in which many females produce broods of size five (or four) and in which dispersal by sons and daughters is common. What distribution of combinations of the sexes within broods of these sizes in this large population might be expected if there is selection for facultative manipulation by females who find themselves in particular ecological circumstances? In diploid species facultative manipulation will be against the backdrop of essential frequency dependence of sex ratio selection. So, it seems obvious that in large populations any facultative effects on the sex ratio may be obscured by the swamping effects of the advantages of frequency dependent selection for the sex ratio. This might be particularly so for birds, especially flying birds that have the potential to sample multiple social situations as adults, where parents will be unable to predict precisely the social or ecological circumstances in which their offspring will breed. In other

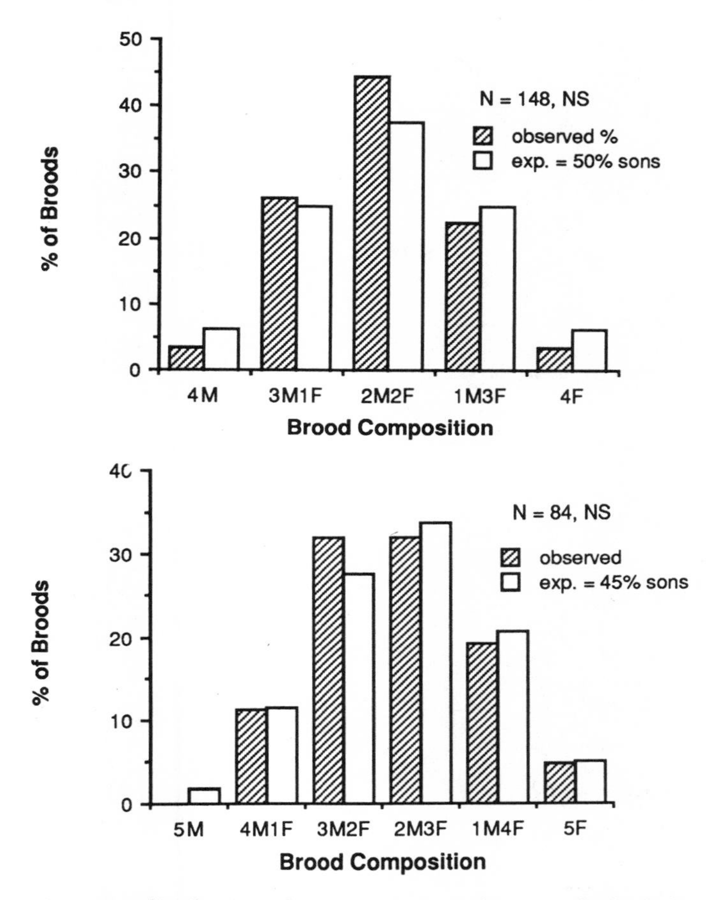

FIGURE 4. Frequency distributions of sex compositions of Eastern Bluebirds from broods of four (top) and five (bottom) nestlings. The expected distributions were computed using the observed sex ratios for broods of four and five.

words, it is likely that tests against the binomial in large populations will be silent about facultative manipulation of the sex ratio unless they are conveniently statistically significant. And, even if they are significant, in some cases the biological significance of nonbinomial distributions might remain obscure.

Selection for the Ability to Adjust Sex Ratios Facultatively?

Snow Geese. Harmsen and Cooke (1983) present a remarkably large data set of sex ratios for Lesser Snow Geese (*Chen caerulescens*

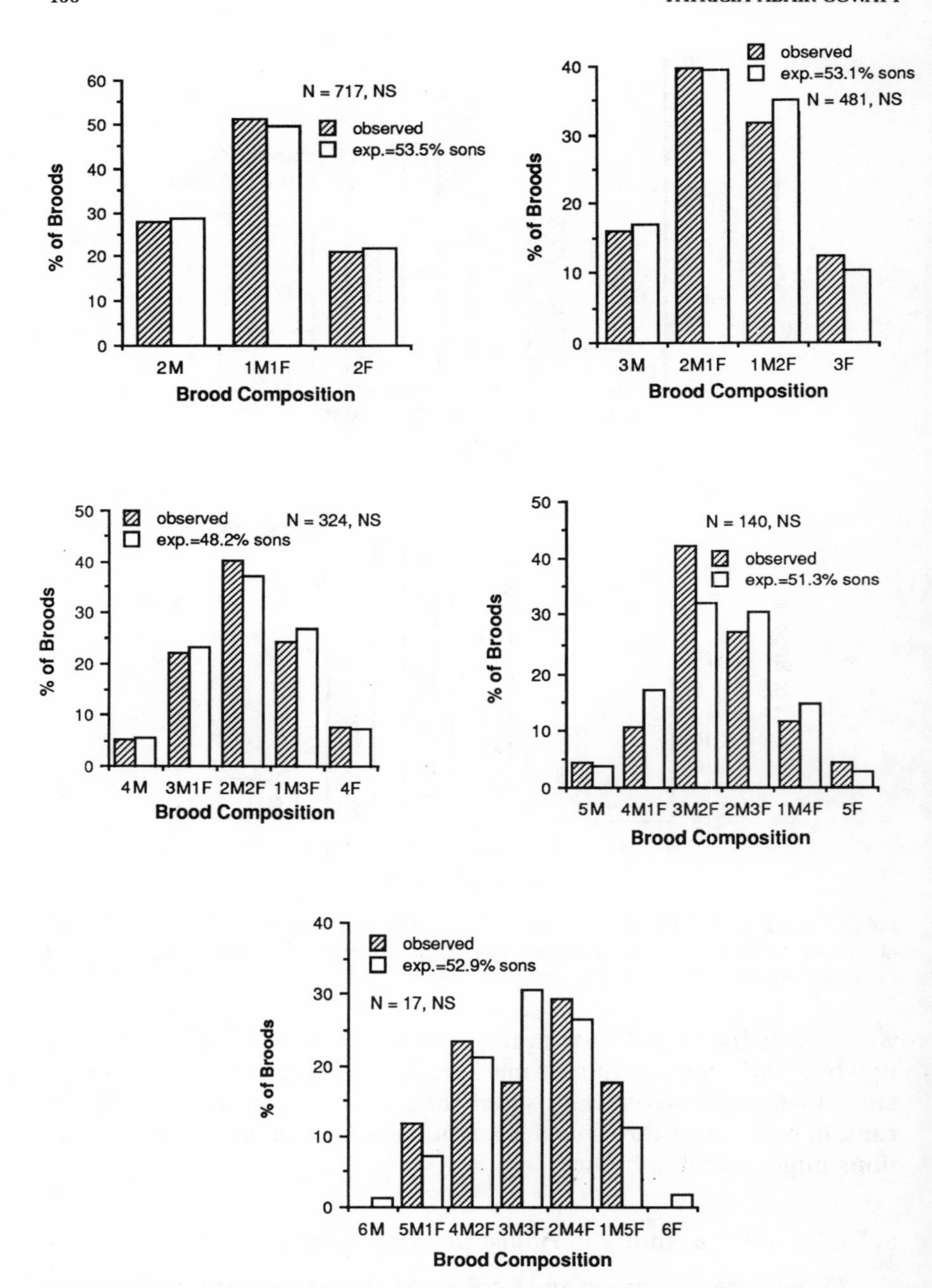

FIGURE 5. Frequency distributions of sex compositions of Lesser Snow Geese for brood sizes two to six, graphed from data in Harmsen and Cooke (1983).

caerulescens). They tested sex ratio distributions within clutch sizes against expected binomial distributions calculated from the observed sex ratios for each clutch size. (They included only completely sexed clutches in their analysis). The distribution of sex ratios for clutches of one, two, three, four, and six were binomial in their original analysis. They considered the significant result they obtained for clutches of five a chance effect rendered likely by the fact that they had performed multiple tests. I calculated a more realistic significance level for their five tests using the sequential Bonferroni technique (Holm 1979; Rice 1989), a method that controls for the increased likelihood of a Type I error when performing multiple tests. Consistent with Harmsen and Cooke's (1983) conclusions, clutches of size five were not distributed differently from the binomial when using adjusted significance levels. Their results are graphed in Figure 5. Harmsen and Cooke (1983) used the binomial to test the prediction that in large, outbreeding populations, there should be no selection against mutants that predominantly produce one sex or another (Kolman, 1960), so that in such large populations there should be greater than binomial variation in clutches of a given size. These data offer no support for this idea, i.e., these data show that there is no selection for an accumulation of all-son or all-daughter producers in this population. Thus, their data contribute to what they termed a theoretical enigma (Harmsen and Cooke, 1983).

The enigma is solved if these data are viewed in another light. If there is selective advantage to being able to adjust the sex ratio of progeny facultatively, there should be selection against a fixed trait, "produce all sons" or "produce all daughters." The Snow Goose data are exactly what one might expect given selection for the ability to manipulate sex ratios in adaptive fashions. Remember they have tested for binomial variation across an apparently undifferentiated cross section of an extremely large population. They did not test the broods females produced in different ecological or social circumstances that one might predict are associated with different sex ratios. Given the size of the population they tested, selection for frequency-dependence of sex ratios may have obscured any effects that arose as a result of facultative manipulation.

4. WHAT DATA ARE NECESSARY AND SUFFICIENT TO SHOW FACULTATIVE MANIPULATION?

One strategy for testing facultative manipulation of sex ratios would be to construct specific hypotheses about factors that would lead to overproduction of sons or daughters by individuals within a population. Individuals would then be grouped within these factors, as, for example, was done by Austad and Sonquist (1988) for opossums. The two groups would then be tested for differences in their sex ratios. Although Austad and Sonquist (1988) did not use separate tests against the binomial for nutritionally augmented and nonaugmented females, they could have. Instead, they used the comparatively simple Student's *t*-test for differences between two groups to examine their data for differences in predicted directions. Had they used tests against the binomial they would have had to repeat the test for each brood size within each factor, thus increasing their probability of a Type I error. Furthermore, tests against the binomial within brood sizes within factors will not allow comparisons between, say, experimental groups. On the other hand, if they could have shown that the two groups of females not only had different sex ratios but that the distribution of sex compositions within broods for augmented females was nonbinomial, that would be especially powerful evidence for facultative manipulation. As their study stands, they have only demonstrated powerful, rather than especially powerful, evidence for facultative manipulation of sex ratios!

4.1. Zebra Finches

Another study with powerful evidence of facultative manipulation of sex ratios is Nancy Burley's (1986) study of captive, color-banded populations of Zebra Finches. To my mind, her paper offers a model of how our continuing investigations of facultative manipulation by birds might be carried out. Burley provides two instances of experimentally manipulated sex ratio. In one, "the banded male experiment," the perceived attractiveness of the males as mates to females was manipulated by varying the color of bands worn by the males. In this experiment only males wore color bands. In the other experiment, "the banded female experiment," the perceived attractiveness of the females as mates for males was manipulated by varying the color of bands worn by the females. Note that these were two separate experiments; no birds used in one experiment were used in the other. In general in both experiments the sex ratio produced favored the sex of the relatively more attractive parent. Figure 6 shows the percent sons produced by females mated to attractive red-banded males, neutral orange-banded males, and unattractive green-banded males. Overall, higher proportions of sons were produced by females mated to attractively banded males. Sex ratios of females mated to red-banded males equaled 61.3% sons; females mated to orange banded males equaled 53.5% sons; and

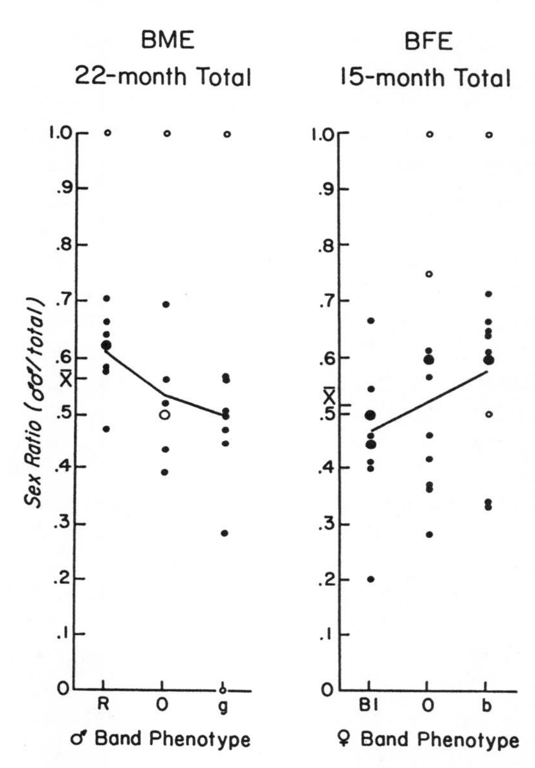

FIGURE 6. Sex ratios of nestling Zebra Finches produced by females paired with males (BME) banded red (R), a preferred color; orange (O), a neutral color; and green (g), an unattractive color; and produced by banded females (BFE) wearing black (Bl), a preferred color; orange (O), a neutral color; and blue (B), an unattractive color. Figures are from Burley (1986) with permission. Note that these data are from two separate experiments, "the banded male experiment" (BME) and "the banded female experiment" (BFE).

females mated to green-banded males equaled 50.2% sons. In the reciprocal experiment in which females but not males were banded, pairs with attractive black-banded females produced more daughters while pairs with unattractively blue-banded females produced relatively fewer daughters. Specifically, black-banded females produced 46.5% sons; orange-banded females 52.2% sons, and blue-banded females 58.2% sons. Figure 6 also demonstrates that even when facultative sex ratio manipulation surely has occurred, it has done so against a backdrop of frequency dependent selection for the sex ratio. In both the banded male experiment and the banded female experiment the aggregate sex ratios are not different from 50% sons. Yet, as illustrated, there are significant correlations between perceived attractiveness of the color-banded sex and the sex ratio produced.

It would be interesting to know whether the sex ratio produced in the banded male experiment (or the banded female experiment) was binomially distributed within broods of a given size. They clearly should be. But what if one tested the distribution of sex combinations within broods of black-banded females, and again within broods of orange-banded females? If Burley had performed these tests and found nonbinomial distributions of sex combinations she would have had especially powerful, rather than just powerful, evidence of facultative manipulation of sex ratios. Some (e.g., Williams, 1979) feel that adequate sample sizes for such tests are at least 65 broods within a given brood size category. Given this sample size constraint in her experimental design, Burley may never have published. I hope this point is clear: if we were to limit our acceptable evidence of facultative manipulation to nonbinomial distributions of sex compositions, data from, say, black-banded females would be tested separately for each brood size Zebra Finches produce; if we further used the 65 brood minimum criterion, this experiment, which already required two years in the case of the banded male experiment, would have taken perhaps decades more to accomplish. The criterion that we test against the binomial within brood sizes using very large sample sizes (for birds, anyway) for each test offers an unrealistic, usually impossible statistical constraint to our searches for facultative manipulation of sex ratios in birds. I advocate the use of less restrictive, though still robust, statistical approaches to this problem.

Among the striking things about Burley's sex ratio investigations is that she performed experiments—something astonishingly rare in the current literature about bird sex ratios. Thus, I must advocate the use of experimental approaches in both the laboratory and the field.

4.2. Yellow-headed Blackbirds

Yellow-headed Blackbirds (*Xanthocephalus xanthocephalus*) breed in western North America. Males are larger than females, and the males are often polygynous. In the late 1970's Cindy Patterson performed experiments on variation in male parental investments to primary, secondary, and tertiary females mated to polygynous males (Patterson *et al.*, 1980). I have regraphed the data from Patterson *et al.* (1980) in Figure 7. The order of settling of females on males' territories is primary, secondary, and tertiary. Males generally contribute parental care only to the offspring of primary females. In most cases, primary,

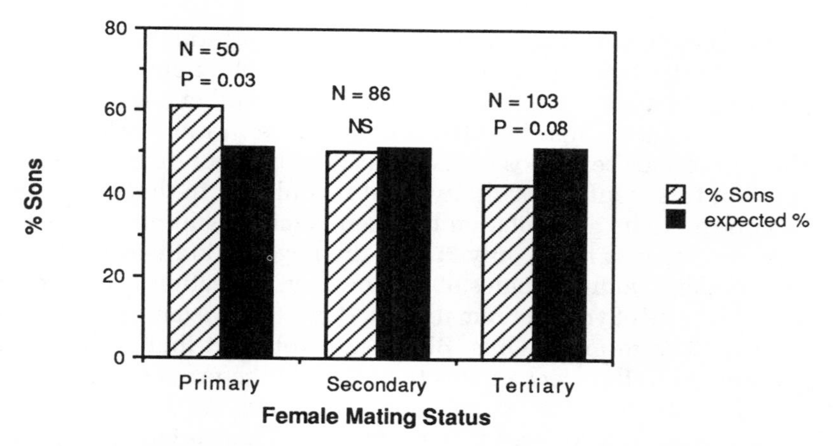

FIGURE 7. Sex ratios produced by Yellow-headed Blackbird females mated first (primary), second (secondary), and third (tertiary) to polygynous males. Sex ratios are graphed from data in Patterson *et al.* (1980).

secondary, and tertiary females were initiating nests within days of one another, such that when offspring of primary females were removed from territories, males re-allocated their parental investments to their secondary females.

Patterson's analysis of sex ratio variation in the broods on unmanipulated territories indicates that sex ratio is dependent on the mating status of particular females, such that primary females produce relatively more sons (61%), while tertiary females produce relatively fewer sons and secondary females produced equal numbers of sons and daughters (Fig. 7). This sex ratio variation is difficult to understand unless females are able to vary the sex ratios of their broods facultatively depending on the ecological and social circumstances in which they find themselves. Remember that these females initiate nests within days of one another so that whatever mediates the sex ratio variation surely must be facultative. If "primary" females are removed before secondary females begin to lay, do "secondary" females modify their sex ratios? Such a manipulation with appropriate controls would offer a relatively simple experimental field test of the possibility.

4.3. Red-winged Blackbirds

Red-winged Blackbirds (*Agelaius phoeniceus*) are also sexually dimorphic birds in which the larger males are often polygynous. Blank and Nolan (1983) examined the sex ratios produced by males of different ages. There was a significant association of sex ratio produced with the age of the mother (Fig. 8). One- and two-year-old mothers produced significantly daughter-biased sex ratios, while mothers aged five and six years produced significantly son-biased sex ratios. Middle-aged mothers produced sex ratios not significantly different from 50 : 50. It is not clear if this result is facultative, but it could be. In this population there appears to be a correlation between hatching success and maternal age: old females hatch only 75% of their eggs, middle-aged females hatch 84%, and young females 93%. However, starvation is most common in the nests of younger females. Of 17 nestlings examined to determine sex, 15 were male. Thus, Blank and Nolan attributed the skews they observed to the mechanisms that affected the production of sons. Young mothers seem to starve sons selectively, while old mothers somehow seem to favor eggs containing sons. This variation may arise through two types of facultative behavior, selective hatching of eggs of older mothers, and less efficient selective allocation to nestlings of younger mothers. Whether these patterns lead to greater numbers of grandchildren for older females who allocated towards sons and younger females that allocated against sons is not yet known, but such an analysis of long-term reproductive success would surely be of interest and should be a more common tactic in our searches for facultative sex ratio variation in birds.

4.4. Red-cockaded Woodpeckers

Red-cockaded Woodpeckers are cooperative breeders in which sons are philopatric and daughters emigrate. The sex ratios of the populations of nestlings and fledglings are equal to 59% sons, a value significantly different from 50 : 50 (Gowaty and Lennartz, 1985). The selection pressure that reduces individual son cost is associated with local resource enhancement by stay-at-home sons (Malcolm and Martin, 1982; Gowaty and Lennartz, 1985; Emlen, Emlen, and Levin, 1986; Lessells and Avery, 1987). Presented in this way these data and the local resource enhancement idea are silent about the possibility of facultatively manipulated sex ratios by Red-cockaded Woodpeckers.

What about the possibility of facultative manipulation of sex ratios by female Red-cockaded Woodpeckers that find themselves in varying ecological and social circumstances? Among the curiosities of Red-cockaded Woodpecker social life is the fact that adult females often move from breeding situation to breeding situation (Lennartz *et al.*, 1987; Walters *et al.*, 1988). Consistent with this female habit, Gowaty and Lennartz (1985) described females as those having never been on

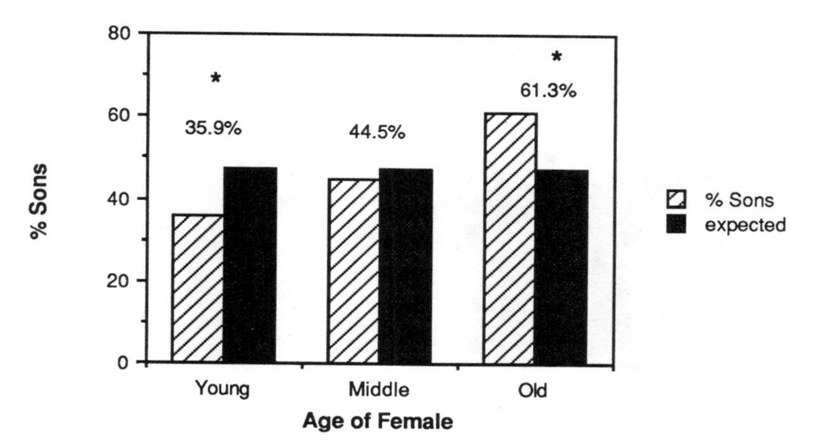

FIGURE 8. Sex ratios produced by old, middle-aged, and young Red-winged Blackbirds. Sex ratios are graphed from data in Blank and Nolan (1983).

the study site before, a class of females with "no tenure"; other females have been observed on the study sites and these females we described as "with tenure." In addition females may breed with or without helpers, so that at least four classes of females can be described: (1) Females with no tenure, but with helpers; (2) females with no tenure and without helpers; (3) females with tenure and no helpers; and (4) females with tenure and with helpers. Because their tenure varies, we infer that their kinship to the males in the groups varies. For example, a female that is new to a territory, but nevertheless has helpers, is likely to be unrelated to either the breeder or helper males. However, a female with tenure who is in a territory with helpers is most likely being helped by her own sons. The percent sons produced by these various classes of females are in Figure 9 (redrawn from Gowaty and Lennartz, 1985, Table 2). Females without and with tenure but with helpers produce 73% and 50% sons, respectively. Pairs in which females lack tenure produce 100% sons; pairs in which females have tenure produce 66% sons. The overall contingency analysis is significant (reported in Gowaty and Lennartz, 1985). However, the presence or absence of helpers is not significantly associated with the sex ratio, while the tenure class of breeding females is. Figure 10 shows that females without tenure produce significantly more sons than females with tenure. I hypothesize that this is an adaptive response to the intersexual behavior of Red-cockaded Woodpeckers and the habit of females who move between breeding opportunities (Gowaty and Lennartz, 1985). It may also reflect tactics on the part of females to reduce competition for

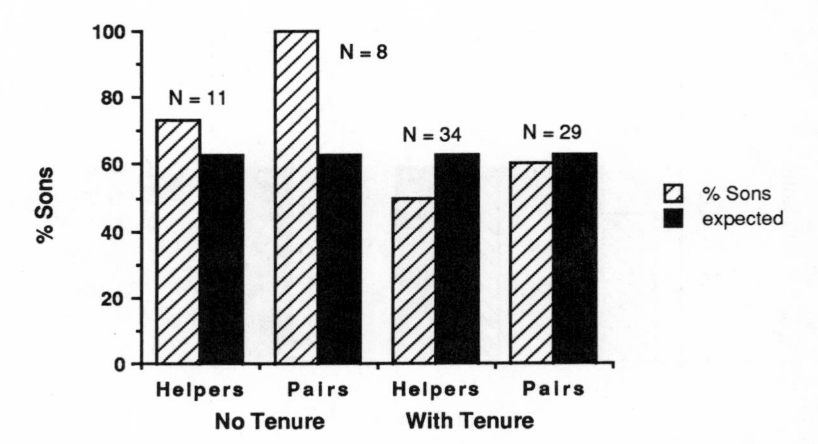

FIGURE 9. Sex ratios produced by Red-cockaded Woodpecker females with helpers and without (pairs) who have (with tenure) and have not (no tenure) been seen on study sites before. Data are regraphed from Table 2 in Gowaty and Lennartz (1985).

mates among their sons. When females have tenure, they are likely already to have genetic representation in a particular place through their sons. To facilitate breeding by these sons, mothers might move; to reduce breeding competition among their sons females may selectively allocate against sons. When they are without tenure, there seems to be a high priority on the production of sons who help guarantee females' genetic representation in particular places or within particular groups.

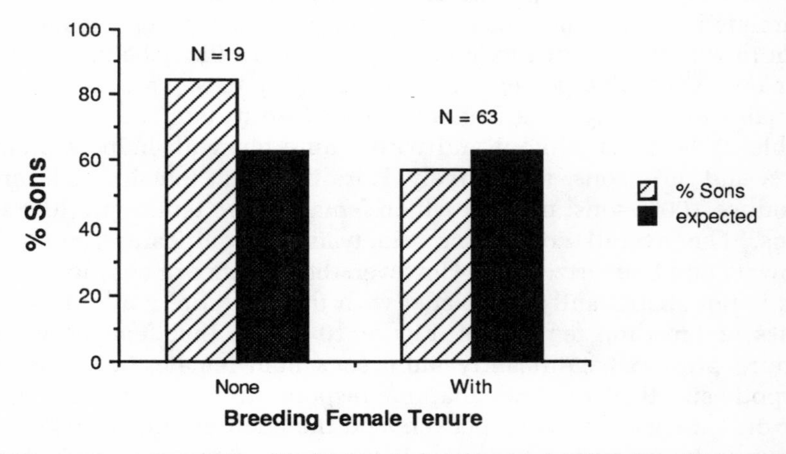

FIGURE 10. Sex ratios produced by Red-cockaded Woodpecker females with and without tenure. Data are regraphed from Table 2 in Gowaty and Lennartz (1985).

This idea is strengthened by closer inspection of the sex ratio variation in Figure 9. Females with no tenure breeding with helpers (who cannot be related to her) and females in pairs without helpers have an excess of sons in comparison to the relatively daughter-biased sex ratios of tenured females who are being helped (probably by their own sons). Remember that the nonbinomial variation in broods of size two in Red-cockaded Woodpeckers comes from the higher than expected frequency of broods with one son and one daughter. These two observations may be related. At any rate, I find it difficult to understand the variation in sex ratios and tenure class without using some notion of facultative manipulation. However, I want to stress that, at this point, this idea of facultative manipulation by Red-cockaded Woodpecker females is an untested hypothesis. There is much that remains to be investigated. Red-cockaded Woodpeckers are an endangered species, thus experimental manipulation of wild populations is unlikely. However, because females move from breeding opportunity to breeding opportunity, we hope to be able to take advantage of the naturally occurring experiments thus created. If facultative manipulation occurs in Red-cockaded Woodpeckers in the way that I have suggested, we should be able to document that the sex ratios produced by given females change in directions predicted by the kinship hypothesis when females move from one group to another. I also suspect that behavioral variation in groups of different kinship will yield important clues about the mechanisms of sex ratio variation in this species.

4.5. Harris' Hawks

Harris' Hawks (*Parabuteo unicinctus*) are cooperative breeders; sons are the philopatric, helping sex, and daughters disperse and are larger than sons. Both the smaller size and the habit of helping by sons have been used to predict an equilibrium sex ratio favoring sons (Bednarz and Hayden, 1990). As predicted, the equilibrium sex ratio equals 57% sons. Bednarz and Hayden (1990) were unable to offer support for or against the maternal condition hypothesis (Trivers and Williard, 1973) or the production-of-helpers hypotheses (Gowaty and Lennartz, 1985). However, their careful data on hatching sequence and sex ratio are especially interesting in relation to potential mechanisms of facultative sex ratio adjustment. First-hatched young were most often sons, so that the sex ratio of first-hatched eggs was significantly biased towards sons. Bednarz and Hayden (1990) offer an adaptive explanation for this observation: Because sons are smaller than daughters, asynchronously first-hatched sons were at a competitive advantage in relation to their larger, faster growing sisters, so that broods with sons hatched first experienced lower mortality and higher success than broods in which daughters hatched first. What is curious about all this is that there is a significant bias in the sex ratio of first-hatched young. How might this happen? Meiotic variation could lead to an overproduction of son-producing ova, faster maturation of son-producing follicles (Ankney, 1982), or differential incubation of eggs containing sons and daughters could guarantee that son eggs hatch first, or eggs that contain sons may develop faster than daughter-containing eggs. At the very least, what this study offers is a potential mechanism for facultative adjustment of sex ratios. Harris' Hawks need only adjust the hatching rate of first-laid versus later-laid eggs when they find themselves in different ecological circumstances that would favor the production of more sons versus more daughters.

4.6. Bald Eagles

What is the brood combination that results in the most grandchildren? If some combinations of offspring by sex result in differential expectations of grandchildren, because some combinations are consistently unlikely to fledge, selection should act against these combinations. Eggs in Bald Eagle (*Haliaeetus leucocephalus*) clutches hatch asynchronously, which results in one offspring being larger than the other. Because females are 25% larger as adults than males, hatching asynchrony in broods of size two could lead to daughters with large competitive advantages over sons. Thus broods with first-hatched daughters and second-hatched sons may be unlikely to fledge more than one young, or so one might reason. Yet, females are 63% of first-hatched eggs. This apparent anomaly is explained by the competitive differences between sibling Bald Eagles which depends not only on the asymmetries produced by hatching asynchrony but also on the growth pattern of male and female chicks (Bortolotti, 1986). In Bald Eagles, sons grow faster than daughters; their growth curves have significantly earlier inflection points than the daughters' curves such that first-hatched sons probably have a relatively greater size advantage over their sisters in the same brood. In an elegant analysis of growth curve variation under different assumptions about growth parameters by sex, Bortolotti showed that in M–F (males first, females second) broods, sons should have a greater size advantage over their sisters during the crucial first week of nestling life than first-hatched offspring in other sex-sequence combinations. Thus, when resources are limited, sibling competition in M–F broods is expected to result in the production of only the first-hatched sons. Thus, if parents are selected not to waste

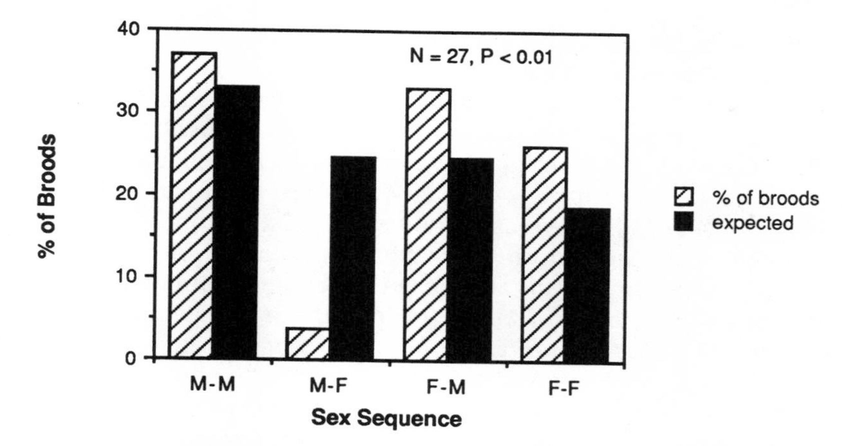

FIGURE 11. Percent of Bald Eagle broods of size two in which males hatched first (M–M and M–F) and in which females hatched first (F–M and F–F). Values are from data in Table 1 of Bortolotti (1986), though, here, expected values are based on a sex ratio of 56% sons.

investment on doomed offspring, one would expect M–F broods to be rare, and indeed, they are significantly rarer than other sex-sequence combinations (Fig. 11). Does this mean that the sex ratio distribution is nonrandom? Figure 12 shows the frequency distribution of sex combinations for 1980–1982 tested against the binomially expected dis-

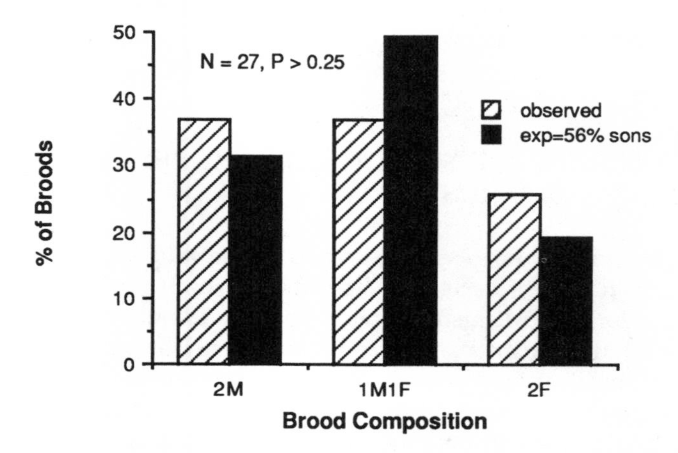

FIGURE 12. Frequency distribution of sex combinations within broods of Bald Eagles. The expected distribution is from the observed sex ratio of 56% sons (Bortolli, 1986).

tribution with 56% sons. (56% sons is the observed sex ratio for broods Bortolotti observed from 1980–1982.) The distribution of sex combinations is not different from the binomially expected distribution. Yet, if one computes the sex ratio that would result from equal production of M–F and F–M broods coupled with 100% loss of daughters in M–F broods, the sex ratio itself would be 60% sons. Thus, it appears that parents are selected to reduce son numbers relative to daughter numbers. Is the underproduction of M–F broods facultative?

The mechanism for this skew need not be non-Mendelian segregation (Godfray and Harvey, 1986); in fact, the evidence suggests that segregation is Mendelian (Fig. 12). It is possible that male and female follicles have different growth rates (Ankney, 1982), with males in this species being the faster sex. If son follicles develop just a bit faster than daughter follicles, one might expect M–M duos to predominate within samples, M–F duos to be next most common, F–M next, and F–F last. I do not think the observed frequencies (Fig. 11) are consistent with faster follicle development among sons. On the other hand, if parents can control development rate of eggs within nests by differentially incubating son and daughter eggs, they can easily adjust the relative frequencies of M–F and F–M broods. In this light, I think it may be no accident that the one M–F brood Bortolotti observed (in his sample from 1980–82), was situated near a superabundance of prey during the time when the second chick would have been at risk of death. I hypothesize that the production of this one M–F brood was a facultative response to high food availability; likewise, I suggest that the failure of some populations of Bald Eagles to produce expected frequencies of M–F broods is a regular facultative response to relatively low food abundance.

5. ARE FACULTATIVE SEX RATIOS RARE OR RARELY OBSERVED IN BIRDS?

Students of sex ratio variation in birds should keep in mind that they are faced with a variety of handicaps which include: (1) sexing young birds is difficult, usually requiring laparotomies; (2) the magnitude of variation in sex ratios, no doubt, will be small due to the constraints built into a system of chromosomal sex determination (Charnov, 1982); (3) field experiments on sex ratio variation are difficult to carry out; and (4) unrealistic statistical constraints are inhibiting.

The review of case studies examined here suggests some strategies for future research. First and foremost, sex ratio studies should be embedded in a matrix of natural history information about particular species in particular environments. Although my bias is clearly in favor of field studies, one should not overlook the startlingly clear conclusions that are possible with laboratory manipulations of resources and social environments. Burley's studies of caged Zebra Finches may have allowed observation of significant factors that otherwise would have been swamped in a field study. Her birds did not have to respond to the vagaries of local environments, including predators. Without control of the extraneous factors, sufficient data to adequately test the effect of perceived attractiveness of partners on offspring sex ratios would have been very long in coming.

Despite the generally pessimistic view of some investigators, I believe evidence for facultative variation in sex ratios of offspring in birds is strong enough to encourage additional energetic searches for facultative variation. Given the difficulties in observing sex ratio variation in birds, one might conclude that evidence for facultative adjustment in birds is relatively strong. Although there are still very few appropriate data on which to base conclusions, I suspect that facultative sex ratio variation in birds is rarely observed rather than truly rare.

ACKNOWLEDGMENTS. I first presented these ideas at a symposium on facultative sex allocation at the American Society of Zoologists meeting in New Orleans in 1987. I thank all the participants of that symposium who gave me constructive criticism of these ideas at that time. I thank M. R. Lennartz for his collegial attitude toward our data on Red-cockaded Woodpeckers. I thank Nancy Burley for our ongoing discussions about avian sex ratios and for her written comments on a previous draft of this manuscript. I also thank William C. Bridges, who read and commented on a previous draft of the manuscript and provided advice about statistical matters, and Luc-Alain Giraldeau, whose comments helped me focus on even more future problems associated with facultative sex ratios in birds. Dale Droge provided me useful comments even though I solicited them at the very last minute. I much appreciate the constructive and insightful criticism of all of these colleagues. I thank especially Gabriel Acebo for financial and emotional support. During part of the preparation of this manuscript, I was supported by a grant from the National Science Foundation.

REFERENCES

Ankney, C. D., 1982, Sex ratio varies with egg sequence in Lesser Snow Geese, *Auk* **99**:662–666.

Austad, S. N., and M. E. Sunquist, 1988, Sex-ratio manipulation in the common opossum, *Nature* **324**:58–60.

Bednarz, J. C., and T. J. Hayden, 1990, Skewed brood sex ratio and sex-biased hatching sequence in Harris' Hawk, *Am. Nat.* (in press).

Blank, J. L., and V. Nolan, Jr., 1983, Offspring sex ratio in Red-winged Blackbirds is dependent on maternal age, *Proc. Natl. Acad. Sci. USA* **80**:6141–6145.

Bortolotti, G. R., 1986, Influence of sibling competition on nestling sex ratios of sexually dimorphic birds, *Am. Nat.* **127**:495–507.

Burley, N., 1982, Facultative sex-ratio manipulation, *Am. Nat.* **115**:223–246.

Burley, N., 1986, Sex-ratio manipulation in color-banded populations of Zebra Finches, *Evolution* **40**:1191–1206.

Burley, N., 1988, The differential allocation hypothesis: An experimental test, *Am. Nat.* **132**:611–628.

Charnov, E. L., 1982, *The Theory of Sex Allocation*, Princeton University Press, Princeton, New Jersey.

Clutton-Brock, T. H., 1986, Sex ratio variation in birds, *Ibis* **128**:318–329.

Cronmiller, J. R., and C. F. Thompson, 1981, Sex ratio adjustment in malnourished Red-winged Blackbird broods. *J. Field Ornithol.* **52**:65–67.

Darwin, C., 1871, *The Descent of Man and Selection in Relation to Sex*, John Murray, London.

Emlen, S. T., J. M. Emlen, and S. A. Levin, 1986, Sex ratio selection in species with helpers-at-the-nest, *Am. Nat.* **127**:1–8.

Fisher, R. A., 1930, *The Genetical Theory of Natural Selection*, Oxford University Press, London.

Gowaty, P. A., 1980, The origin of mating system variability and behavioral and demographic correlates of the mating system of eastern bluebirds (*Sialia sialis*), Ph.D. dissertation, Clemson University, Clemson, South Carolina. xxi–123

Gowaty, P. A., and M. R. Lennartz, 1985, Sex ratios of nestling and fledgling Red-cockaded Woodpeckers (*Picoides borealis*) favor males, *Am. Nat.* **126**:347–353.

Godfray, H. C. J., and P. H. Harvey, 1986, Bald eagle sex ratios: Ladies come first, *TREE* **1**:56–57.

Hamilton, W. D., 1967, Extraordinary sex ratios, *Science* **156**:477–488.

Harmsen, R. and F. Cooke, 1983, Binomial sex-ratio distribution in the Lesser Snow Goose: A theoretical enigma, *Am. Nat.* **121**:1–8.

Holm, S., 1979, A simple sequentially rejective multiple test procedure, *Scand. J. Stat.* **6**:65–70.

Howe, H. F., 1977, Sex ratio adjustment in the Common Grackle, *Science* **198**:744–747.

Koenig, W. D., and R. L. Mumme, 1987, *Population Ecology of the Cooperatively Breeding Acorn Worrdpecker*, Princeton University Press, Princeton, New Jersey.

Kolman, W. A., 1960, The mechanism of natural selection for the sex ratio, *Am. Nat.* **878**:373–377.

Lennartz, M. R., Hooper, R. G., and R. F. Harlow, 1987, Sociality and cooperative breeding of Red-cockaded Woodpeckers, Picoides borealis, *Behav. Ecol. Sociobiol.* **20**:77–88.

Lessells, C. M., and M. I. Avery, 1987, Sex ratio selection in species with helpers at the nest: Some extensions of the repayment model, *Am. Nat.* **129**:610–620.

Lombardo, M. P., 1982, Sex ratios in the Eastern Bluebird, *Evolution* **36**:615–617.

Malcolm, J. R., and K. Marten, 1982, Natural selection and the communal rearing of pups in African wild dogs (*Lycaon pictus*), *Behav. Ecol. Sociobiol.* **10**:1–13.

Packer, C., and Anne E. Pusey, 1987, Intrasexual cooperation and the sex ratio in African lions, *Am. Nat.* **130**:636–642.

Parker, G. A., 1984, Evolutionarily stable strategies, in: *Behavioural Ecology* (J. R. Krebs and N. B. Davies, eds), Sinauer Associates, Sunderland, Massachusetts, pp. 30–61.

Patterson, C. B., W. J. Erckman, and G. H. Orians, 1980, An experimental study of parental investment and polygyny in male blackbirds, *Am. Nat.* **116**:757–769.
Rice, W. R., 1989, Analyzing tables of statistical tests, *Evolution* **43**:223–225.
Sokal, R. R., and F. J. Rohlf, 1981, *Biometry*, 2nd ed., W. H. Freeman and Co., New York.
Stamps, J., A. Clark, P. Arrowood, and B. Kus, 1987, The effects of parent and offspring gender on food allocation in Budgerigars, *Behaviour* **101**:177–199.
Stamps, J., 1990, When should parents differentially provision sons and daughters? *Am. Nat.* (in press).
Trivers, R. L., and D. E. Williard, 1973, Natural selection of parental ability to vary the sex ratio of offspring, *Science* **179**:90–92.
Walters, J. R., 1990, Red-cockaded Woodpeckers: A 'primitive' cooperative breeder, in: *Cooperative Breeding in Birds: Long-term Studies of Ecology and Behavior* (P. B. Stacey and W. D. Koenig, eds.), Cambridge University Press, Cambridge, England, pp. 67–102.
Walters, J. R., Doerr, P. D., and Carter, J. H. III, 1988, The cooperative breeding system of the Red-cockaded Woodpecker, *Ethology* **78**:275–305.
Williams, G. C., 1979, The question of adaptive sex ratio in outcrossed vertebrates, *Proc. Roy. Soc. London* (B) **205**:567–580.

CHAPTER 4

ENEMY RECOGNITION AND RESPONSE IN BIRDS

IAN G. McLEAN and GILLIAN RHODES

1. INTRODUCTION

Traditionally, defense in animals has been discussed in relation to concepts such as coloration, crypsis, mimicry, flight, aggressive behavior, and grouping (e.g., Edmunds, 1974; Alcock, 1984). At issue has been the evolutionary origins of the relevant structures or behaviors. Tinbergen and Lorenz's classic studies on enemy recognition by young birds underlined the importance of development as a factor in the bird/enemy relationship (e.g., Tinbergen, 1951), but adaptiveness remained the primary theoretical framework underpinning research paradigms. The idea that response to enemies is "species-specific" pervades the ethological literature (e.g., Greig-Smith, 1981; Drickamer and Vessey, 1982), due, in part, to the influence of the innate releasing mechanism concept (see Curio, 1975, for a discussion). Clear demonstrations that enemy recognition can be based on experience (e.g., Schleidt, 1961) were not influential.

In birds, studies of recognition have focused primarily on individual variability in song (e.g., Weedon and Falls, 1959; and many subsequent publications) or on parent/chick recognition in colonial breeders (review in Beecher, 1982). Studies of enemy recognition have

IAN G. McLEAN • Department of Zoology, University of Canterbury, Christchurch, New Zealand. GILLIAN RHODES • Department of Psychology, University of Canterbury, Christchurch, New Zealand.

looked primarily at the functional significance of mobbing behavior (see review of seminal papers in Montgomerie and Weatherhead, 1988; also Altmann, 1956; Curio, 1978), particularly its effectiveness in different contexts such as in colonial versus solitary breeding (Wittenberger and Hunt, 1985; Brown and Hoogland, 1986). Some emphasis has been placed on the significance for communication of different alarm or distress calls given in different contexts (Marler, 1955; Thorpe, 1961; Shalter and Schleidt, 1977; Hogstedt, 1983; Greig-Smith, 1984; Klump and Shalter, 1984). Although data are limited, birds are apparently capable of fine-scale discrimination among different enemies (Kruuk, 1964; Curio, 1975; Patterson *et al.*, 1980; Buitron, 1983), as are some mammals (Seyfarth *et al.*, 1980; Owings *et al.*, 1986) and may use displays and vocalizations to communicate the kind of enemy being faced or their likely future behavior in relation to that enemy (Gottfried, 1979; Greig-Smith, 1980; East, 1981; Gottfried *et al.*, 1985).

Optimality theory involves study of the costs and benefits of a trait, preferably measured in terms of fitness (Alcock, 1984). The popularity of optimality theory has resulted in a plethora of studies investigating tradeoffs in behavior. For example researchers may measure variation in vigilance in relation to size or species composition of groups, dominance status, or breeding status (e.g., Coraco *et al.*, 1980; Goldman, 1980; Greig-Smith, 1981; Hogstad, 1988; see also a considerable literature on fish, e.g., Godin and Smith, 1988; Huntingford and Wright, 1989). Parental investment theory (which is derived from optimality theory) has led to the hypothesis that parents should determine level of acceptable risk (i.e., cost) when defending their young by referring to the relative difference in cost of rearing the current clutch to completion versus starting again (Andersson *et al.* 1980; as the young approach independence parents should take more risks). As with many predictions derived from optimality theory (Gray, 1987), available field data on this hypothesis are equivocal (e.g., Buitron, 1983; Regelmann and Curio, 1983; Knight and Temple, 1986a; McLean *et al.*, 1986; Hobson *et al.*, 1988; see Montgomerie and Weatherhead, 1988, and below, for an extensive discussion).

Generalizations about how birds respond to enemies are difficult to make. Different species clearly respond to the same or similar enemies in many different ways (Edwards *et al.*, 1949; Altmann, 1956; Robertson and Norman, 1977; Barash, 1980; Brown and Hoogland, 1986). However, different species may respond in similar ways (Stefanski and Falls, 1972; Greig-Smith, 1981; Dean and McLean, ms; see below), particularly if they are morphologically similar, or are group-living (e.g., mixed-species flocks of forest birds). It is also clear that the response of birds to enemies is highly dependent on context (Barash, 1975; Curio, 1975; Shalter, 1978a,b; Conover and Perito, 1981; Frankenberg, 1981; Shields, 1984; Curio and Regelmann, 1985; Cully and Ligon, 1986; Hill, 1986), and that the notion of a "species-specific" response is too simplistic.

Even when tests are conducted under identical conditions, within-species variation may range from no response whatsoever to extreme mobbing involving physical attack and loud continuous alarm calling (e.g. McLean *et al.*, 1986). Response varies with season (Shedd, 1982, 1983; Cully and Ligon, 1986), dominance status (Hogstad, 1988), group size (Greig-Smith, 1981), environmental factors (Lombardi and Curio, 1985a), age (Smith *et al.*, 1984), sex (Curio, 1975, 1980; McLean 1987), social context (Lombardi and Curio, 1985b; Gyger *et al.*, 1986), behavior of the enemy (Frankenburg, 1981; Buitron, 1983; Eden, 1987), and previous experience (Shalter, 1978b; McPherson and Brown, 1981; Knight *et al.*, 1987). Birds are clearly remarkably flexible in their response to enemies, suggesting that interaction with enemies involves assessment as well as recognition and response.

In order to study recognition, one must first define a response that is measurable within the constraints of the experimental design. Monitoring different brain receptors (conducted successfully with chickens, Holst and Saint-Paul, 1960 *in* Eibl-Eibesfeldt, 1970), or heart rate (Mueller and Parker, 1980) is not usually an option for field biologists, or even for many laboratory researchers. The measurement of response is generally at a fairly gross level (e.g., number of alarm calls) and many subtleties in behavior are often lost or ignored. The question of what constitutes recognition has not been seriously addressed in the bird literature. In Section 2, we outline a possible model of the cognitive processes involved in recognizing and responding to enemies, including the influence of assessment on cognitive processing.

In Section 3, we look in more detail at mobbing behavior. Mobbing represents the most widely distributed response to enemies among birds, and has been the subject of numerous studies in the recent literature. The major issues and hypotheses underlying research on mobbing have been reviewed by Curio (1978) and will not be dealt with in detail here. Rather, we will focus on the contribution of ultimate and proximate factors to the expression of mobbing behavior during the nesting cycle, reciprocity in mobbing assemblages, and the possible significance of one hypothesized function of mobbing behavior (cultural transmission) for conservation biology. These three aspects were chosen because we believe they offer possibilities for future research, and because they reflect current research at the University of Canterbury. Specifically, we challenge two commonly accepted hypotheses: (1) that optimality reasoning allows us to understand mobbing responses of parent birds, and (2) that mobbing birds do not exhibit reciprocal altruism.

We define "enemy" to mean any organism which may harm or injure a bird, its offspring, or members of its flock or breeding group. Conspecifics could be included under this definition, but will be excluded here. In Section 4 we explore some of the issues raised in Sections 2 and 3 by outlining our recent work on the relationship between a small passerine, the Grey Gerygone (*Gerygone igata*), and its brood parasite enemy, the Shining Bronze Cuckoo (*Chrysococcyx lucidus*). Investigations of the relationship between hosts and brood parasites have focused on apparent mimicry between eggs, juvenile appearance (particularly gape), and vocalizations (Hamilton and Orians, 1965; Payne, 1977; Rothstein, 1982a; McLean and Waas, 1987). Most research on behavioral responses (i.e., recognition) has looked at tactics used by different host species when faced with the egg of a brood parasite (Clark and Robertson, 1981; Rothstein, 1982b; Graham, 1988). Many hosts which regularly encounter adult brood parasites recognize and respond to them as enemies, usually by mobbing (Edwards *et al.*, 1950; Robertson and Norman, 1976, 1977; Smith *et al.*, 1984; Payne *et al.*, 1985; McLean, 1987). Why host birds continue to raise parasite chicks appearing in their nest, despite their ability to recognize the adult parasite as an enemy, is a mystery upon which we hope to shed some light.

2. A COGNITIVE MODEL OF RECOGNITION AND RESPONSE

Recognition mechanisms vary tremendously in complexity, ranging from simple reflexive systems based on key stimuli, such as the red bills used by male Zebra Finches (*Taeniopygia guttata*) to recognize rivals (Immelmann, 1959), to complex and flexible cognitive systems in which perceptual input is interpreted in relation to stored knowledge (see below). The type of recognition system used by a particular organism for a particular task will depend on the ecological and evolutionary constraints associated with the task. A reflexive system would be appropriate when speed is critical, when the class of objects to be recognized is fixed, when simple cues signal the presence of a member of that class, when the cost of failure to respond rapidly is high, when an invariant response is usually successful and when the risk associated with making that response is low and fixed. A more flexible cognitive system would be appropriate when the organism must be able to respond to new stimuli, when the members of the class cannot be recognized using simple, physically specifiable cues, when a variety of responses (whose utility depends on context) are possible, and when the risks associated with each response are either high or variable.

The essence of a cognitive recognition model is that stored knowledge of the world (either innately given or based on experience) is used to interpret perceptual input, and to choose an appropriate response. The main advantage of such a system is its flexibility. New knowledge can be acquired, either through direct experience or cultural transmission of information, so that the organism can learn and new ways of coping can be developed. Because recognition is not neurally committed to any particular response, the choice of response can be sensitive to contextual factors such as the availability of conspecifics (e.g., for mobbing) or the stage of the breeding cycle. These situational factors may affect response choice by influencing the perceived threat and risks associated with possible responses.

Dennett (1983, 1988) has argued that animal behavior can best be characterized, predicted, and explained using intentional idioms such as beliefs and desires, i.e., by adopting an "intentional stance." In essence, Dennett imputes onto animals a desire to achieve an end. This teleology is expressed as behaviors such as displays or vocalizations which are likely to alter the behavior of others. The object of the exercise is to provide a language which will ". . . describe, in predictive, fruitful, and illuminating ways, the cognitive prowess of creatures in their environments" (Dennett, 1988, p. 350).

An analysis of the cognitive processes underlying recognition and control of response for enemy recognition is shown in Figure 1. First, an input (e.g., visual or auditory) is encoded to yield a *perceptual representation* of the stimulus. This perceptual representation may itself be the result of a complex series of information processing stages, including edge detection, and the use of shading, texture gradients, motion, stereopsis, and other depth cues to determine the three-dimensional form of the stimulus (Marr, 1982). The resulting representation is then compared with *stored representations*. If a match occurs, i.e., if the perceptual representation is sufficiently similar to one of the stored representations, then the stimulus is recognized as being familiar, and *semantic associations* are triggered that provide information about it, e.g., that it is an enemy.

The stimulus may be recognized, or classified, at different *levels of specificity*, for example, it may be recognized as a particular individual, a particular type of enemy, a general type of predator, or simply an

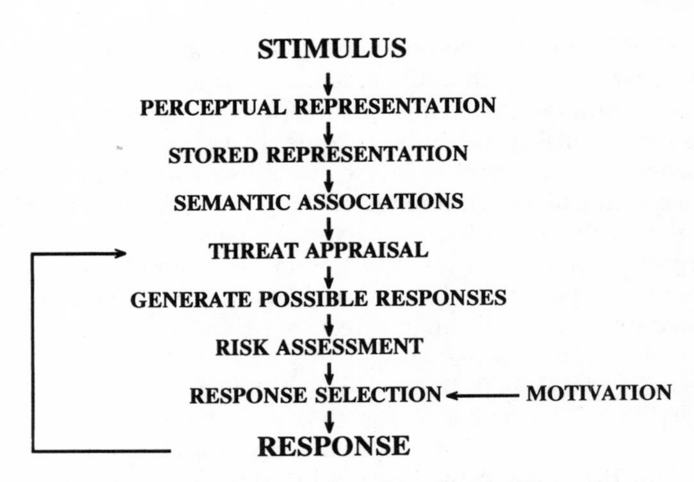

FIGURE 1. A simple cognitive model of enemy recognition.

enemy. Which level is *appropriate* depends on the extent to which different enemies pose different threats, and on whether different responses are appropriate for different types of enemy. If all types of enemy pose the same threat and require the same response, then there is no need to distinguish different types of enemy, and recognition at the most general level, "enemy," is appropriate. However, when different types of enemy represent different levels or kinds of threat (e.g., cat versus cuckoo), then selection of the appropriate response requires recognition at a more specific level.

Enemy recognition at all levels of specificity appears to occur in birds, although the extent to which these abilities are used is less clear. The ability to recognize types of enemies (e.g. ground predator versus flying predator) is likely to be universal. Demonstration that finer scale differentiation occurs (e.g., among species of predators) is dependent on birds giving measurably different responses to those enemies, and may be more difficult to demonstrate (examples in Patterson *et al.*, 1980; Greig-Smith, 1980; Buitron, 1983). Some birds are capable of distinguishing among individual humans (Herrnstein *et al.*, 1976, demonstrated this experimentally for pigeons) and one gull researcher we know insists that he is persecuted even away from the colony (J. Mills, personal communication). Identification of individuals of other enemy species seems likely, but whether birds use this ability to adjust their response in the presence of those individuals has not been established.

Recognition is followed by an *appraisal of threat,* based on experience with the predator and on details of context, including the animal's current state or motivation (Houston and McNamara, 1988), resulting in a belief that a certain level and type of threat exists. Neither the present model, nor Dennett's intentional stance, assumes that such beliefs are *consciously held.* Rather, they predict that the resulting behavior (e.g., the response to the predator) can best be predicted and understood within an intentional framework.

Following threat appraisal, a set of *possible responses* is generated (e.g., mobbing, alarm calling, hiding, distraction behavior) and the appropriateness and risk associated with each is assessed before a response occurs. The complexity of the behavioral repertoire will determine the extent to which stages such as threat appraisal and risk assessment are relevant: "the more flexible the action, the more complex must be the computational resources for monitoring, planning and scheduling different types of activity" (Boden, 1984, p. 155). Motivation, or internal state, of the animal will strongly influence response choice. Also, an enemy may remain present for some time, and the response to it may be an ongoing event. Therefore, continuous monitoring of the threat may be carried out, and the response can be modified. This monitoring process is indicated by the arrow from *response* back to *threat appraisal* in the model.

The decision-making process described here may involve a trade-off between speed and flexibility of response. Processes such as threat appraisal and risk assessment may allow the animal greater flexibility of response in novel situations, but they are relatively time consuming. However, studies of human information processing suggest that repeated exposure to a classification task can result in responses becoming highly automated and very fast (Schneider and Shiffrin, 1977; Shiffrin and Schneider, 1977), presumably through the development of simple decision rules. Humans may even develop rapid responses in relatively novel situations, for example when an experienced musician sight-reads music. Therefore, for commonly experienced enemies, recognition and response selection may occur very quickly. In less routine situations, or with less commonly encountered predators, more time may be required.

"Animal cognition," involving questions about how animals gather and process information from their environment, is a rapidly growing area in psychology (Walker, 1983; Roitblat *et al.*, 1984; Roitblat, 1987). To date, the approach is experimental and little contact has been made with ethological data. For example, in a recent paper on concept learning in pigeons Roberts and Mazmanian (1988) showed that pigeons learned to distinguish pictures of kingfishers from pictures of other birds. The pigeons had great difficulty distinguishing pictures of animals from nonanimals (which they eventually learned to do), and pictures of birds from pictures of other kinds of animals (which they never learned to do). The authors found this result puzzling, because the birds could learn the most concrete or specific category (kingfisher) and the least specific category (animal), but not one at an intermediate level of abstractness (bird).

Yet, ethological data clearly show that birds can distinguish between bird and animal enemies (mammals, for example; references above). Roberts and Mazmanian make no reference to this literature, which should have made their results even more puzzling. We suggest that abstract categories such as "birds" or "animals" have little meaning for pigeons. Using the language in Figure 1, they may have neither stored representations nor semantic associations for these stimuli. For pigeons, what counts are members of one's social group (who may be recognized individually), members of one's species who are not members of one's social group (who may be recognized as members of a class of unknown individuals), and other animals of significance such as kinds of enemies. It may have been fortunate that Roberts and Mazmanian chose a kingfisher (a possible, although unlikely, enemy) rather than some nonthreatening species for their specific bird category.

The important points are (1) that greater consultation of the biological literature could help psychologists to ascertain the important factors influencing the behavior of their subjects, and (2) that biologists can learn much from the theoretical frameworks used by psychologists. In the case of enemy recognition, much recent theory has focused on ultimate influences on behavior, and few attempts have been made to determine what actually triggers observed responses, or whether there are any connections between those triggers and the hypothesized ultimate causation. Dennett's intentional stance, and the cognitive model outlined here, provide frameworks for discussing response patterns of animals at a proximate level, and deserve intensive scrutiny from biologists.

3. MOBBING BEHAVIOR

As noted by Curio *et al.* (1978, p. 899): "There can be little doubt that mobbing has survival value, because it can be potentially dangerous for the mobber or its brood, is time-consuming, and is extremely widespread in vertebrates." Numerous anecdotal reports are available of mobbing birds, or birds giving distraction displays, being taken by predators (Smith, 1969; Myers, 1978; Denson, 1979; England, 1986; Brunton, 1986). Several behavioral studies of brood parasites and predators have suggested that mobbing responses may be provoked by the parasite or predator in order to attract or obtain information about the host or prey (Smith, 1969; Robertson and Norman, 1977; Smith *et al.*, 1984; McLean, 1987). Clearly, mobbing behavior is commonly experienced by both birds and their enemies, and each probably behaves in ways that maximize the effectiveness of mobbing for the mobber or minimize it for the enemy.

Functions of mobbing include providing direct benefits to the mobber, providing benefits to members of the mobber's group or society (including offspring), and discouraging or driving away the enemy (Curio, 1978; also see Caro, 1986a,b). Most hypotheses include immediate or short-term benefits (e.g., the enemy is encouraged to give up), whereas a few involve longer term components (e.g., cultural transmission of enemy recognition). Presumably several functions can be operating at one time, and establishing experimental rationales for demonstrating the operation of a particular function will often be difficult.

Our aim here is to challenge two hypotheses which are virtual dogma in the literature on mobbing behavior of birds. The first is that optimality reasoning provides a useful framework for understanding the patterns of mobbing behavior exhibited by birds in natural situations (Andersson *et al.*, 1980; Montgomerie and Weatherhead, 1988; for a more general approach to optimality reasoning, including comments on the cost of coping with predators, see Houston and McNamara, 1988; for a philosophical analysis see Dupré, 1987).

Briefly, optimality reasoning involves an assessment of function, preferably measured in terms of fitness (Houston and McNamara, 1988). Any currency may be used as the assessment involves a cost/benefit analysis. Optimality studies often use more proximate measures than fitness, such as food intake or survival of individuals. In the case of breeding birds responding to enemies, measurement is usually of behaviors associated with mobbing. These behaviors are related to nest success (a crude measure of fitness) and the analyses involve the assumption that there are optimal behavioral responses to enemies which birds have been selected to exhibit.

Here, we restrict ourselves to breeding birds, and argue that the variation in responses exhibited by birds to different enemies, or to the same enemies at different times, is influenced primarily by the interaction between parents and their nest and young. We call this the *feedback hypothesis*. We argue that hypotheses such as this, which are based on proximate mechanisms, provide more fertile ground for studying many questions about enemy recognition than do functional analyses based on optimality theory.

The second challenge is to the hypothesis that mobbing birds do not exhibit reciprocal altruism. Although this hypothesis is rarely stated explicitly, the combination of the common occurrence of kin in the immediate vicinity of a mobber, the many arguments for selfish benefits derived from mobbing, and the undirected nature of the display, have resulted in the reciprocal altruism hypothesis being neglected. Curio (1978) expressly denied the possibility of reciprocal altruism in mobbing assemblages, Klump and Shalter (1984) offered tentative support, and most papers do not comment and focus on hypotheses which are more easily tested. Curio (1978) noted that a mobber appears to be displaying to an "anonymous community," and cannot be protected against cheaters who would swamp any benefits derived from undirected apparent altruism. In contrast, we agree with Smith (1986), who argued that territorial neighbors do not constitute an anonymous community. Rather, neighboring resident nonconspecifics are the most likely individuals to come to a mobber's assistance at a time when conspecifics have been excluded from the mobber's immediate vicinity (Dean and McLean, ms). Such individuals need not be known individually as long as they are resident locally for enough time for a reasonable number of mobbing events to occur (Trivers, 1971; Axelrod and Hamilton, 1981). Assistance with mobbing may even cross major phyletic boundaries (Rasa, 1981).

Lastly in this section, we offer some comments on the utility of cultural transmission for conservation biology. In the context of enemy recognition, cultural transmission involves transfer of an ability to recognize enemies from one individual to another by behavioral means. Transfer of appropriate kinds of response may also occur, although this has not been demonstrated to our knowledge.

3.1. Predictions Derived from Optimality Theory

The allure of optimality arguments is that they are eminently suited to mathematical analysis and they fit directly into the framework of natural selection (Maynard Smith, 1978). It is difficult, therefore, to suggest alternatives to optimality reasoning without also appearing to criticize natural selection (Myers, 1983). Currently, despite some dissatisfaction with optimization as a valuable heuristic in biology (e.g., Gray,

1987; Emlen, 1987; Jamieson, 1989), there are no alternative frameworks available which offer predictions on the same scale, or with the same elegance. The recent review by Houston and McNamara (1988) indicates that optimality reasoning is still enjoying strong support in biology. One reason for this may be that those who find support for optimality work with relatively short-term behaviors where animals can use simple decision rules to mimic predictions of theory, whereas those who question the utility of the theory work with systems subject to historical constraints (see discussion in Krebs *et al.* 1983).

Montgomerie and Weatherhead (1988) provide the latest summary of predicted patterns of nest defense derived from optimality theory. They predict that the intensity of nest defense (actually exposure to risk by parents) should increase through the nesting cycle either because renesting potential declines or because the probability of offspring survival increases relatively rapidly compared to that of parents. In this general representation, intensity of defense increases continuously from egg laying to a peak at departure of the chicks from the nest, after which it declines because chicks become less vulnerable. A version of the curve specific to Great Tits (*Parus major*) is in Regelmann and Curio (1986). Montgomerie and Weatherhead predict that intensity of nest defense will also vary with parental experience and confidence of parenthood; offspring number, quality and vulnerability; nest accessibility and conspicuousness; and a host of other factors. To all these factors must be added an individual's "state" at the time of the enemy interaction (Houston and McNamara, 1988), although this may contribute to variance rather than to any predictable patterns. We call this model the *optimality hypothesis,* and depict its prediction in Figure 2a, which is taken from Figure 2 in Montgomerie and Weatherhead (1988, p.178).

Our difficulty with the optimality hypothesis is that it requires a level of control of behavior by genes which we consider to be unrealistic (see especially Heyman, 1988, critiquing Houston and McNamara, 1988; also several chapters in Dupré, 1987). It seems unlikely to us that selection will operate on the host of factors which influence reproductive success listed by Montgomerie and Weatherhead (1988) in relation to a complex suite of defensive behaviors and cognitive abilities in any way which directly links those behaviors and genes. Also, individual birds are unlikely to gain enough experience with enemies during the breeding cycle to allow them to develop simple decision rules which would mimic predictions of theory. Selection has presumably encouraged defensive behavior by birds. However, we hypothesize that the details of the defensive response exhibited by each bird each time an enemy is encountered will be influenced by proximate factors (see Fantino, 1988, critiquing Houston and McNamara, 1988).

A small digression by way of a simple analogy may help elucidate this point. Let us agree that driving a car is adaptive because cars allow quick and efficient transfer from one point to another. There are many types of car and many ways of driving each one. Even when driving the same car different drivers will use different amounts of fuel and will cause different amounts of wear and tear. However, availability of cars, driving styles, and routes will change little over several years, during which each individual will drive many times. It will therefore be possible for drivers to develop strategies (or decision rules or Darwinian algorithms, see below) for determining the best way to drive. Presumably these will be based on relative costs, comfort, the types of car available, the required route for going with certain friends, the weather, the personality of the driver, and a host of other factors. However, none of these factors strongly influence the adaptiveness of driving and all are functionally equivalent in terms of achieving that adaptive end. If we, as researchers, measure the details of routes, costs, and so on, in order to establish which decision rules our study drivers are using, we will learn nothing about why it is adaptive to drive, although we will learn about the processes which influence the development of decision rules.

In this analogy, driving options, although reasonably constant, are not invariant. Car models change, as do routes and the kinds of fuel available. Drivers earn more, develop new friendships, change jobs, feel tired, a storm is brewing. It seems unlikely that anyone would argue that decisions made in response to these details are entrained into genetic structures with selection constantly acting to update our response to them as they vary. Yet this hypothesis underlies the optimality model for response to enemies during the nesting cycle. Driving is analagous to defensive behavior in the general sense. The options available for driving from A to B are equivalent to different sorts of mobbing responses. Predictions about the most likely way for an individual to behave in a particular context will depend on how that individual developed its decision rules. We suggest that the cognitive abilities of birds allow for context and experience to be the most important influential factors in determining patterns of response through the breeding cycle.

Cosmides and Tooby (1987) have developed this argument by referring to what they term "psychological mechanisms," which they hypothesize provide the "missing link" between adaptive behavior, and the expression of behavior in a particular context. They argue that selection has given animals information processing machinery which provides for the development of "Darwinian algorithms" (i.e., decision rules). We believe that further attention to the cognitive abilities of animals that exhibit complex behaviors, and the influence of experience and context on the expression of those behaviors, may provide more fruitful predictions about how animals will behave than elegant mathematics can offer on how they should behave.

3.2. An Alternative Scenario: The Feedback Hypothesis

The cognitive model of enemy recognition describes a system that is sensitive to proximal factors such as context and experience. If, as published data suggest, birds show predictable patterns of response to enemies during the breeding cycle, and if response patterns are influenced by proximal factors, then we must search for environmental factors that could influence observed responses. We were encouraged by the result that defensive behavior of birds when faced with an enemy is strongly dependent on context (see references in Section 1). However, the details of each interaction primarily influence *variance* rather than pattern of response (see, for example, scatter in Figure 14 in Curio (1975, p. 30)), and do not predict response curves. We suggest that the proximate factor that influences the pattern of response is the continual interaction between parents and young.

In our scenario, current young are viewed as having two influences on adult behavior. First, they will necessarily be perceived by the parent as being of *greater immediate value than any other clutch* (past or future). Thus they should be defended at all times in ways that are appropriate to the context of the interaction with the enemy (rather than to the optimal "value" of the young). This suggestion is derived from the notion that the proximate stimulus of existing young reinforces parental care (including defensive behavior). We therefore predict that, all other things being equal (which they are not, at least for birds with altricial young, see below), *within-parent* variation in response through the breeding cycle will be small. We note that *between-parent* variation may still be large due to factors such as parent age and experience, nest-site characteristics, or differences between the sexes.

Second, the pattern of parental response will be influenced by the *feedback signals provided by current young*. It is variation in the signals provided by current young which explains why all things are not equal, as mentioned above. The nest and eggs provide a static visual signal which maintains parental activities at the nest. It is difficult to assess the influence of the nest without eggs, but this is an object with which parents interact and is an intrinsic component of rearing young; thus it should be defended, although possibly with a lower intensity than for a nest containing eggs. After hatching, the chicks provide a stronger signal due to their active behaviors which encourage a still higher level of parental activity.

We depict this pattern in Figure 2b. Here, the intensity of defense rises rapidly from zero to a plateau appropriate to defense of a nest and clutch of eggs as the nest is built and/or eggs are laid. Rate of increase and attainment of the plateau for each bird will depend on when incubation is initiated, how often the nest is visited during laying, and which birds are attending the nest. At hatching, a step function occurs because the parents considerably increase their interest in and interaction with the nest contents, and the static eggs are replaced by active mobile chicks.

Shape of the curve after hatching is more difficult to predict because it depends on the contribution to reinforcement by at least three aspects of chick development which do not follow similar developmental pathways, nor are they the same between species. These are chick size, vocalizations, and the irruption of feathers which change the visual appearance and overall size of chicks almost instantaneously (personal observation). Representations of these three components of development are provided in the three bottom graphs in Figure 2. In general, growth is maintained at a relatively constant rate throughout time in the nest, although there is usually some sigmoidal tendency (Ricklefs, 1983; the line is taken from his Figure 2, p. 14). Increase in vocal volume is strongly sigmoidal, being initially quiet, then increasing in volume over a few days and peaking well before departure from the nest [Knight and Temple, 1986c; the line is taken from their Figure 2, p. 889, (Greig-Smith, 1980, represents this as a straight line, but provides too few data points to properly assess the shape of the curve)]. We summarize these three curves into a monotonic increase containing a second step function in the nestling phase (Fig. 2b).

Once chicks depart the nest, we predict that defense will remain initially high, but will decline as parents reduce their contribution to chick maintenance and their interactions with chicks. Response should remain somewhat higher initially than predicted by Montgomerie and Weatherhead who suggested a rapid decline because fledglings become more capable of escaping danger and predators are unlikely to destroy an entire brood. This difference may be rather subtle and difficult to detect.

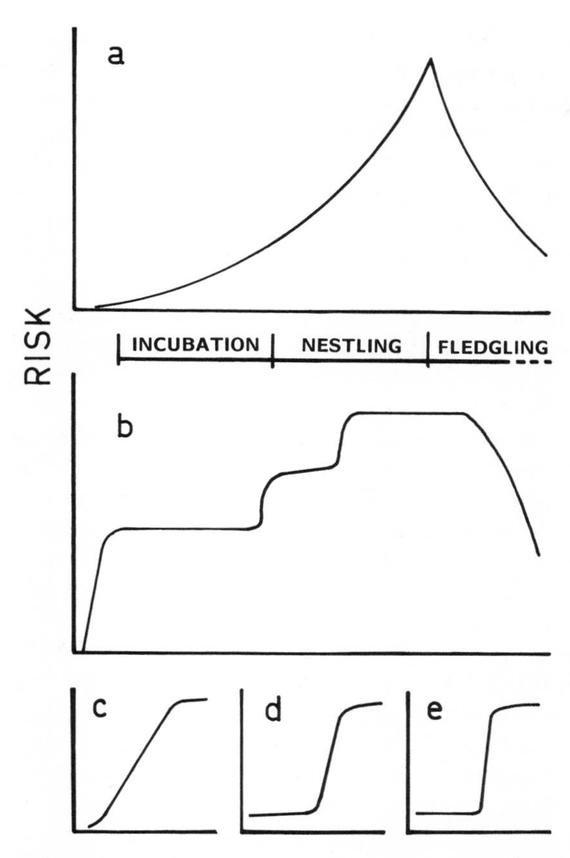

FIGURE 2. The response curves for exposure to risk by parents during the breeding cycle predicted by optimality theory (a; from Montgomerie and Weatherhead, 1988), and predicted by the feedback hypothesis (b). The nestling component of the feedback hypothesis curve is made up of a combination of the curves for growth (c; from Ricklefs, 1983), volume of nestling begging calls (d; from Knight and Temple, 1986c), and first appearance of juvenile plumage (e).

3.3. Evidence for the Feedback Hypothesis

A relatively small number of studies have been conducted intensively enough for the shape of the response curve through the nesting cycle to be available in the required detail. Methodological problems with human approaches raised by Knight and Temple (1986a) and Hobson *et al.* (1988) decrease confidence in at least some of the studies which are available. Further investigation of these problems by Redondo and Carranza (1989) and Weatherhead (1989) suggest that a series of human approaches may be an acceptable research technique. Here we limit discussion to those studies which provide details on the shape of the response curve through the breeding cycle and ignore the methodological issues. The most detailed studies are those of Curio and coworkers (Curio, 1975; Regelmann and Curio 1983; Curio *et al.*, 1984; Curio *et al.*, 1985; Regelmann and Curio, 1986), Greig-Smith (1980), Andersson *et al.* (1980), East (1981), Redondo and Carranza (1989) and Weatherhead (1989).

The patterns from Greig-Smith (1980) for Stonechats (*Saxicola torquata*) clearly fit predictions of the feedback hypothesis. Response during incubation is flat and low. There is an increase in response through the nestling period, and this response is maintained into the early period after departure from the nest, after which it declines. Particularly remarkable is the rapid jump in call rates for one alarm call (chacks) part-way through the nestling period (determined by combining data in his Figs. 9a and 10, pp. 614 and 615). The other call (whits), which does not show this step function (although it does flatten out at the top end), is used specifically to warn chicks to remain quiet, which may influence its pattern of use in subtle ways.

Data from Curio and co-workers on the Great Tit (*Parus major*) fit the feedback hypothesis almost as well as those of Greig-Smith. The following summary is taken from data in the papers listed above. Response during nest building is higher than during incubation, drops early in incubation, then increases (although not statistically significantly) through the second part of incubation. The slope of the response curve increases markedly during the nestling period with a jump at about nestling age 12, after which it flattens out. The jump at day 12 was found in two different studies (Curio, 1975; Regelmann and Curio, 1983), and strongly supports the feedback hypothesis as it represents the second step in the nestling period in Figure 2b. Neither hypothesis predicts a drop in response between nest building and incubation.

East (1981), working with European Robins (*Erithacus rubecula*), found a flat response during incubation and a rapid increase in response through the nestling period, with the peak maintained into the initial period after departure of chicks from the nest. There is a suggestion of a step function during the nestling period (although this is unlikely to be statistically significant). These data strongly support the feedback hypothesis.

Andersson *et al.* (1980) compared colonial and solitarily breeding Fieldfares (*Turdus pilaris*) and found both types responded before completion of the clutch at a level slightly lower than after incubation began. There was a slight (but not statistically significant) increase in response through the incubation period and a step upwards at hatching. Response was flat after hatching. This pattern was also found by Blancher and Robertson (1982), although there is a suggestion of a step function in the response curve for the nestling period in their data. These data support the feedback hypothesis better than the optimality hypothesis, but neither model predicts a flat response post-hatching.

Weatherhead (1989) presented a series of analyses of data taken from Song Sparrows (*Melospiza melodia*). Overall, he found a flat response during incubation, and a steady increase in response through the nestling period. Some graphs appear to contain step functions at or near hatching and/or during the nestling period, but the significance of these cannot be assessed from the data presented. No data were obtained after chicks left the nest. These data support the feedback hypothesis more closely than the optimality hypothesis.

Redondo and Carranza (1989) present data from Magpies (*Pica pica*) taken during the nestling period only. The response curves are initially flat, then are followed by a rapid rise about two thirds of the way through the nestling period. Although suggestive of the step function predicted by the feedback hypothesis, these results do not contain data after chicks leave the nest and it is not clear if there is a plateau after the increase in response. However, the data fit the feedback hypothesis more closely than the optimality hypothesis. Redondo and Carranza interpret their results in relation to another proximate factor which may influence feedback: the behavior of Magpie chicks in the presence of an enemy changes from crouching, to calling and abandoning the nest, at about the age that adults rapidly increase their response.

These data, although limited, provide enough support for the feedback hypothesis to warrant its being given further attention. We encourage other researchers to consider the influence of proximate factors on their data, and perhaps to extend our model to incorporate factors that we have not considered.

The feedback hypothesis is simpler than the optimality hypothesis, provides different predictions, and is based on assumptions about the cognitive abilities of birds rather than assumptions about the operation of natural selection. However, the difference in levels at which these two hypotheses operate (proximate versus ultimate) means that they cannot be directly compared. In fact, feedback may be the best mechanism currently available to selection in the evolutionary attempt to provide birds with an optimal pattern of response to enemies (a point belabored by numerous reviewers of this chapter). Our primary purpose in developing the feedback hypothesis is to point out that questions about mechanisms are usually ignored in the current literature because of emphasis on the search for a fit between data and predictions of optimality theory under the guise of testing that theory. Tests of predictions of optimality theory may be possible under controlled conditions using short-term behaviors such as foraging. Testing optimality using patterns of defensive behavior during breeding is probably impossible, in part because a lack of fit between data and theory is unlikely to result in rejection of the theory (Gray, 1987). We suggest that attention to the mechanisms involved in the interaction between birds and their enemies will provide a richer source of information about the behavioral abilities of birds than will studies of optimal response curves.

3.4. Other Factors Influencing Feedback, the Response Curve, and Variance in Response

We cannot predict the influence of clutch or brood size on feedback without further information. Although it seems likely that larger numbers of eggs or chicks will provide a greater stimulus than one or two, we doubt that the relationship is linear and suspect that the effects of additional young added to the first will be relatively minor, especially when this feedback is expressed as nest defense.

Differences in levels of activity by parents at the nest (e.g., between sexes) may provide a test of differences in predictions between the models. In species where females do all nest building and incubation, we predict higher levels of response for females than for males; whereas male response should increase once he begins feeding nestlings. Most studies in which sexes have been assessed separately have found differences in response (review in Regelmann and Curio, 1986). We caution that in relation to the feedback hypothesis careful attempts must be made to quantify the interaction of the male with the nest even if he does not build or incubate. In many species males accompany (or guard) the female during nest-building and laying and attend her at the nest either by feeding her or by accompanying her back to the nest after a feeding sortie. Males may also use a song post within view of the nest or may remain near the nest or check it frequently while the female is incubating. His experience of the nest may be little different to hers despite his lack of direct involvement.

Two studies have found that birds have a specific call for silencing young when an enemy is near (Greig-Smith, 1980; Knight and Temple, 1986c). This is presumably an adaptive behavior which has been selected for independently of response curves. However, once a bird starts calling, it has notified the enemy of its position and that it is concerned.

It may then be constrained to continue calling because mobbing behavior, and the associated attraction of others to assist, may be the only practical response. Also, if calling by chicks will resume soon after adult calling stops (Greig-Smith, 1980), then the parents may be constrained to continue calling anyway. They therefore lose the low risk option of not calling or mobbing that is available before eggs hatch. Although we doubt that silencing nestlings is the only reason that birds call when an enemy is near the nest, the simple rule of "chicks in nest, must call; versus eggs in nest, calling may not be required" may be influencing the patterns seen, and are entirely consistent with the cognitive model in which assessment is an intrinsic part of response choice.

Like researchers, animals must often make decisions without obtaining complete information about the situation being faced (see Section 2). They must often respond immediately, without sampling, considering alternatives, developing strategies, or adjusting their behavior in relation to their internal state. Also, birds sometimes appear to adjust their response during a mobbing episode as they obtain more information about the enemy. Such context-dependent adjustments are not usually considered in optimality analyses of defense behavior, although they appear in studies of optimal foraging (Houston and McNamara, 1988). The difficulty for testing any theory is that an initial response may commit the bird to a set of response options of which none would have been the preferred (best, optimal, or most appropriate to feedback level), had all necessary information been available when the initial response was made. Such birds will appear to researchers to have behaved inappropriately, since even if they achieve the best response by the end of the interaction event, it is their entire response which enters the data. For both the optimality and feedback hypotheses, such situations will add increased variance to the data.

Two other factors whose primary influence will be to increase variance in response are that birds may respond erratically, or different responses may be equally effective. These generate difficulties for the researcher attempting to find a meaningful variable to measure and are unlikely to help in differentiating between competing hypotheses. Although a few studies have found a relationship between level of nest defense and probability of nest success, others have not (review in Montgomerie and Weatherhead, 1988), and it is still not at all clear that the variables we measure are highly correlated with the birds' view of their interaction with the enemy. Measuring everything is not always the best option because some variables are correlated either negatively or positively (e.g., Redondo and Carranza, 1989, Fig. 4, p. 373). We desperately need an independent technique for determining which variables are most important to the bird, in terms of effectiveness of nest defense and degree of risk.

3.5. Do Mobbing Birds Exhibit Reciprocal Altruism?

We know of only one study which has specifically set out to determine if reciprocal altruism occurs in mobbing assemblages of birds (Dean and McLean, ms). The difficulty here is that it is virtually impossible to design an experiment which excludes all likely competing hypotheses concerning why birds mob. We believe that Dean and McLean have come close to achieving this objective in a study on response to a standard enemy by mixed-species flocks (MSF) of birds in New Zealand, and summarize this study below.

New Zealand birds do not form the spectacular MSF seen in the tropics (Munn and Terborgh, 1979). However, several forest bird species form MSF during the nonbreeding part of the year (McLean *et al.*, 1987; Dean, 1990). The species which join flocks include endemic, native, recently self-introduced, and human-introduced birds. In the study area at Kowhai Bush, Kaikoura, on the northeastern seaboard of South Island, six species join MSF, including the native Grey Gerygone, New Zealand Brown Creeper (*Mohoua novaeseelandiae*, Grey Fantail *Rhipidura fuliginosa*, and Silvereye *Zosterops lateralis*, and the introduced Chaffinch *Fringilla coelebs* and Common Redpoll *Carduelis flammea*) (Dean, 1990). Other bird species resident locally include three native and five introduced species which do not join MSF in forest.

Of the species that join MSF, our observations and banding records indicate that New Zealand Brown Creepers and Silvereyes regularly flock. Brown Creepers remain in small family groups throughout the year, occasionally forming larger amalgamations in winter. Silvereyes form flocks of up to 30 individuals which may contain small family groups. Fantails occasionally form flocks of up to 15 birds along forest margins or in clearings, but do not remain in family groups and are usually solitary unless in MSF. Grey Gerygones do not form monospecific flocks and are usually solitary. The two finches join large MSF in forest in the fall only, and move onto adjacent farmland in the winter, where they often join large MSF of finches. Some Chaffinches occur in forest in winter, but Common Redpolls do not. One other species remains in family groups in the study area (Rifleman *Acanthisitta chloris*), but does not join MSF. All native species are relatively long-lived, often reaching five to six years of age. Life history information on introduced species is unavailable.

All species, except the nonflocking Shining Bronze Cuckoo, are resident at Kaikoura throughout the year. Populations of some (Grey Gerygone, Grey Fantail, Silvereye and probably most finches) increase in the winter due to local movements to the coast by inland birds. However, observations of marked birds confirm that at least some individuals which join MSF are likely to be neighbors during the breeding season. These birds appear to fulfill the conditions required for the evolution of reciprocal altruism; of lengthy associations relative to the frequency of the altruistic act (Trivers, 1971; Axelrod and Hamilton, 1981; Rothstein and Pierotti, 1988).

A suite of predators occur at Kowhai Bush (Moors, 1983), including introduced mammals (mustelids, cats, rats) and birds (Pacific Marsh Harrier *Circus approximans*, New Zealand Falcon *Falco novaeseelandiae*, Little Owl *Athene noctua*, the Morepork Owl *Ninox novaeseelandiae*, and Kingfisher *Halcyon sancta*). These predators are mobbed by resident forest birds using calls and displays similar to those used when an experimental predator (a stuffed Little Owl) was presented. Mobbing events are reasonably frequent; we note about one per week. In contrast to some views, New Zealand was not depauperate of bird predators prior to human arrival (Holdaway, 1989) although there were no mammalian predators (Moors, 1983), and mobbing birds use behaviors typical of small birds anywhere when faced with an enemy (Curio, 1978); they are likely to have been mobbing enemies for millenia.

In order to investigate whether birds which join MSF exhibit reciprocity, Dean and McLean (ms) conducted a series of tests in which alarm calls of one of six bird species were played back in the presence of the stuffed Little Owl. The aim in this study was to investigate which birds joined mobbing flocks that formed when mobbing was "initiated" using playback. Two of the variables tested are relevant here: (i) response in relation to whether the call played back was a flocking or nonflocking species, and (ii) whether patterns of species joining mobbing assemblages differed between winter and spring. Measured responses included proportion of tests responded to by each species, number of individuals responding, number of alarm calls given by each species, and proportion of time close to the model.

The results were unambiguous. First, the four native species which join MSF responded most strongly to the alarm calls of three native MSF species played back in the presence of the owl; they responded less strongly to the Chaffinch, and least strongly to alarm calls of two nonflocking species (introduced Blackbird *Turdus merula* and a control, the Red-breasted Flycatcher *Ficedula parva*, which does not occur in New Zealand). Of the two introduced species that join MSF, Chaffinches responded about equally to alarm calls of the three native species and Chaffinch alarm, and less to Blackbird and Red-breasted Flycatcher alarm. Common Redpolls, which only responded in spring (they do not occur in forest in winter), did not show significant variation in response patterns across alarm calls, in part because of small sample size of response overall. Three species which did not join MSF responded; of these, the native New Zealand Bellbird (*Anthornis melanura*) and introduced Dunnock (*Prunella modularis*) responded about equally to all alarm calls, whereas the Blackbird showed some tendency to respond significantly more to Blackbird alarm. Five species did not respond to any alarm calls.

Second, there was no difference in patterns of response between winter and spring for species which joined MSF. All native species which joined MSF responded about equally strongly to calls of the three native MSF species in spring as in winter, even though flocks do not form in spring. Chaffinches responded more in spring than in winter, and Common Redpolls only responded in spring. Bellbird and Dunnock both responded significantly more in winter than in spring. Blackbirds responded about equally in both seasons (but primarily to Blackbird alarm).

These results are interpreted to mean that birds which join MSF respond reciprocally to each others' alarm calls. They appear to be operating by the simple rule: "if it joins my mixed-species flock and it is alarm calling, I should join the mobbing assemblage." This rule applies at all times of year, even when flocks are not forming, and suggests a complexity to community structure which has not been previously suspected. These birds are not members of an anonymous community (Smith, 1986) and are assisting each other independently of kin associations or parental care.

Two likely driving forces of this reciprocity are, first, that during the breeding season, the members of ones' own species are the least likely individuals to come to assist when an enemy is threatening because they have been excluded by territorial behavior. Thus, the only birds available to assist when mobbing is needed to discourage an enemy are members of other species. This problem is faced by all breeding birds (even including the New Zealand Brown Creeper whose flock sizes in summer are limited to small family groups). Assuming that mobbing really is effective at discouraging enemies, all individuals living in a small community should benefit from assisting each other when an enemy is near. Second, a preference for responding to each others' alarm calls can be established while birds are in MSF when they are likely to both hear alarm calls of other species and see the cause of the alarm (in contrast, species which do not associate with the flocks will hear the alarm because they live in the vicinity but may not see its cause). They therefore can learn which enemies the other members of MSF respond to. All species which join MSF at Kowhai Bush build cryptic nests in the canopy or subcanopy, and feed in all parts of the forest including on the ground, although they spend different proportions of time at different heights and using different substrates (I. G. McLean and S. M. Dean, unpublished data). They are thus likely to be threatened in similar ways by the resident predators, and benefit in similar ways from mobbing them.

Does the reciprocity of response among birds which join MSF indicate that reciprocal altruism is occurring in this community? The primary condition of reciprocal altruism, that X suffers a cost to Y's benefit, and Y suffers a cost to X's benefit at some later time (Trivers, 1971), was not demonstrated in this study. Discussion of the problem of individuals cheating in a system of reciprocity may be found in Rothstein and Pierotti (1988). Unfortunately, it is impossible to show that a mobbing individual gains nothing from a mobbing event, although it presumably suffers some cost in terms of time spent and a slight possibility of being attacked by the enemy. Mobbing behavior in this situation appears to be a form of simultaneous cooperation in which there is an asymmetry in the initiator of the mobbing event (see Rothstein and Pierotti, 1988, for a more detailed discussion of terms and of cost/benefit tradeoffs). Other asymmetries will derive from how much benefit each individual accrues from the mobbing event (e.g., is there a nest nearby, does the bird have dependent young, or is it currently spending a lot of time feeding on the ground). We suggest that these asymmetries, and the reciprocity which occurs in this system, mean that some form of reciprocal altruism is occurring among New Zealand forest birds.

3.6. The Significance of Cultural Transmission

Cultural transmission of enemy recognition has now been demonstrated in both the laboratory (Curio *et al.*, 1978; Vieth *et al.*, 1980) and the field (Conover, 1987). Limitations on the ability of birds to learn about enemies have not yet been explored, and we believe that there are many exciting areas of research in this area awaiting attention. For example, are there analogies with imprinting, where the object on which young birds will imprint must be within proscribed dimensions (Johnson, 1988); and do birds learn both about which organisms should be treated as enemies, and how to respond to each one? We suspect that a considerable proportion of the variance normally encountered in studies of response to enemies (e.g., McLean *et al.*, 1986) can be explained in terms of each bird's experience with that enemy, and observation of encounters between the enemy and other members of its species.

We are particularly excited about possible applications of cultural transmission of enemy recognition for conservation biology. New Zealand contains possibly the largest proportion of endangered bird species of any country in the world. The decline of most of these species is due to the joint problems of habitat loss and introduced predators and competitors. Although establishment of reserves can limit habitat destruction, control of predators and competitors is expensive, limited in scope, and subject to vagaries of funding. An important solution to predation is transfer of endangered species to habitat islands where predators and other ecological competitors have been eradicated or were never introduced. However, such programs may only be a short-term solution if few large islands with appropriate habitat are available (as is the case in New Zealand, e.g., Williams, 1986), or if the habitat required by the species is unavailable on islands.

We believe that it may be possible to train at least some endangered populations to cope with introduced enemies, and therefore to assist their survival in the long term. If the species does not recognize the new enemy, or alternatively does not have the behavioral ability to cope with that enemy when it appears nearby, then it may be possible to provide those skills using a training regime. Once established in the population, transfer of the skills could be achieved by cultural transmission, perhaps assisted by people involved in the management program at critical times (e.g., when chicks are becoming independent). We are currently testing such a program with island-endemic members of the genus *Petroica* (I. G. McLean and R. Maloney, unpublished data).

McLean and Maloney (1989) have shown that a subspecies of New Zealand Tits (*P. macrocephala dannefaerdi*), which are endemic to the Snares Islands, recognize some natural enemies of tits on mainland New Zealand. This recognition ability is presumably genetic as these enemies are not resident on the Snares. However, response levels were considerably lower than for mainland tits and the aim is to improve the responses of island birds to levels exhibited by mainland birds. The research design involves using models and playback of alarm vocalizations to raise the level of response of the island birds to that shown by mainland birds and also to train them to recognize and respond to mammals such as rats which are not natural enemies. Although it is not known if such recognition and response abilities will save the tits in the presence of rats, we believe that there are species (such as the Black

Stilt *Himantopus novaezealandiae*) whose survival could improve significantly using such a program. We particularly encourage researchers attempting to use this technique to ensure that young birds are exposed to the training regime, either directly or by watching their parents giving appropriate responses.

4. A CASE STUDY OF RECOGNITION: THE WARBLER AND THE CUCKOO

The Grey Gerygone is a small (6.5 g) territorial song bird which is parasitized by the Shining Bronze Cuckoo. These cuckoos are host-specific on Grey Gerygones in New Zealand, perhaps because this is the only species which builds a domed hanging nest (Gill, 1983). Many details of the interaction between the gerygone and cuckoo have been documented by Gill (1982a,b; 1983) and the summary presented here is a review of his work and our own unpublished observations.

Cuckoos are the only forest species which are long-distance migrants in New Zealand (Oliver, 1955). Shining Bronze Cuckoos arrive in late September or early October, when eggs in some first warbler nests have already hatched. Thus they generally parasitize second nests of gerygones, most of which are initiated in November. Second nests may be initiated in October after breeding failures and it is likely that cuckoos cause breeding failures by gerygones (we have one observation of a cuckoo removing the last egg from a nest). However, gerygones suffer only low direct fitness cost from nest parasitism of second nests, because, although all chicks are expelled from the nest by the cuckoo chick, survival of chicks from unparasitized second nests is low. Parents provide food for a cuckoo chick at a rate equivalent to that required for about 2.5 gerygone chicks (the typical clutch is three or four). Thus the work required to raise a cuckoo chick is moderate rather than high.

Cuckoos of the genus *Chrysococcyx* parasitize many Australasian species which build domed nests (Gill, 1983; Payne *et al.*, 1985), and it is likely that this parasitism has been going on for millenia. Why, then, have warblers not developed tactics for preventing parasitism by cuckoos, for example by rejecting the egg or chick from the nest when they appear? In other brood parasite/host interactions some species have the ability to reject cuckoo eggs, either through removal of the egg or abandonment of the nest (Payne, 1977; Rothstein, 1982a; Rohwer and Spaw, 1988). Rothstein (1982a; personal communication) has shown that rejector species do not recognize parasite eggs; rather they recognize their own eggs and reject any egg that is sufficiently different, an ability that does not limit the potential host to targetting specific brood parasites and may even help in preventing intraspecific parasitism. This ability may have led to selection for mimicry by cuckoos of host eggs, and even to the development of subpopulations of cuckoos called "gens" which specialize on particular host species (Wyllie, 1981; Brooke, 1989). However, there is no evidence for any host recognizing the cuckoo chick once it appears in the nest, although there are suggestions of chicks mimicking morphology or vocalizations of host chicks (Payne, 1977; McLean and Waas, 1987; Davies and Brooke, 1988).

The egg of the Shining Bronze Cuckoo is slightly larger, rounder, and greener than the geryone egg, which is white with brownish speckling. As far as we can tell, gerygones do not reject cuckoo eggs, nor do cuckoos need to time egg-laying to coincide with host laying. We know of three cases in which a cuckoo egg was laid about a week after incubation was initiated; in two of these the cuckoo hatched successfully and was raised. The cuckoo may be assisted by the domed nest (and therefore dark interior) of the host. Gerygones cannot build over a clutch of eggs, their ability to see details of eggs within the nest is limited, and we have no records of nests being abandoned after a cuckoo egg appeared in the nest.

We have been investigating two recognition questions in the warbler/cuckoo interaction. First, we asked if cuckoos were able to parasitize gerygones because the hosts did not recognize cuckoos as enemies. Apart from tail size, the plumage of juvenile cuckoos is virtually identical to that of adults, so cuckoo chicks become very adultlike before they leave the nest. For the last three weeks of dependency, cuckoo chicks look identical to adults. Second, McLean and Waas (1987) had shown that cuckoo chicks do not sound like adult cuckoos. Rather, they give calls virtually identical to those given by gerygone chicks. We asked if these hosts are "fooled" by cuckoo begging calls. Perhaps it is the auditory rather than the visual stimuli that are most important to gerygones.

4.1. Recognition of Adult Cuckoos as an Enemy by Grey Gerygones

We tested whether gerygones recognize adult cuckoos by placing a stuffed cuckoo near gerygone nests during the first week of incubation. At this time, cuckoos can still successfully parasitize the nests. As a control, we placed a Greenfinch *Carduelis chloris* (a species that also breeds in this forest) in the same posture and location as the cuckoo.

Order of presentation at each nest was alternated and models were presented one or two days apart. Greenfinches and Shining Bronze Cuckoos are of similar size and coloration. The models were placed while the female was away on a feeding trip (males do not incubate) so that on her return she found the model sitting staring at her nest from about 1.5 m away. Models were left for one minute from when the female looked directly at them. Observations of naturally occurring incubation periods and foraging trips were made either the day before or immediately before the first model test.

Twelve nests were tested. On her return to the nest with no models present (a natural return), a female gerygone flew rapidly to a perch just below or to the side of the nest entrance from 5 to 10 m away. She immediately entered the nest and settled on the eggs. If one of the models was present, her response on seeing the model—sometimes from a perch within 1 to 2 m of the model and the nest, and more usually from immediately below or beside the nest—was to move away from the nest and to inspect the model carefully. If the model was the Greenfinch, she visibly became less agitated during the one minute test period; in five trials she fed, and in five trials she entered the nest before one minute had passed. If the model was the cuckoo, she remained agitated, did not feed, and usually moved to a location where she was not between the model and the nest. In two trials she rapidly entered the nest within a few seconds but used flight paths and perches which she had not used during natural returns or while the Greenfinch was present. Her total time off the nest and the delay from when she first sighted the model, were recorded, as were all behaviors exhibited during the test period. Responses of males were included in the data if they were present.

The response of females to the cuckoo was rather subtle. Most clearly recognized the cuckoo as an enemy, but they did not respond to it using intensive mobbing behavior. Males arrived with the female for five tests and mobbed the cuckoo (using alarm calls and swoops on the model) in four of these, but only one female alarm called (once) and no females swooped on the model. The Greenfinch was mobbed (by a male) for one test only, and this was a case in which the Greenfinch was presented second, the cuckoo having been mobbed on that perch by the male the previous day; the swoop was part of an initial mobbing response which declined rapidly to cautious appraisal (the birds sat nearby and watched the model). One female appeared to go into a catatonic state when she saw the cuckoo; she "froze" on a perch 2 m from the cuckoo and 0.5 m from the nest and did not move for 16 minutes (the cuckoo was removed after one minute). After 16 minutes she moved her head slightly, but remained on the perch until 25 minutes when the male arrived and chased her off.

There was no significant difference in response to the two models in number of hops (movements < 10 cm) or flights (perch changes > 10 cm), the number of alarm calls or songs, or the nearest approach distance to the model (Table 1, $p > 0.05$, Wilcoxon signed-ranks tests on each behavior). However, the appearance of the cuckoo near the nest for one minute significantly delayed the females' return to the nest relative to her time off the nest during a natural foraging episode or the appearance of a Greenfinch near the nest (Table 1; Friedmans test; $p = 0.004$) with the mean delay caused by the cuckoo (14.8 minutes) being seven times the delay caused by the Greenfinch (2.0 minutes). During the delay period some females constantly inspected the perch where the cuckoo had been sitting, sitting on it and occasionally alarm calling or giving short songs. Also, the females' incubation and feeding regime changed dramatically after a cuckoo appeared near the nest (these observations were made only on the later nests to be tested, because initially we did not realize the importance of monitoring behavior for

TABLE I
Response of Grey Gerygones to a Model Shining Bronze Cuckoo and a Greenfinch Placed Near the Nest Early in Incubation, When Brood Parasitism Could Occur ($N = 12$, unless otherwise indicated)

	Response to				
	Cuckoo		Greenfinch		
	Mean	SD	Mean	SD	p
# hops	17.0	12.0	12.0	8.5	NS*
# flights	6.7	5.9	5.0	3.6	NS*
# alarm calls	3.8	7.1	0.3	0.9	NS*
# songs	0.3	0.7	0	0	NS*
nearest distance (m)	1.0	1.0	1.2	0.7	NS*

	Return to nest			
	Natural	After cuckoo	After Greenfinch	p
Delay (min) ($N = 12$)	—	14.8 ± 13.7	2.0 ± 2.0	0.003*
Time on nest (min) ($N = 7$)	21.6 ± 7.7	52.0 ± 23.0	21.8 ± 11.8	0.007+
Time off nest (min) ($N = 6$)	14.8 ± 5.7	7.8 ± 3.9	11.1 ± 2.4	0.006+

*Wilcoxon test for matched pairs.
+Friedman's test (only matched data were used, hence N used for natural observations was the same as for model tests, although a full set of natural observations was available).

several hours after the test). On their first incubation after appearance of the cuckoo, females remained on the nest significantly longer than after appearance of the Greenfinch or for natural incubating episodes, and the time spent off the nest after that incubation period was significantly shorter for cuckoo tests than for Greenfinch tests or natural foraging episodes. Also, when females did eventually come off the nest after the lengthy incubation periods following cuckoo tests, they remained close to the nest, rarely moving more than 15 m away. In natural situations, and after Greenfinch tests, females almost always left the nest vicinity immediately and were out of sight until they returned rapidly to the nest entrance.

Thus, although females did not mob or hide from the cuckoo in ways different from their treatment of the Greenfinch, they took longer to return to the nest after cuckoo tests, they remained on the nest for long periods (one for 90+ minutes), and they stayed near the nest during their first foraging trip. Some reinspected the cuckoo perch, and they gave every appearance of being concerned that the cuckoo might appear again. As has been argued previously (McLean, 1987), mobbing of cuckoos near nests at a time when brood parasitism might occur may not be the most effective tactic for coping with cuckoos, even though the cuckoo is recognized as an enemy.

4.2. Vocal Mimicry of Begging Calls by Shining Bronze Cuckoo Chicks

In the second experiment, we asked if gerygones were "fooled" by the apparent mimicry of gerygone begging calls by cuckoo chicks. Here, we offered the gerygones a choice of cuckoo and gerygone begging calls near nests containing chicks (cuckoo, $N = 6$; gerygone, $N = 7$) close to fledging age. Tests were also conducted using four newly fledged cuckoo chicks. Cuckoo chicks remain still for long periods and it was possible to set up the apparatus and run the test without disturbing them. We removed chicks from some nests before running the test, but presence or absence of chicks in the nest did not appear to influence response of parents to the speakers, hence our use of newly fledged cuckoo chicks. As chicks were close to fledging when the test was conducted, we reasoned that parents should investigate chicks calling nearby as if they were from the nest. Parents did not usually go to the nest before investigating the played-back calls, and whether or not there was calling from the nest did not appear to influence their response.

Two sets of calls were played back. For tests conducted in summer

1987/8 recordings were of newly fledged gerygones and cuckoo chicks recorded in summer 1986/7. For tests conducted in summer 1988/9 recordings were of a cuckoo and gerygone chick raised in the same nest in 1987/8. Both chicks were near fledging age when recorded, but the gerygone had been pushed out of the nest by the cuckoo and was being fed in a nearby tree. All recordings were made as a parent approached with food, hence begging calls were loud and continuous. About 20 seconds of tape was dubbed repeatedly to give the necessary two minutes of playback tape.

Two speakers were placed 8 m on each side of the nest or cuckoo chick. Our criterion for initiation of playback was that the parent(s) be equidistant between the speakers and at least one of them be within 1 m of the nest or cuckoo chick. Playback began through both speakers at the same instant. After one minute, playback stopped and speakers were switched during a delay of 30 seconds, allowing enough time for the parents to move back to the nest or to cease hopping about the speaker. Then playback resumed for a further minute, but with each begging call coming from the other speaker. During tests conducted in 1987/8, some birds remained at the speaker chosen on the first trial for the 30 second break and were therefore beside that speaker for the second trial. At least some of these birds did not appear to hear the other speaker during the second trial. As a result, in 1988/9 the design was altered slightly: criterion for initiating the second trial was the same as for the first trial. In practice, the delay between trials was less than 1.5 minutes in all tests except for one where the birds left the area and came back 10 minutes later.

We predicted that gerygones feeding gerygone chicks would go to the "gerygone" speaker, and those feeding cuckoo chicks would go to the "cuckoo" speaker, because the procedure was designed to suggest that a chick had just fledged from the nest. In the analysis we defined a response as a bird moving to within 4 m of a speaker (i.e., more than half way towards it) and clearly orienting towards it. A "first response" was usually to the first trial, but for one female and one male with gerygone chicks, and two females and two males with cuckoo chicks, the first response was to the second trial.

As predicted, at nests containing gerygone chicks five of the six females and six of the seven males present went to the gerygone speaker on their first response. The other female and male remained near the nest. For the second trial one female and three males switched to the other speaker (now playing gerygone), two females and one male remained at the first speaker (now playing cuckoo), and one female chose the cuckoo speaker. Gerygones feeding gerygone chicks clearly preferred gerygone chick calls on their first response. Of those who made a true choice on the second trial, four went to the gerygone and one went to the cuckoo.

Gerygones feeding cuckoo chicks also preferred gerygone chick calls. Nine of the 10 females and seven of the eight males present went to the gerygone speaker on their first response. One female did not respond and one male chose the cuckoo speaker (this way the only one of 28 birds which chose the cuckoo on its first response). For the second trial, four females and three males switched to the other (gerygone) speaker, two females and one male remained at the first speaker (now playing cuckoo), and no birds chose the cuckoo speaker. Thus, of those birds given a true choice on the second trial, all chose the gerygone calls.

Statistical testing of these data is unnecessary. If we use only the first response of the stronger responder within a pair, of pairs with gerygone chicks seven chose the gerygone calls and none chose the cuckoo. Of pairs with cuckoo chicks 10 chose the gerygone calls and one chose the cuckoo. Of pairs given a true choice on the second trial, 11 chose the gerygone calls and one (a female with gerygone chicks) chose the cuckoo. Parent gerygones were clearly not fooled by the apparent mimicry of begging calls by cuckoos.

5. GENERAL DISCUSSION

Two important factors influencing the recognition process are, first, ultimate factors related to the adaptive nature of biological systems, and second, the proximate factors which influence each recognition event. We have argued that birds are complex organisms which employ cognitive processes well above the level of reflex pathways. The advantage of such complexity is that it allows for flexibility in any situation where alternative behavioral responses are both available and appropriate, and where contextual factors can be incorporated into decisionmaking. A further advantage is that experience can contribute significantly to the expression of responses.

The primary disadvantage is that decisionmaking may be time consuming. In the case of enemy recognition and response, many situations will arise in which time for assessment will not be available. Two consequences of this are, first, that animals will not always appear to behave optimally, for example because the response given may have been an instantaneous reaction which was not particularly appropriate to the situation. Second, they are likely to respond using simple decision rules (Gray, 1987; Barnard, 1988). This is certainly a procedure adopted by humans to simplify decisionmaking (Glass and Holyoake, 1986). Although it may be possible for selection to build those rules into the genetic structure of a population, for birds at least we predict that many rules will be developed by individuals as a result of experience. It is the ability to develop decision rules which is adaptive; the rules themselves stem from many sources. Thus responses observed by researchers studying enemy recognition may well be uncoupled from selection in the sense that the primary factors influencing observed responses are immediate context and recent experience.

A part of recent experience is the interaction between birds, their young, and the nest. The existence of the nest and contents presumably reinforces parental care, and we have suggested that this feedback process will influence the expression of parental care as defense of young. As with the ability to develop decision rules, the ability to be reinforced via feedback may be a product of selection. These sorts of abilities are entirely consistent with the notion that birds process information from their environment in sophisticated ways, a notion which is fundamental to the cognitive model. We are encouraged by our preliminary review of available data which gives some support to the feedback model.

If feedback patterns experienced by birds working at nests strongly influence the expression of parental behavior, then the determination of host birds to rear brood parasites once the egg of the parasite has been accepted becomes unsurprising. In the example we studied, the host species clearly recognized that the brood parasite was an enemy, and was not fooled by apparent mimicry of its own young by the brood parasite. The most successful defense developed by many bird species to brood parasites is rejection of parasite eggs, apparently by recognizing those eggs as being different from one's own (Rothstein, 1982a). Selection apparently has operated on a simple accept or reject ability. However, once the egg is accepted, it becomes "one's own" and should be nurtured in the same way as all nest contents.

Direct behavioral defenses against adult brood parasites have not been shown to successfully prevent brood parasitism in any study, and may even be initiated by the brood parasite if it uses information gained from the defensive event to determine quality of potential hosts (Robertson and Norman, 1977; Smith, *et al.*, 1984; McLean, 1987). We suggest that brood parasites exploit the hypothesized decoupling of selection and the details of defensive behavior exhibited by nesting birds. By providing hosts with a powerful feedback signal in the nest, they overcome any more general tendency of the hosts to perceive the signaller as an enemy. As long as host species do not develop the ability to reject eggs of brood parasites, brood parasitism should remain a successful strategy. If hosts do develop such an ability, selection may well encourage egg mimicry (and presumably host specificity), but the provision of parental care for nest contents by hosts will remain an overriding force which brood parasite chicks strongly reinforce.

Enemy recognition in birds is clearly an area that offers exciting potential for future research. We have pointed to areas of both theoretical and applied significance, and have offered a proximate framework for viewing one aspect of enemy recognition. We emphasize that a careful analysis of the interaction between defensive behavior and context (e.g., type of enemy, breeding stage) is needed if the complexity of enemy recognition and response behavior is to be understood. Researchers have only just begun to plumb the cognitive abilities of birds in this area; a marriage of psychology and ethology may help to point the way forward.

ACKNOWLEDGMENTS. Many colleagues have discussed with us the ideas in this review, especially C. L. Cassady St. Clair, J. L. Craig, S. M. Dawson, I. G. Jamieson, R. Maloney, E. Minot, R. D. Montgomerie, E. Slooten, J. N. M. Smith, J. R. Waas, and P. J. Weatherhead. For assistance in the field we thank H. Cameron, S. Dean, R. de Hamel, S. Fegley, and J. van Berkel. S. Dean made a major contribution toward compiling the literature and T. Robinson assisted with manuscript preparation. For reviewing the manuscript, we thank G. J. Fletcher, R. N. Hughes, C. M. Miskelly, S. I. Rothstein, and J. N. M. Smith. D. M. Power's editorial assistance and tolerance was exceptional. Our research is funded by the New Zealand University Grants Committee, the University of Canterbury, the Department of Conservation, the Earthwatch Organization, and the New Zealand Lottery Board.

REFERENCES

Alcock, J., 1984, *Animal Behavior, an Evolutionary Approach*, Sinauer Associates, Sunderland, Massachusetts, 596 pp.

Altmann, S. A., 1956, Avian mobbing behavior and predator recognition, *Condor* **58**:241–253.

Andersson, M., Wiklund, C. G., and Rundgren, H., 1980, Parental defense of offspring: A model and an example, *Anim. Behav.*, **28**:536–542.

Axelrod, R., and Hamilton, W. D., 1981, The evolution of cooperation, *Science* **211**:1390–1396.

Barash, D. P., 1975, Mobbing behavior by crows: The effect of the "crow-in-distress" model, *Condor* **74**:120.

Barash, D. P., 1980, Predictive sociobiology: mate selection in Damselfishes and brood defense in White-crowned Sparrows, in: *Sociobiology: Beyond Nature/Nurture?* (G. W. Barlow and J. Silverberg, eds.), AAAS, Washington, pp. 209–226.

Barnard, C. J., 1988, Policy-making for survival: Reading the rules and small print, *Beh. Br. Sci.* **11**:130–131.

Beecher, M. D., 1982, Signature systems and kin recognition, *Amer. Zool.* **22**:477–490.

Blancher P. J., and Robertson, R. J., 1982, Kingbird aggression: Does it deter predation?, *Anim. Behav.* **30**:929–945.

Boden, M. A., 1984, Animal perception from an artificial intelligence viewpoint, in: *Minds, Machines and Evolution* (C. Hookway, ed.), Cambridge University Press, Cambridge, pp. 153–174.

Brooke, M., 1989, Tricks of the egg trade, *Nat. Hist.* **89(4)**:50–54.

Brown, C. R., and Hoogland, J. L., 1986, Risk in mobbing for solitary and colonial swallows, *Anim. Behav.* **34**:1319–1323.

Brunton, D. H., 1986, Fatal antipredator behavior of a Killdeer, *Wilson Bull.* **98**:605–607.

Buitron, D., 1983, Variability in the responses of Black-billed Magpies to natural predators, *Behaviour* **88**:209–235.

Caro, T. M., 1986a, The functions of stotting: A review of the hypotheses, *Anim. Behav.* **34**:649–662.

Caro, T. M., 1986b, The functions of stotting in Thompson's Gazelles: Some tests of the predictions, *Anim. Behav.* **34**:663–684.

Clark, K. L., and Robertson, R. J., 1981, Cowbird parasitism and evolution of anti-parasite strategies in the Yellow Warbler, *Wilson Bull.* **93**:249–258.

Conover, M. R., 1987, Acquisition of predator information by active and passive mobbers in Ring-billed Gull colonies, *Behaviour* **102**:41–57.

Conover, M. R., and Perito, J. J., 1981, Response of Starlings to distress calls and predator models holding conspecific prey, *Z. Tierpsychol.* **57**:163–172.

Coraco, T., Martindale, S., and Pulliam, H. R., 1980, Avian flocking in the presence of a predator, *Nature* **285**:400–401.

Cosmides, L., and Tooby, J., 1987, From evolution to behavior: evolutionary psychology as the missing link, in: *The Latest on the Best* (J. Dupré, ed.), MIT Press, Massachusetts, pp. 277–306.

Cully, J. F., and Ligon, J. D., 1986, Seasonality of mobbing intensity in the Pinyon Jay, *Ethology* **71**:333–339.

Curio, E., 1975, The functional organization of anti-predator behavior in the Pied Flycatcher: A study of avian visual perception, *Anim. Behav.* **23**:1–115.

Curio, E., 1978, The adaptive significance of avian mobbing. I. Teleonomic hypotheses and predictions, *Z. Tierpsychol.* **48**:175–183.

Curio, E., 1980, An unknown determinant of a sex-specific altruism, *Z. Tierpsychol.* **53**:139–152.

Curio, E., and Regelmann, K., 1985, The behavioral dynamics of Great Tits (*Parus major*) approaching a predator, *Z. Tierpsychol.* **69**:3–18.

Curio, E., Ernst, U., and Vieth, W., 1978, Cultural transmission of enemy recognition: one function of mobbing, *Science* **202**:899–901.

Curio, E., Regelmann, K., and Zimmermann, U., 1984, The defence of first and second broods by Great Tit (*Parus major*) parents: A test of predictive sociobiology, *Z. Tierpsychol.* **66**:101–127.

Curio, E., Regelmann, K., and Zimmerman, U., 1985, Brood defense in the Great Tit (*Parus major*): The influence of life-history and habitat, *Behav. Ecol. Sociobiol.* **16**:273–283.

Davies, N. B., and Brooke, M. de L., 1988, Cuckoos versus reed warblers: Adaptations and counteradaptations, *Anim. Behav.* **36**:262–284.

Dean, S. M., 1990, Composition and seasonality of mixed-species flocks of insectivorous birds at Kaikoura, New Zealand, *Notornis* **37**:27–36.

Dean, S. M., and McLean, I. G., Mobbing and reciprocity in a forest bird community, (unpubl. manus.)

Dennett, D. C., 1983, Intentional systems in cognitive ethology: The "Panglossian paradigm" defended, *Beh. Br. Sci.* **6**:343–390.

Dennett, D. C., 1988, Précis of The Intentional Stance, *Beh. Br. Sci.* **11**:495–546.

Denson, R. D., 1979, Owl predation on a mobbing crow, *Wils. Bull.* **91**:133.

Drickamer, L. C. and Vessey, S. H., 1982, *Animal Behavior: Concepts, Processes, and Methods*, Prindle, Weber & Schmidt, Boston, 510 pp.

Dupré, J., (ed.), 1987, *The Latest on the Best*, MIT Press, Massachusetts, 359 pp.

East, M., 1981, Alarm calling and parental investment in the Robin *Erithacus rubecula*, *Ibis* **123**:223–229.

Eden, S. F., 1987, The influence of Carrion Crows on the foraging behavior of Magpies, *Anim. Behav.* **35**:608–610.

Edmunds, M., 1974, *Defence in Animals*, Longman Group Ltd., Essex, England, 357 pp.

Edwards, G., Hosking, E., and Smith, S., 1949, Reactions of some passerine birds to a stuffed cuckoo, *Brit. Birds* **62**:13–19.

Edwards, G., Hosking, E., and Smith, S., 1950, Reactions of some passerine birds to a stuffed cuckoo. II. A detailed study of the Willow-Warbler, *Brit. Birds* **63**:144–150.

Eibl-Eibesfeldt, I. 1970, *Ethology, the Biology of Behavior*, Holt, Rinehart and Winston, New York, 530 pp.

Emlen, J. M., 1987, Evolutionary ecology and the optimality assumption, in: *The Latest on the Best* (J. Dupré, ed.), MIT Press, Cambridge, Massachusetts, pp. 163–178.

England, M. E., 1986, Harrier kills mobbing Willet, *Raptor Research* **20**:78–79.

Fantino, E., 1988, Conditioned reinforcement and reproductive success, *Beh. Br. Sci.* **11**:135.

Frankenberg, E., 1981, The adaptive significance of avian mobbing. IV. "Alerting others" and "perception advertisement" in Blackbirds facing an owl, *Z. Tierpsychol.* **55**:97–118.

Gill, B. J., 1982a, Notes on the Shining Cuckoo (*Chrysococcyx lucidus*) in New Zealand, *Notornis* **29**:215–227.

Gill, B. J., 1982b, The Grey Warbler's care of nestlings: A comparison between unparasitised broods and those comprising a Shining Bronze-Cuckoo, *Emu* **82**:177–181.

Gill, B. J., 1983, Brood-parasitism by the Shining Cuckoo *Chrysococcyx lucidus* at Kaikoura, New Zealand, *Ibis* **125**:40–55.

Glass, A. L., and Holyoake, K. J., 1986, *Cognition*, 2nd ed., Random House, New York, 570 pp.

Goldman, P., 1980, Flocking as a possible predator defense in Dark-eyed Juncos, *Wilson Bull.* **92**:88–95.

Godin, J-G. J., and Smith, S. A., 1988, A fitness cost of foraging in the guppy, *Nature* **333**:69–71.

Gottfried, B., 1979, An experimental analysis of the interrelationship between nest density and predation in old-field habitats, *Condor* **81**:251–257.

Gottfried, B. M., Andrews, K., and Haug, M., 1985, Breeding Robins and nest predators: Effect of predator type and defense strategy on initial vocalization patterns, *Wilson Bull.* **97**:183–190.

Graham, D. S., 1988, Responses of five host species to cowbird parasitism, *Condor* **90**:588–591.

Gray, R. D., 1987, Faith and foraging: A critique of the "paradigm argument from design", in: *Foraging Behavior* (A. C. Kamil, J. R. Krebs, and H. R. Pulliam, eds.), Plenum Press, New York, pp. 69–140.

Greig-Smith, P. W., 1980, Parental investment in nest defense by Stonechats (*Saxicola torquata*), *Anim. Behav.* **28**:604–619.

Greig-Smith, P. W., 1981, Responses to disturbance in relation to flock size in foraging groups of Barred Ground Doves *Geopelia striata*, *Ibis* **123**:103–106.

Greig-Smith, P. W., 1984, Distress calling by woodland birds: Seasonal patterns, individual consistency and the presence of conspecifics, *Z. Tierpsychol.* **66**:1–10.

Gyger, M., Karakashian, S. J., and Marler, P., 1986, Avian alarm calling: Is there an audience effect?, *Anim. Behav.* **34**:1570–1572.

Hamilton, W. J., and Orians, G. H., 1965, Evolution of brood parasitism in altricial birds, *Condor* **67**:361–382.

Herrnstein, R. J., Loveland, D. H., and Cable, C., 1976, Natural concepts in pigeons, *Journ. Exp. Psychol.: Anim. Behav. Proc.* **2**:285–302.

Heyman, G. M., 1988, Optimization theory: A too narrow path, *Beh. Br. Sci.* **11**:136–137.

Hill, G. E., 1986, The function of distress calls given by Tufted Titmice (*Parus bicolor*): An experimental approach, *Anim. Behav.* **34**:590–598.

Hobson, K., Bouchart, M. L., and Sealy, S. G., 1988, Responses of naive Yellow Warblers to a novel nest predator, *Anim. Behav.* **36**:1823–1830.

Hogstad, O., 1988, Social rank and antipredator behavior of Willow Tits *Parus montanus* in winter flocks, *Ibis* **130**:45–56.

Hogstedt, G., 1983, Adaptation unto death: Function of fear screams, *Amer. Natur.* **121**:562–569.

Holdaway, R. N., 1989, New Zealand's pre-human avifauna and its vulnerability, *N. Z. Journ. Ecol. (supplement)* **12**:11–25.

Houston, A. D., and McNamara, J. M., 1988, A framework for the functional analysis of behavior, *Beh. Br. Sci.* **11**:117–130.

Huntingford, F. A., and Wright, P. J., 1989, How sticklebacks learn to avoid dangerous feeding patches, *Behav. Proc.* **19**:181–189.

Immelmann, K., 1959, Experimentelle untersuchungen uber die biologische bedeutung artspezifischer merkmale beim Zebrafinken (*Taeniopygia castanotis* Gould), *Zool. Jahrb., Abt. Syst. Okol. Geogr.* **86**:437–592.

Jamieson, I. G., 1989, Behavioral heterochrony and the evolution of birds helping at the nest: An unselected consequence of communal breeding? *Amer. Natur.* **133**:394–406.

Johnson, M., 1988, Memories of mother, *New Scientist* **117**:60–62.

Klump, G. M., and Shalter, M. D., 1984, Acoustic behavior of birds and mammals in the predator context. I. Factors affecting the structure of alarm signals; II. The functional significance and evolution of alarm signals, *Z. Tierpsychol.* **66**:189–226.

Knight, R. L., and Temple, S. A., 1986a, Why does intensity of avian nest defense increase during the nesting cycle?, *Auk* **103**:318–327.

Knight, R. L., and Temple, S. A., 1986b, Methodological problems in studies of avian nest defense, *Anim. Behav.* **34**:561–566.

Knight, R. L., and Temple, S. A., 1986c, Nest defense in the American Goldfinch, *Anim. Behav.* **34**:887–897.

Knight, R. L., Grout, O. J., and Temple, S. A., 1987, Nest-defense behavior of the American Crow in urban and rural areas, *Condor* **89**:175–177.

Krebs, J. R., Stephens, D. W., and Sutherland, W. J., 1983, Perspectives in optimal foraging, in: *Perspectives in Ornithology* (A. H. Brush and G. A. Clark, eds.), Cambridge Univ. Press, Cambridge, pp. 165–216.

Kruuk, H., 1964, Predators and anti-predator behavior of the Black-headed Gull (*Larus ridibundus* L.), *Behav. Suppl.* **11**:1–129.

Lombardi, C. M., and Curio, E., 1985a, Influence of environment on mobbing by Zebra Finches, *Bird Behaviour* **6**:28–33.

Lombardi, C. M., and Curio, E., 1985b, Social facilitation of mobbing in the Zebra Finch *Taeniopygia guttata, Bird Behaviour* **6**:34–40.

Marler, P., 1955, Characteristics of some animal calls, *Nature* **176**:6–8.

Marr, D., 1982, *Vision*, W. H. Freeman & Co., San Francisco, 397 pp.

Maynard Smith, J., 1978, Optimization theory in evolution, *Ann. Rev. Ecol. Syst.* **9**:31–56.

McLean, I. G., 1987, Response to a dangerous enemy: Should a brood parasite be mobbed?, *Ethology* **75**:235–245.

McLean, I. G., and Maloney, R., 1989, Genetic and cultural factors in enemy recognition by an island-endemic bird, Paper pres. at 21st International Ethological Congress, Utrecht, The Netherlands.

McLean, I. G., Smith, J. N. M., and Stewart, K. G., 1986, Mobbing behavior, nest exposure, and breeding success in the American Robin, *Behaviour* **96**:171–186.

McLean, I. G., and Waas, J. R., 1987, Do cuckoo chicks mimic the begging calls of their hosts? *Anim. Behav.* **35**:1896–1898.

McLean, I. G., Wells, M. S., Brown, R., Creswell, P., McKenzie, J., and Musgrove, R., 1987, Mixed-species flocking of forest birds on Little Barrier Island, *N. Z. Journ. Zool.* **14**:143–147.

McPherson, R. J., and Brown, R. D., 1981, Mobbing responses of some passerines to the calls and location of the Screech Owl, *Raptor Research* **15**:23–30.

Montgomerie, R. D., and Weatherhead, P. J., 1988, Risks and rewards of nest defense by parent birds, *Quart. Rev. Biol.* **63**:167–187.

Moors, P. J., 1983, Predation by mustelids and rodents on the eggs and chicks of native and introduced birds in Kowhai Bush, New Zealand, *Ibis* **125**:137–154.

Mueller, H. C., and Parker, P. G., 1980, Naive ducklings show different cardiac response to hawk than to goose models, *Behaviour* **74**:101–113.

Myers, J. P., 1978, One deleterious effect of mobbing in the Southern Lapwing (*Vanellus chilensis*), *Auk* **95**:419–420.

Myers, J. P., 1983, Commentary, in: *Perspectives in Ornithology* (A. H. Brush and G. A. Clark, eds.), Cambridge Univ. Press,, Cambridge, England, pp. 216–221.

Munn, C. A., Terborgh, J. W., 1979, Multispecies territoriality in neotropical foraging flocks, *Condor* **81**:338–347.

Oliver, W. R. B., 1955, *New Zealand Birds*, Reed, Wellington.

Owings, D. H., Hennessy, D. F., Leger, D. W., and Gladney, A. B., 1986, Different functions of "alarm" calling for different time scales: A preliminary report on ground squirrels, *Behaviour* **99**:101–116.

Patterson, T. L., Petrinovich, L., and James D. K., 1980, Reproductive value and appropriateness of response to predators by White-crowned Sparrows, *Behav. Ecol. Sociobiol.* **7**:227–231.

Payne, R. B., 1977, The ecology of brood parasitism in birds, *Ann. Rev. Ecol. Syst.* **8**:1–28.

Payne, R. B., Payne, L. K., and Rowley, I., 1985, Splendid Wren *Malurus splendens* response to cuckoos: An experimental test of social organization in a communal bird, *Behaviour* **94**:108–127.

Rasa, O. A. E., 1981, Raptor recognition: an interspecific tradition? *Naturwiss.* **68**:151–152.

Redondo, T., and Carranza, J., 1989, Offspring reproductive value and nest defense in the magpie (*Pica pica*), *Behav. Ecol. Sociobiol.* **25**:369–378.

Regelmann, K., and Curio, E., 1983, Determinants of brood defense in the Great Tit *Parus major* L., *Behav. Ecol. Sociobiol.* **13**:131–145.

Regelmann, K., and Curio, E., 1986, How do Great Tit (*Parus major*) pair mates cooperate in brood defense?, *Behaviour* **97**:10–36.

Ricklefs, R. E., 1983, Avian postnatal development, in: *Avian Biology*, Vol. III (D. S. Farner, J. R. King, and K. C. Parkes, eds.), Academic Press, New York, pp. 2–83.

Roberts, W. A., and Mazmanian, D. S., 1988, Concept learning at different levels of abstraction by pigeons, monkeys and people, *Journ. Exper. Psych.: Anim. Behav. Proc.* **14**:247–260.

Robertson, R. J., and Norman, R. F., 1976, Behavioral defenses to brood parasitism by potential hosts of the Brown-headed Cowbird, *Condor* **78**:166–173.

Robertson, R. J., and Norman, R. F., 1977, The function and evolution of aggressive host behavior towards the Brown-headed Cowbird (*Molothrus ater*), *Can. J. Zool.* **55**:508–518.

Rohwer, S., and Spaw, C. D., 1988, Evolutionary lag versus bill-size constraints: A comparative study of the acceptance of cowbird eggs by old hosts, *Evol. Ecol.* **2**:27–36.

Roitblat, H. L. 1987, *Introduction to Comparative Cognition*, W. J. Freeman & Co., New York.

Roitblat, H. L., Bever, T. G., and Terrace, H. S., 1984, *Animal Cognition*, Erlbaum, Hillsdale, New Jersey.

Rothstein, S. I., 1982a, Successes and failure in avian egg and nestling recognition with comments on the utility of optimality reasoning, *Amer. Zool.* **22**:547–560.

Rothstein, S. I., 1982b, Mechanisms of avian egg recognition: Which egg parameters elicit responses by rejector species? *Behav. Ecol. Sociobiol.* **11**:229–239.

Rothstein, S. I., and Pierotti, R., 1988, Distinctions among reciprocal altruism, kin selection, and cooperation and a model for the initial evolution of beneficient behavior, *Ethol. Sociobiol.* **9**:189–209.

Schleidt, W. M., 1961, Uber die auslosung der flucht vor raubvogeln bei truthuhnern, *Naturwiss.* **48**:141–142.

Schneider, W., and Shiffrin, R. M., 1977, Controlled and automatic human information processing: I. Detection search and attention, *Psych. Rev.* **84**:1–66.

Seyfarth, R., Cheney, D. L., and Marler, P., 1980, Monkey responses to three different alarm calls: Evidence of predator classification and semantic communication, *Science* **210**:801–803.

Shalter, M. D., 1978a, Localization of passerine seet and mobbing calls by Goshawks and Pygmy Owls, *Z. Tierpsychol.* **46**:260–267.

Shalter, M. D., 1978b, Effect of spatial context on the mobbing reaction of Pied Flycatchers to a predator model, *Anim. Behav.* **26**:1219–1221.

Shalter, M. D., and Schleidt, W. M., 1977, The ability of Barn Owls, *Tyto alba*, to discriminate and localize avian alarm calls, *Ibis* **119**:22–27.

Shedd, D. H., 1982, Seasonal variation and function of mobbing and related antipredator behaviors of the American Robin (*Turdus migratorius*), *Auk* **99**:342–346.

Shedd, D. H., 1983, Seasonal variation in mobbing intensity in the Black-capped Chickadee, *Wilson Bull.* **95**:343–348.

Shields, W. M., 1984, Barn Swallow mobbing: Self-defense, collateral kin defense, group defense, or parental care?, *Anim. Behav.* **32**:132–148.

Shiffrin, R. M., and Schneider, W., 1977, Controlled and automatic human information

processing: II. Perceptual learning, automatic attending and a general theory, *Psych. Review* **84**:127–190.
Smith, J. N. M., Arcese, P., and McLean, I. G., 1984, Age, experience, and enemy recognition by wild Song Sparrows, *Behav. Ecol. Sociobiol.* **14**:101–106.
Smith, N. G., 1969, Provoked release of mobbing—a hunting technique of *Micrastur* falcons, *Ibis* **111**:241–243.
Smith, R. J. F., 1986, Evolution of alarm signals: Role of benefits of retaining group members of territorial neighbors, *Amer. Natur.* **128**:604–610.
Stefanski, R. A., and Falls, J. B., 1972, A study of distress calls of Song, Swamp, and White-throated Sparrows (Aves: Fringillidae). I. Intraspecific responses and functions, *Can. J. Zool.* **50**:1501–1512.
Thorpe, W. H., 1961, *The Biology of Vocal Communication and Expression in Birds*, Cambridge University Press, Cambridge, England, 142 pp.
Tinbergen, N., 1951, *The Study of Instinct*, Clarendon Press, Oxford, England, 288 pp.
Trivers, R. L., 1971, The evolution of reciprocal altruism, *Quart. Rev. Biol.* **46**:35–57.
Vieth, W., Curio, E. and Ulrich, E., 1980, The adaptive significance of avian mobbing. III. Cultural transmission of enemy recognition in Blackbirds: Cross-species tutoring and properties of learning, *Anim. Behav.* **28**:1217–1229.
Walker, S., 1983, *Animal Thought*, Routledge & Kegan Paul, London, 452 pp.
Weatherhead, P. J., 1989, Nest defense by song sparrows: Methodological and life history considerations, *Behav. Ecol. Sociobiol.* **25**:129–136.
Weedon, J. S., and Falls, J. B., 1959, Differential responses of male Ovenbirds to recorded songs of neighboring and more distant individuals, *Auk* **76**:343–351.
Williams, M., 1986, Native bird management, *Forest and Bird* **17**:7–9.
Wittenberger, J. F., and Hunt, G. L., 1985, The adaptive significance of coloniality in birds, in: *Avian Biology*, Vol. VIII (D. S. Farner, J. R. King, and K. C. Parkes, eds.), Academic Press, New York, pp. 1–78.
Wyllie, I., 1981, *The Cuckoo*, Universe Books, New York, 223 pp.

CHAPTER 5

PARASITES AND SEXUAL SELECTION IN A NEW GUINEA AVIFAUNA

STEPHEN G. PRUETT-JONES, MELINDA A. PRUETT-JONES, and HUGH I. JONES

1. INTRODUCTION

How and why animals make the mating decisions that they do has long puzzled evolutionary biologists. The evolution of exaggerated secondary sexual characters in males is believed to result, in part, from a preference in females for such traits, but both the mechanisms of perception and assessment of such traits (cf. Janetos, 1980; Bradbury *et al.*, 1985) and the question of why females should have such preferences (Borgia, 1979; Lande, 1981; Kirkpatrick, 1982, 1987) remain obscure and matters of controversy. Hamilton and Zuk (1982; also Hamilton, 1982) proposed that females prefer males with elaborate traits because those traits provide information about the male's relative resistance to

STEPHEN G. PRUETT-JONES • Department of Biology, University of California at San Diego, La Jolla, California 92093. MELINDA A. PRUETT-JONES • Natural Reserve System, Scripps Institution of Oceanography, University of California at San Diego, La Jolla, California 92093. HUGH I. JONES • Department of Zoology, University of Western Australia, Nedlands, Western Australia 6009, Australia. *Present address for S. G. P-J.:* Department of Ecology and Evolution, University of Chicago, Chicago, Illinois 60637. *Present address for M. A. P-J.:* Field Museum of Natural History, Chicago, Illinois 60605. *Present address for H. I. J.:* Gwynedd Health Authority, Coed Mawr, Bangor, Gwynedd LL57 4TP, Wales.

parasites. This hypothesis can be described more fully as follows: if parasite resistance in host individuals is heritable, and if parasites negatively affect the expression of secondary sexual character in males, then an individual male's genetic fitness in terms of resistance would be discernable through its external phenotype. If females, by comparing males on the basis of traits affected by parasites, could identify the fittest individuals they would receive direct genetic benefits (i.e., increase the probability that their offspring would inherit genes for parasite resistance) by preferentially mating with those individuals. Elaboration of the traits in males that allow females to discriminate among individuals would be favored by sexual selection.

The parasite hypothesis is complex because it involves both genetic and phenotypic processes. It concerns the nature of coevolutionary cycles between parasites and their hosts (Hamilton, 1982; May and Anderson, 1983; Kirkpatrick, 1986; Seger and Hamilton, 1988), the joint evolution of female preferences and male traits (Lande, 1981; Kirkpatrick, 1982, 1987; Pomiankowski, 1988), and the adaptive significance of female choice (Borgia, 1979; Lande, 1981; Kirkpatrick, 1982; Pomiankowski, 1987, 1988).

The specific genetic mechanism proposed by Hamilton and Zuk for the maintenance in hosts of heritable variation in resistance involves coadaptational genotypic cycling between parasites and hosts (cf. Kirkpatrick, 1986; Pomiankowski, 1987, 1988; Seger and Hamilton, 1988). Within such populational cycles, hosts are viewed as never fully reaching an equilibrium with respect to resistance. Other genetic mechanisms are also feasible. For example, in monogamous species, females may receive direct parental care benefits from making an active choice of mates, and parasites may influence the traits in males that indicate general abilities at parental care (Read, 1990).

The parasite hypothesis makes predictions at both the intraspecific and interspecific level. Hamilton and Zuk (1982) stated: "Our hypothesis is contradicted if *within* a species preferred mates have most parasites; it is supported if *among* species those with most evident sexual selection are most subject to attack by debilitating parasites" (italics theirs; citations left out). That is, the model predicts that within a species there exists a negative relationship between parasite load (any measure of parasite burden; see below) and the fitness of males, and that across species there is a positive correlation between parasite load and the result or effect of sexual selection. The sexual characters Hamilton and Zuk analyzed as indicators of the extent of sexual selection across species were plumage brightness and "showiness," and song complexity in passerine birds.

The intraspecific prediction of the parasite hypothesis concerns how parasites affect host fitness. A negative relationship between parasites and the expression of some phenotypic trait in males changes an otherwise positive relationship between that trait and fitness (i.e., reproductive success) to a negative correlation between parasites and fitness. If the phenotypic trait affects mating success, it will be the nature of the relationship between parasites and the phenotypic trait that proximally determines the relationship between parasites and fitness.

The prediction of the hypothesis at the interspecific level assumes that differences in selection within species will result in differences in male phenotypic traits across species. Although Hamilton and Zuk did not make use of the concept of "opportunity for selection" (Crow, 1958; Wade and Arnold, 1980; Arnold and Wade, 1984; Wade, 1987) it is appropriate to discuss it as a means of explaining their interspecific prediction (Clayton, Pruett-Jones, and Lande, submitted manuscript). The index I, the opportunity for selection, is defined as the variance in a trait divided by the square of the mean value for that trait. Selection on some trait is only possible when there is variance in that trait. Noting that variance as a statistic often increases with the mean, the opportunity for selection is an appropriate index to use when comparing species as it is a measure of the relative variance in the trait under consideration. A large relative variance is a necessary but not sufficient condition for there to be selection (Crow, 1958; Wade, 1979; Wade and Arnold, 1980; Arnold and Wade, 1984; Wade, 1987). For selection to be realized there needs to be a positive covariance between fitness and one or more of the phenotypic traits in question. For evolution to occur the variation in the phenotypic traits must be heritable.

The Hamilton and Zuk hypothesis assumes that an increasing parasite load increases the opportunity for selection and that as the opportunity for selection increases so does the magnitude of the evolutionary change; i.e., that differences across species in parasite load translate into differences across species in phenotypic traits. These assumptions are the basis for the prediction that across species there should be a positive correlation between parasite loads and effects of sexual selection (Clayton et al., submitted). Within a species, an individual male does not have an "opportunity for selection"; he either mates or he doesn't. The opportunity for selection is the outcome of factors affecting variance in parasite loads and female choice for the traits indicative of an individual's relative parasite burden. Hamilton and Zuk envisioned selection on two traits: selection for parasite resistance and selection for secondary sexual characters which allow females to assess a male's relative parasite load. They implicitly assume an initial or resultant pleiotropy (or linkage disequilibrium) between the genes for resistance and the genes controlling the expression of secondary sexual characters.

In support of the interspecific prediction, Hamilton and Zuk (1982) documented a positive correlation between plumage showiness and parasite loads in North American passerine birds, and subsequent studies by Read (1987, 1988), Ward (1988), Pruett-Jones *et al.* (1990), and Zuk (in press) have shown similar correlations for other avifaunas or fish species (but see Read and Harvey, 1989; Harvey et al., in press). Hamilton and Zuk (1982) also demonstrated a positive covariance between parasites and song complexity in passerines, results recently questioned by Read and Weary (1990).

In this chapter we examine the interspecific prediction of the Hamilton and Zuk hypothesis with respect to a tropical, montane avifauna in Papua New Guinea. Our concern is, first, the distribution of parasites across host individuals and species, and with respect to altitude and seasons, and second, how variation in species' patterns of plumage brightness and showiness vary with parasite load. We show that although parasite burdens do correlate with aspects of plumage elaboration in passerines, whether the mechanism proposed by Hamilton and Zuk underlies these correlations is not yet resolved.

2. STUDY AREA

The data we present were gathered in two separate but complementary studies on Mount Kaindi and Mount Missim, Morobe Province, Papua New Guinea. Mount Kaindi rises from the eastern, and Mount Missim the western, side of the Wau Valley. H. I. J. worked on Mount Missim and Mount Kaindi from December 1981 to January 1982 (Jones, 1985). S. G. P-J. and M. A. P-J. worked on Mount Missim throughout 1982 and 1983 and again from September to December each year from 1985 to 1987.

The ecology of both field sites has been previously described (Beehler, 1982; Gressitt and Nadkarni, 1978; Pratt and Stiles, 1985; Pruett-Jones and Pruett-Jones, 1982, 1986, in press), although some details of the Mount Missim study area, where the majority of samples were collected, are relevant to point out here. The area was approximately 1000 ha in size and ranged from 1300 m to 2200 m altitude. It was characterized by continuous, montane rainforest with an annual rainfall of 190 to 220 cm, increasing with altitude. On Mount Missim, as across much of New Guinea, May to October represents the dry season and November to April the wet season (McAlpine *et al.* 1983; Pruett-Jones and Pruett-Jones, 1988). The seasonality of rainfall was not extreme, and rain generally fell every week of the year.

The avifauna at 1300 m on Mount Missim consists of approximately 130 species, decreasing to 90 species at 2200 m altitude (Beehler, 1982, 1983; Pratt, 1983; personal observation). At all altitudes, it is dominated by fruit pigeons (*Columbidae*), parrots (*Psittacidae*), honeyeaters (*Meliphagidae*), and birds of paradise (*Paradisaeidae*). The majority of species, excluding fruit doves, shared an important aspect of ecology in that they were resident in the rainforest the year round and did not migrate. Some species may have shown altitudinal migrations, if not across seasons then with age of the individual (Diamond, 1972; Pruett-Jones and Pruett-Jones, 1986).

3. FIELD METHODS

3.1. Collection and Examination of Blood Smears

Blood samples were taken from clipped toenails of birds caught in mist nets. All birds were banded prior to release. The blood samples were collected as thin smears and were fixed in 95–100% ethanol within 24 hours and stained with Giemsa between three and five weeks after collection. All smears were examined microscopically by H. I. J. for 15 min with a 40 × objective and for a minimum of 15 min under oil immersion. Samples were examined for blood protozoa as well as microfilarial larvae of filarid nematodes. *Haemoproteus* spp. and *Plasmodium* spp. (hereafter abbreviated as H/P) were counted as the number of protozoa per 10,000 red blood cells (RBC). *Leucocytozoon* spp., *Trypanosoma* spp., and microfilaria were counted as the total number seen on the entire slide. Approximately one-half of the smears were examined without knowledge of the host species.

Duplicate blood smears were taken from birds recaptured at later dates. We analyze these duplicate samples for temporal changes in intensity, however, for all comparative analyses only data from the first sample for each individual were used.

3.2. Measures of Parasitemia

We refer to three measures of parasite infections (cf. Margolis *et al.*, 1982; Hamilton and Zuk, 1982; Duffy, 1983): 1, parasite diversity, the number of different types of parasites present (H/P were counted as one type, yielding a rank scale of diversity from 0–4); 2, parasite prevalence, the proportion of individuals infected with at least one parasite of any type; and 3, parasite intensity, the mean density of H/P protozoa (number per 10,000 RBC) across all individuals sampled. In a comparison of parasite prevalence versus intensity (see Section 4), we calculate prevalence considering only H/P protozoa and intensity as the mean number of protozoa for just infected individuals; otherwise, prevalence and intensity were calculated as described above across all individuals sampled. Intensity values were highly skewed and were log transformed prior to analysis. We also calculate the variance in relative intensity and the index I, the opportunity for selection, as the variance in intensity divided by the squared mean intensity (see Section 1).

3.3. Host Plumage Brightness and Showiness

Objectively defining and quantifying species' patterns of plumage brightness or evaluating the extent of development of secondary sexual characters is difficult. The practical problems involved have led researchers to rely on subjective evaluation by independent observers as the only realistic solution. Hamilton and Zuk (1982), as well as subsequent investigators (Read, 1987, 1988; Read and Harvey, 1989; Zuk, in press) visually scored species from illustrations in standard field guides. A six point scale was used, with 1 representing very dull species and 6, very bright species. Hamilton and Zuk (1982; Zuk, 1989) refer to both plumage brightness and overall "showiness" of appearance in the species.

The difference between brightness and showiness has caused, we believe, some of the confusion over testing this hypothesis (Read and Harvey, 1989; Zuk, 1989). If parasites do influence sexual selection, there are several aspects of the phenotype of males, excluding behavior, which could be affected by parasites and thus act as the focus of female choice, e.g., plumage brightness, the color of brightness of legs, skin patches or wattles, iris color, the contrast between colors, etcetera. Whereas plumage brightness reflects one measurable trait or set of traits, showiness encompasses both plumage brightness and other potentially important characters, such as those above. The differences between scores for brightness and showiness can be significant (see Section 4) and can greatly affect the conclusions drawn in any study of association between parasites and secondary sexual characters (Read and Harvey, 1989; Zuk, 1989).

We thought it appropriate to score species for both brightness and showiness, but we did not score the species ourselves. We asked five colleagues (four men and one woman) to score the male and female of each species, again using a six point scale, for "1) plumage brightness and 2) showiness of appearance, which would include brightness and elaborateness of plumage and appearance." This definition of showiness reflects the basis on which Hamilton and Zuk did their original scoring (Zuk, 1989). Each observer used the same field guide (Beehler *et al.*, 1986), and they did not see each other's scores or talk together about the scoring. The observers knew of the Hamilton and Zuk hypothesis and our reasons for asking them to do the scoring, but they had not seen any of the parasite data or sample sizes for any species at the time they assisted us.

Separate from having species scored for plumage brightness and showiness, one of the authors (S. G. P-J.) categorized each species as either sexually dichromatic or sexually monochromatic, again using the field guide of Beehler *et al.* (1986). Species scored as dichromatic were those showing any color differences between males and females.

3.4. Host Ecology

For each species we summarized the following aspects of ecology: body size (total body length, from Beehler *et al.*, 1986); foraging strata, the strata within the forest where the species foraged [four levels were used: 1, ground (< 0.5 m), 2, understorey (0.5–2 m), 3, lower canopy (2–10 m), and 4, upper canopy (> 10 m); mean values were calculated for species foraging over more than one level]; foraging levels, the number of different forest strata the species utilized; and diet [the categories used were 1, nectarivorous, 2, frugivorous, 3, invertebrate eaters, 4, predatory (species eating terrestrial vertebrates), 5, omnivorous (combination of nectarivorous, frugivorous, and insectivorous), and 6, general predators (combination of insectivorous and predatory); list modified from Bennett (1986) as cited in Read (1987) and A. F. Read (unpublished manuscript)].

As part of another study (Pruett-Jones and Pruett-Jones, in press) data were available for each species on prevalence of tick parasites. These data were gathered separately from the sampling for blood parasites.

3.5. Statistical Analysis

In our analysis, we examine relevant correlations across all species, across species within families, and across families. The latter two comparisons are made in order to reduce phylogenetic artifacts which can influence comparative studies involving species from related taxonomic groups (cf. Ridley, 1983; Felsenstein, 1985; Pagel and Harvey, 1988). Family-level values were calculated as the grand mean of the individual species' means within each family. For all variables under study, family-level values calculated as grand means were highly correlated with values weighted for sample sizes ($p < 0.0001$ in each regression). The majority of our analyses focus on passerines specifically, as this order encompasses the majority of species studied and previous comparative tests of the parasite hypothesis (Hamilton and Zuk, 1982; Read, 1987, 1988; Read and Harvey, 1989; Read and Weary, 1990, Zuk, in press) have concentrated on this group.

The probability values reported are based on two-tailed tests (Read and Harvey, 1989).

4. RESULTS

The data set comprised 141 blood samples from 45 bird species collected by H. I. J. (Jones, 1985) and 658 samples from 66 species collected by S. G. P-J. and M. A. P-J. (Table I; Appendix).

4.1. Distribution of Parasites across Species, Altitude, and Seasons

Approximately two-thirds of all species (51 of 79, 64.6%) and one-half of all individuals (448 of 799, 56.1%) examined were infected with blood parasites. Species were, on average, infected by just one type of parasite (mean diversity for all species combined was 1.08, $SD = 1.07$, range = 0–4; for passerines specifically, mean = 1.25, $SD = 1.10$, range = 0–4). Only two species, both birds of paradise, were infected with all four parasite types. Species harboring a greater diversity of parasites

TABLE I
Blood Samples Included in Analysis

Group	Individuals	Species	Families
Nonpasserines (6 orders)	43	15	6
Passerines	756	64	17
Total	799	79	23

TABLE II
Comparative Parasite Prevalence in Passerines and Nonpasserines

Group	Parasite prevalence		Parasite intensity	
	Mean	SD	Mean	SD
Nonpasserines	19.1	31.1	17.7	43.7
Passerines	44.6	40.7	12.9	28.7
Total	39.7	40.2	13.8	31.8

suffered a greater overall prevalence as well ($N = 79$, Spearman rank correlation (r_S) $= 0.844$, $p \ll 0.001$).

Nonpasserine species showed significantly lower prevalences ($df = 77$, $t = -2.267$, $p = 0.026$) but similar intensities ($df = 77$, $t = 0.524$, $p = 0.602$) to passerine species (Table II). Nonpasserines were infected only with H/P protozoa; *Trypanosoma*, *Leucocytozoon*, or microfilaria were not found in that group.

Across passerines, the prevalence of H/P, *Trypanosoma*, *Leucocytozoon*, and microfilaria parasites averaged 38.3%, 5.5%, 5.4%, and 13.2%, respectively, with little or no correlation of prevalences between each parasite type across species (Table III; similar correlation coefficients were obtained when non-passerines were included). Parasite intensities covaried with parasite prevalences; species in which a greater proportion of individuals were infected also showed greater levels of infection (considering intensity in infected species, $df = 37$, $R^2 =$

TABLE III
Correlation Matrix for Prevalences of Different Parasites in Passerines[a]

	Haemoproteus/ Plasmodium[b]	Trypanosoma	Leucocytozoon	Microfilaria
Haemoproteus/ Plasmodium	1			
Trypanosoma	−0.150	1		
Leucocytozoon	0.222	0.016	1	
Microfilaria	0.160	0.125	0.363*	1

[a]Sample size is 52 species of passerines.
[b]*Haemoproteus/Plasmodium* protozoa.
*$p < 0.01$.

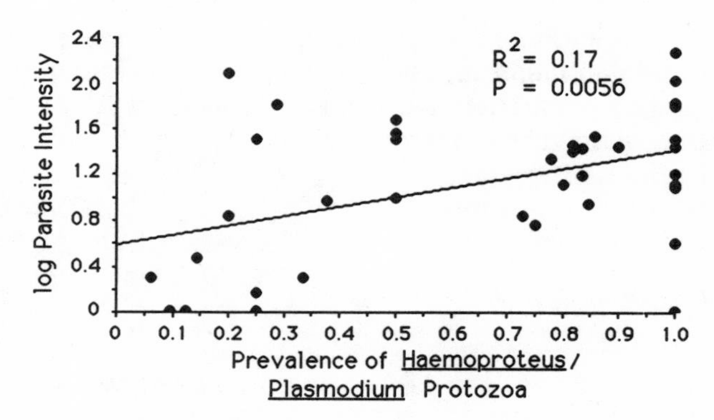

FIGURE 1. Regression of parasite prevalence versus parasite intensity for 38 species of passerines showing some level of parasitemia. Intensity in this analysis is the mean number of protozoa across infected individuals.

0.172, $F = 8.691$, $p = 0.005$, Fig. 1; considering intensity across all species sampled, $df = 63$, $R^2 = 0.738$, $F = 178.74$, $p = 0.0001$). Examining the relationship between prevalence and intensity within each of seven passerine families for which we sampled at least three species (see Section 4.4), positive relationships were observed in all seven families, with significant correlations in five of the seven.

Passerines showed similar parasite prevalences at all altitudes (Fig. 2), whereas there was evidence of a seasonal trend with lower prevalences observed in the dry season (Fig. 3).

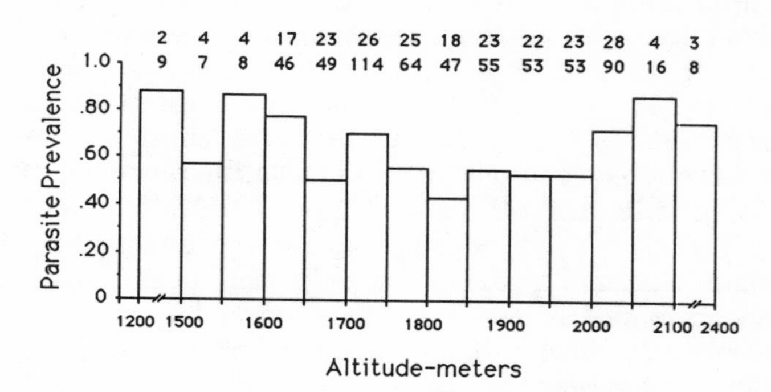

FIGURE 2. Altitudinal variation in parasite prevalence in 52 species of passerines. Shown for each altitudinal interval is the prevalence value for all individuals netted within that interval. Sample sizes, in terms of the number of species and individuals examined within each interval, are listed across the top. Data from all years were combined.

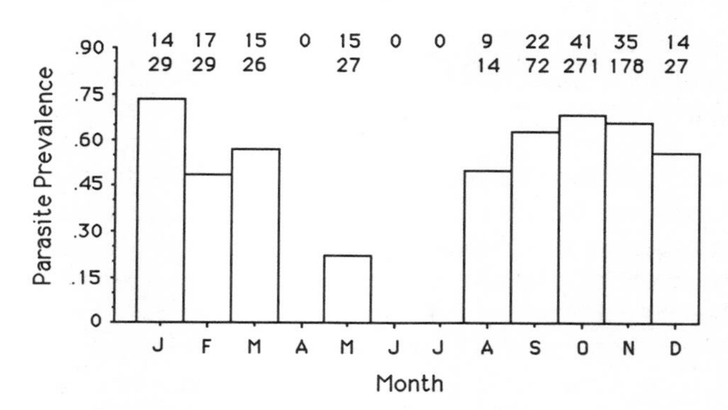

FIGURE 3. Seasonal variation in parasite prevalence in 52 species of passerines. Shown for each month is the prevalence value for all individuals netted during that month. Sample sizes, in terms of the number of species and individuals examined each month, are listed across the top for each interval. No samples were collected during April, June, and July. Data from all years were combined.

4.2. Distribution of Parasites across Individuals

All analyses that follow are restricted to the 64 species of passerines. For each of 19 species we sampled at least 10 individuals, which we considered a minimum sample size in order to examine the distribution of parasites across host individuals. For six of these species, there was no infection by H/P protozoa and they were excluded. Across individuals of 11 of the remaining 13 species, the distribution of H/P protozoa was approximated by the negative binomial distribution (Table IV; Fig. 4). For one species, *Rhagologus leucostigma,* mean intensity was larger than the variance in intensity and comparison with the negative binomial was inappropriate.

The negative binomial distribution (Johnson and Kotz, 1969) is defined by parameters N and P with the frequency distribution of y given by

$$f(y) = \binom{N + y - 1}{N - 1}\left(\frac{P}{Q}\right)^{y} Q^{-N}$$

in which $Q = 1 + P$. We estimated P and N for each species (Table IV) using the method described in Williamson and Bretherton (1963).

For six species for which the sex of individuals could be reliably determined in the field we sampled at least five males and five females and can ask whether parasite prevalence varied with sex. The six spe-

TABLE IV
Parameters of the Negative Binomial Distributions Describing the Distribution of Parasites across Host Individuals in 12 Species of Passerines and Comparison with the Observed Distributions[a]

	Distribution parameters		Statistical test[b]		
Species	*N*	*P*	*df*	X^2	p
Poecilodryas albispecularis	0.420	56.47	1	1.697	> 0.10
Melanocharis versteri	0.631	106.53	2	0.098	> 0.90
Melanocharis striativentris	1.144	53.95	3	0.335	> 0.90
Toxorhamphus poliopterus	0.621	128.87	3	3.771	> 0.10
Ptiloprora guisea	0.992	5.04	1	0.032	> 0.50
Melipotes fumigatus	0.338	63.52	2	1.487	> 0.10
Erythrura papuana	0.223	72.53	3	2.355	> 0.50
Amblyornis macgregoriae	0.260	79.65	9	11.496	> 0.10
Cnemophilus loriae	0.120	1.00	1	0.065	> 0.50
Epimachus meyeri	0.233	107.70	1	0.604	> 0.10
Lophorina superba	0.298	35.63	4	6.560	> 0.10
Parotia lawesii	0.205	36.59	7	18.060	> 0.03

[a]Species of passerines for which we sampled at least 10 individuals and which showed some level of parasite infection.
[b]Statistical test comparing observed distribution with that expected by the negative binomial distribution.

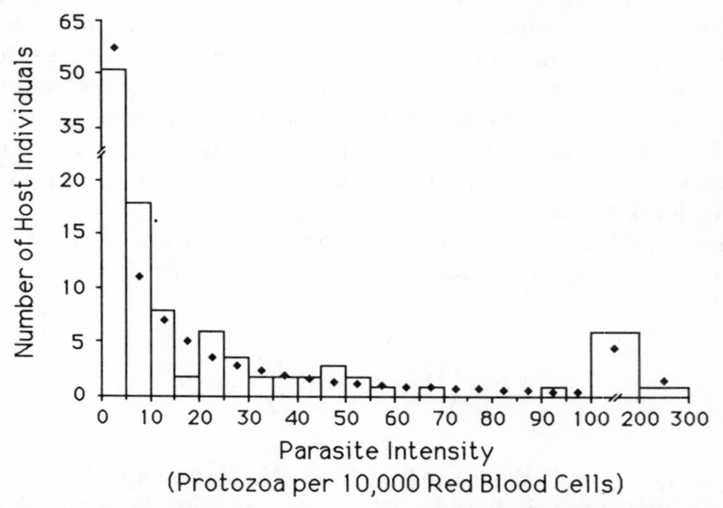

FIGURE 4. Frequency distribution of *Haemoproteus* and *Plasmodium* blood protozoa across 123 individuals MacGregor's bower bird (open bars) compared with expected frequencies (solid diamonds) assuming a negative binomial distribution. The observed distribution did not differ significantly from the expected distribution ($p > 0.10$).

cies were: *Pachycephala soror, Cnemophilus loriae, Lophorina superba, Parotia lawesii, Amblyornis macgregoriae*, and *Toxorhamphus poliopterus*. All individual *P. soror* were parasite free (Appendix). For three (*C. loriae, L. superba*, and *P. lawesii*) of the remaining five species, males and females showed similar parasite prevalences (contingency table analysis; $p > 0.07$ for *P. lawesii*; $p > 0.50$ for the other two species). For both *A. macgregoriae* and *T. poliopterus*, males suffered significantly greater prevalences than did females ($p < 0.05$ in each analysis).

Duplicate samples were available from 70 individuals of 15 species which were captured at least twice. Comparing parasite intensity on the second capture to that on the first, there was no significant association ($df = 69$, $R^2 = 0.026$, $F = 2.822$, $p = 0.0976$; Fig. 5).

4.3. Plumage Brightness versus Showiness

The scores of the five observers for plumage brightness and showiness were fairly consistent, but by no means exactly the same. For male showiness, for example, the average coefficient in the correlation matrix of observers' scores was 0.824 ($R^2 = 0.679$).

Bright species were, by definition, also showy species and considering means of the scores of the five observers, male plumage bright-

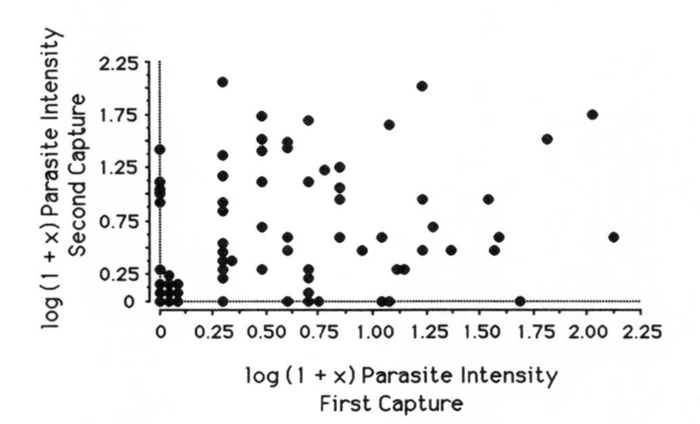

FIGURE 5. Temporal variation in parasite intensity for 70 individuals of 15 species of passerines. Because of zero values for intensity, 1.0 was added to all values prior to log transformation. The relationship is not significant; see text.

ness was highly correlated with male showiness ($df = 78$, $R^2 = 0.768$, $F = 259.602$, $p = 0.0001$; a similar relationship existed between female plumage brightness and female showiness. There was also a consistent trend in that species in which males were showy were characterized as having showy females ($df = 78$, $R^2 = 0.728$, $F = 209.407$, $p < 0.0001$).

Scores for plumage brightness were, however, significantly lower on average than scores for showiness (considering mean values, the average difference in the two scores for each species was 0.44 points; $t = 6.038$, $p = 0.0001$). Considered separately, each of the five observers gave species significantly higher scores, on average, for showiness than plumage brightness ($p < 0.01$ in each comparison).

4.4. Parasitism and Showiness

Significant, positive correlations between parasite loads and measures of brightness were observed for parasite diversity, male brightness, and male and female showiness, and parasite prevalence and male brightness and showiness (Table V; Fig. 6). The variance in intensity and the index *I*, the opportunity for selection, did not correlate with any measure of brightness. Considering the scores of each observer separately, significant correlations across species between parasite prevalence and male showiness were observed in all five cases, and across families in three of the five cases.

The significance of the correlation between parasite prevalence

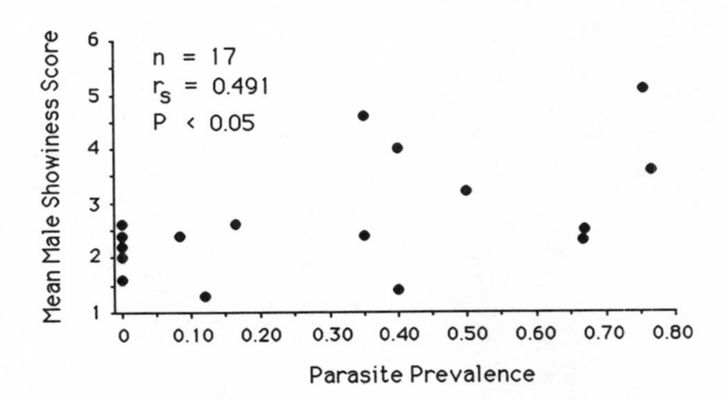

FIGURE 6. Correlation between parasite prevalence and male showiness in 17 families of New Guinea passerines.

TABLE V
Rank Correlations between Measures of Parasite Load and Plumage Elaboration[a]

Measure of parasite load/ measure of brightness	Species level[b]		Family level[c]	
	r_s	p	r_s	p
Parasite diversity				
Male brightness	**0.251**	0.047	0.486	0.052
Male showiness	**0.428**	0.001	**0.523**	0.036
Female brightness	0.070	0.582	0.310	0.215
Female showiness	**0.249**	0.049	0.338	0.177
Parasite prevalence				
Male brightness	0.185	0.144	**0.500**	0.046
Male showiness	**0.293**	0.020	**0.491**	0.050
Female brightness	0.095	0.453	0.259	0.303
Female showiness	0.216	0.087	0.274	0.276
Relative parasite intensity				
Male brightness	0.144	0.254	0.479	0.054
Male showiness	0.201	0.112	0.386	0.120
Female brightness	0.019	0.881	0.072	0.760
Female showiness	0.124	0.327	0.151	0.540
Variance in relative parasite intensity				
Male brightness	0.055	0.682	0.042	0.904
Male showiness	0.091	0.503	−0.292	0.384
Female brightness	−0.084	0.535	−0.091	0.787
Female showiness	0.013	0.928	−0.237	0.441
Index I—the opportunity for selection				
Male brightness	−0.158	0.347	−0.030	0.928
Male showiness	−0.162	0.332	0.085	0.803
Female brightness	−0.093	0.582	−0.066	0.992
Female showiness	−0.124	0.459	−0.188	0.575

[a]Significant correlation coefficients ($P < 0.05$) are highlighted.
[b]Sample size is 37–64 species.
[c]Sample size is 10–17 families.

and male showiness varied with the set of species included in the analysis (Table VI). The strength of the overall correlation appeared to be due to the set of species for which five to ten individuals were sampled. The relationship became nonsignificant when the analysis was restricted to those species in which larger samples were obtained (Table VI).

TABLE VI
Correlations between Parasite Prevalence and Male Showiness in Relation to Sample Size[a]

Category	Number of species	r_s	p
All species combined	64	**0.293**	0.020
Species for which the number of individuals sampled was:			
1–5	28	0.229	0.238
5–10	22	**0.442**	0.043
≥ 10	19	0.183	0.441
≥ 15	11	0.135	0.674
≥ 20	9	0.122	0.734

[a]Probability values < 0.05 are highlighted.

Our data were not sufficient to examine the parasite/showiness association across species within genera, but within-family comparisons could be made for each of seven families for which we sampled at least three species. In all cases, the relationship between parasite prevalence and male showiness was non significant, and in two of the five cases the sign of the correlation was negative (Table VII).

In comparison with sexually monochromatic species, dichromatic species were significantly showier with respect to males ($t = 3.569$, $p = 0.0007$) but not females ($t = 0.633$, $p = 0.529$) and they were not more heavily parasitized ($p > 0.07$ each for diversity, prevalence, and intensity).

TABLE VII
Within-Family Comparisons of Parasite Prevalence and Male Showiness

Family	Number of species	r_s	p
Acanthizidae	3	0.866	0.222
Ripiduridae	4	0.544	0.347
Eopsaltriidae	8	0.152	0.682
Pachycephalidae	8	−0.200	0.596
Dicaeidae	6	−0.621	0.165
Meliphagidae	12	0.539	0.075
Paradisaeidae	10	0.463	0.165

4.5. Parasites and Ecology

In Table VIII are listed rank correlations between parasite prevalence, male showiness, and the ecological variables quantified. Body size and foraging strata showed positive correlations with male showiness, but only foraging strata was separately correlated with parasite prevalence (Table VIII). To examine the join correlations between foraging strata and parasite prevalence and male showiness in greater detail, we corrected parasite prevalence for foraging strata and then did a regression analysis of prevalence versus male showiness; the correlation remained significant ($df = 63$, $R^2 = 0.052$, $F = 4.442$, $p = 0.039$) although the amount of variation explained was small. Sample size again emerged as an important variable as both the number of species and individuals that were sampled showed strong correlations with parasite prevalence (Table VIII).

Considering the ten species of bird of paradise specifically, parasite intensity correlated with the degree of sexual dimorphism (ratio of male/female wing length; $r_s = 0.664$, $p < 0.05$) and parasite prevalence

TABLE VIII
Rank Correlations between Measures of Parasite Prevalence, Male Showiness, Sample Size, and Aspects of Ecology[a]

Prevalence/showiness/ aspect of ecology	Species level[b]		Family level[c]	
	r_s	p	r_s	p
Parasite prevalence				
No. species sampled	—		**0.591**	0.018
No. individuals sampled	0.189	0.136	**0.623**	0.013
Body size	0.192	0.136	0.202	0.424
Foraging strata	**0.445**	0.001	0.383	0.126
Foraging levels	−0.070	0.582	0.225	0.368
Male Showiness				
No. species sampled	—		0.291	0.246
No. individuals sampled	0.137	0.280	0.378	0.134
Body size	**0.382**	0.003	0.467	0.063
Foraging strata	**0.362**	0.004	0.062	0.810
Foraging levels	−0.008	0.952	0.053	0.834

[a]Probability values < 0.05 are highlighted.
[b]Sample size is 62–64 species.
[c]Sample size is 17 families.

varied with the mating system: monogamous species ($N = 2$, mean prevalence = 46.9) harbored lower parasite loads than promiscuous species ($N = 8$, mean prevalence = 82.7; Mann-Whitney $U = 14$, $p < 0.028$). Monogamous species were also less showy than the promiscuous species with mean male showiness scores of 3.6 and 5.4, respectively, for the two groups (Mann-Whitney $U = 16$, $p < 0.035$; Pruett-Jones *et al.*, 1990).

The dietary groups frugivores, invertebrate eaters, and omnivores were observed among the 64 species and significant differences across groups were found with respect to both parasite prevalence ($df = 63$, $F = 19.236$, $p = 0.0001$; Table IX) and male showiness ($df = 63$, $F = 13.197$, $p = 0.0001$). Within each dietary group, however, the correlation between parasite prevalence and male showiness was not significant (for frugivores, $r_s = 0.706$, $p > 0.08$; for invertebrate eaters, $r_s = 0.082$, $p > 0.638$; for omnivores, $r_s = 0.152$, $p > 0.478$).

In comparing prevalences of tick parasites and blood parasites, the data set is restricted to passerine species for which at least 10 individuals were examined for ticks at this was thought to be a minimum sample size for calculation of ectoparasite prevalence values (Pruett-Jones and Pruett-Jones, in press). In rank correlations, neither tick prevalence nor tick intensity correlated with blood parasite prevalence, parasite intensity, or male showiness (comparing tick prevalence and tick intensity with blood parasite prevalence, parasite intensity, and male showiness, respectively, $r_s = 0.25$, $p > 0.11$, $r_s = 0.174$, $p > 0.26$, $r_s = 0.237$, $p > 0.12$, $r_s = 0.134$, $p > 0.39$, $r_s = 0.093$, $p > 0.55$, $r_s = 0.03$, $p > 0.85$).

Comparing passerine and nonpasserines species with respect to the ecological variables examined, nonpasserines were significantly larger ($t = 3.635$, $p = 0.0005$), they foraged over fewer levels in the forest ($t = -2.655$, $p = 0.0096$), but they foraged at similar average levels in the forest ($t = -0.788$, $p = 0.4333$).

TABLE IX
Male Showiness and Parasite Prevalence across Dietary Groups in Passerines

Dietary group	Number of species	Male showiness[a]	Parasite prevalence[a]
Frugivores	7	4.8 (1)	59.7 (4)
Invertebrate Eaters	34	2.4 (2)	21.3 (5)
Omnivores	23	3.0 (3)	74.3 (4)

[a]Shown are the mean values across species in each group. The numbers in parentheses indicate statistical significance; values followed by the same numbers did not differ significantly from one another ($p > 0.05$).

4.6. Parasites and Phylogenetic Order

To examine one possible phylogenetic association which might influence the observed correlations between parasites and the elaboration of secondary sexual characters, we examined the relationship between parasite load and presumed phylogenetic order of the species and subfamilies. By phylogenetic order we refer to the ordering of species or families in standard taxonomic treatments. Whether such ordering accurately reflects the evolutionary sequence in terms of divergence of the respective taxa is debatable, however it nevertheless provides the only readily available approximation of such sequences. We considered the two most commonly used taxonomies, that based on morphology (Peters, 1951–1986) and that based on DNA–DNA hybridization studies (Sibley and Ahlquist, 1985; Sibley *et al.*, 1988). The 17 families of passerines that we studied comprised 15 and 18 subfamilies in the classification schemes of Sibley and Ahlquist and Peters, respectively. In ordering the species, the presumed oldest species was given the score of 1 and the presumed most recent species the scores of 15 or 18 according to the classification.

Phylogenetic order in the scheme proposed by Peters predicted rank order for both parasite prevalence and male showiness at both the species and subfamily levels (Table X; Fig. 7). Despite the extreme differences in the classifications (e.g., Turdinae was number 1 in one list and number 11 in the other), phylogenetic order in the scheme of Sibley and Ahlquist similarly correlated with male showiness (Table X).

TABLE X
Parasite Prevalence, Male Showiness, and Phylogenetic Order[a]

	Ordering by Peters[b]				Ordering by Sibley and Ahlquist[b]			
	Species[c]		Subamilies[d]		Species[c]		Subfamilies[b]	
Category	r_s	p	r_s	p	r_s	p	r_s	p
Parasite prevalence	**0.536**	0.0001	**0.654**	0.007	0.037	0.764	−0.165	0.535
Male showiness	**0.538**	0.0001	**0.617**	0.011	**0.254**	0.044	0.176	0.509

[a]Significant correlation coefficients ($p < 0.05$) are in bold.
[b]References are Peters (1951), and Sibley and Ahlquist (1985).
[c]Sample size is 64 species.
[d]Sample size is 18 subfamilies.
[e]Sample size is 15 subfamilies.

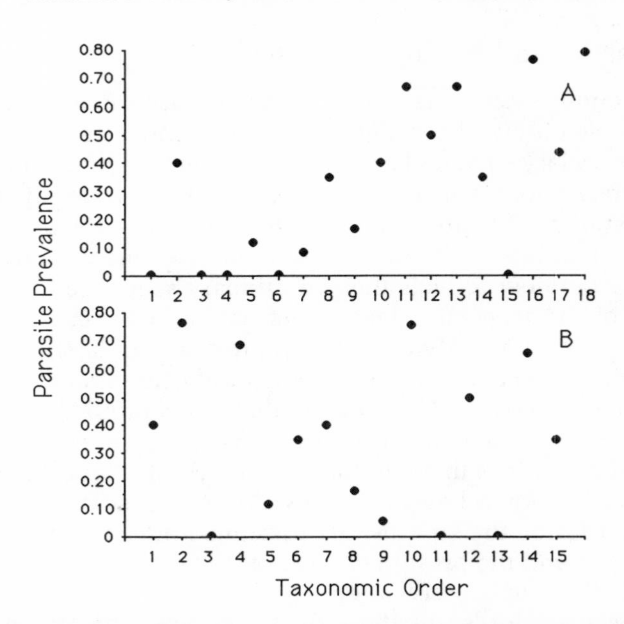

FIGURE 7. Correlation between parasite prevalence and phylogenetic ordering of passerines subfamilies according to the taxonomic scheme of (A) Peters (1951) and (B) Sibley and Ahlquist (1985).

5. DISCUSSION

In this study we considered the distribution of blood parasites across host individuals and species as it relates to ecology and morphological traits in a montane avifauna in Papua New Guinea. Separate from a description of the dynamics of blood parasitism in this avifauna, our central motivation was determining whether the patterns of parasite distribution support the interspecific prediction of the hypothesis by Hamilton and Zuk (1982) on parasites and sexual selection. This data set represents the fourth independent interspecific test of this hypothesis, and complements the earlier work on North American (Hamilton and Zuk, 1982), European (Read, 1987, 1988; Read and Harvey, 1989) and Neotropical (Zuk, in press) passerine birds.

In the avifauna on Mount Missim, *Haemoproteus/Plasmodium* protozoa were the most common blood parasites, infecting slightly over half of all individuals sampled. Blood protozoa were nonrandomly distributed across hosts, with most individuals of each species showing little or no infection but some individuals harboring high intensities.

Such patterns of overdispersion were approximated by a negative binomial distribution (Table IV) which is the most frequently observed pattern of dispersion of parasites across hosts, both with respect to ecto- and endoparasites (Crofton, 1971a, b; Randolph, 1975; Anderson and May, 1978, 1982; Hudson et al., 1985; May, 1985), although random distributions are also known (Pruett-Jones and Pruett-Jones, in press).

A negative binomial distribution of parasites across hosts illustrates one of the numerous difficulties in assessing a species' true parasite burden from low samples sizes. With most individuals having low intensities, large samples are necessary to accurately determine both the mean and variance in parasite intensity for a given species. The same can be said for accurately determining the parasite burden for a given individual as well. We showed that there was no correlation between parasite intensities in individuals sampled on separate occasions (Fig. 5) suggesting that a single sample represents no more than a random sample from an individual's true parasite burden over time. This difficulty does not reflect on the Hamilton and Zuk hypothesis per se, but it illustrates the caution needed in interpreting data on species-specific parasite burdens or any resultant comparisons based on these data. This problem is common to all data sets (references above) used to test the parasite hypothesis.

Passerine species suffered significantly higher parasite burdens than did nonpasserines (Table II), a difference observed in other avifaunas as well, at least with respect to malarial parasites (Garnham, 1966). The reasons for such nonuniformity in parasite burdens across major avian taxa are, at present, unclear. Parrots were uninfected by all blood parasites, and similarly did not show any infestation by ticks (Pruett-Jones and Pruett-Jones, in press). Parrots are certainly brightly colored and strictly in terms of the Hamilton and Zuk model, one might predict that they should suffer high parasite burdens. Parrots are also, however, very cryptic in the habitats where they occur and if the focus of female choice is relative rather than absolute plumage showiness, then a plausible argument could also be made that parrots should have low parasite loads (Endler and Lyles, 1989). The obvious potential discrepancy between how we perceive plumage brightness in birds and how birds themselves perceive and react to brightness is a second difficulty with objectively testing the parasite model (Endler and Lyles, 1989).

Across passerines on Mount Missim, parasite diversity and parasite prevalence correlated with male showiness (Table V). The correlation of showiness with prevalence was observed across species as well as across families but not within families. The predicted association of the Hamilton and Zuk hypothesis between parasites and showiness was thus observed considering some levels of analysis, but not all levels, and the significance of association varied considerably for different measures of parasitism (Table V). The difference in results with different measures of parasitemia highlights the confusion over exactly which measure of parasitemia is the appropriate one. This confusion exists, in part, because the interspecific prediction of the Hamilton and Zuk hypothesis has never been fully discussed with respect to the dynamics of parasite–host coevolution. The interspecific prediction originally focused on mean parasite load across species, but the mean as a statistic does not necessarily express the variance in parasite load on which selection could act. The mean is an appropriate measure to use only to the extent that it correlates with variance in parasite burden and relative variance—the opportunity for selection (Clayton et al., submitted). We believe that the most appropriate variable to use is the opportunity for selection as it represents the relative variance in parasite load. We examined the opportunity for selection based on parasite load, but our analysis is subject to criticism because we have not specified how parasite load relates to parasite resistance in individuals or to individual fitness. Predictions concerning selection will depend on the nature of the relationship between parasites and fitness, e.g., whether it is linear or geometric, as well as the distribution of parasites across hosts.

The significance of the correlations between parasite prevalence and male showiness varied with sample size in terms of the number of individuals of each species that were sampled (Table VI). A strikingly similar dependence of statistical significance on sample size was shown in the data on European passerines (Read and Harvey, 1989). In both analyses, the significance of the overall correlation appeared to depend on those species for which just a few individuals were sampled. The reasons for this relationship are not clear and given that it has appeared in the data from two entirely different avifaunas, it seems unlikely that it is an artifact of one particular study. Cox (1989), in a contradictory criticism of the Hamilton and Zuk model, suggested that this dependence of the correlation between parasites and showiness on sample size was the result of more careful scrutiny by observers of samples from rare species (i.e., those for which few individuals are sampled). He suggests that it is the samples from rare species that are more accurately sampled in comparison with those from common species. As noted by Hamilton and Zuk (1989), however, it is within this group of species that the correlation between showiness and parasite load is strongest. Cox's argument, although offered as a criticism, lends support to the general parasite hypothesis. Nevertheless, we think Cox is wrong. The number of different parasitologists who scored the blood smears analyzed in the data sets used by Hamilton and Zuk (1982), Read (1987), and in this study is large and it is unlikely that all observers showed the same suggested pattern of giving samples from rare species more attention. In our analysis, about half of the blood smears were scored blind to the host. We expect that any variation in accuracy of scoring samples is random with respect to individual species. Lastly, in our data set, it was not the rarest of the species for which the significant correlation was observed, but rather those species for which five to ten individuals were sampled (Table VI).

Earlier (Pruett-Jones *et al.*, 1990) we suggested that one possible explanation for this dependence on statistical significance on sample size was greater random variation in the records for species for which few samples were collected. We now doubt this explanation given the similar patterns between our data set and that analyzed by Read and Harvey (1989). Zuk (1989) and Read and Harvey (1989) offer other points regarding this variation with sample sizes, but until the underlying factors responsible for the association are known, the patterns themselves cannot be viewed as either supporting or refuting the original hypothesis (Zuk, 1989).

Our results also show that rarity or commonness is related in other ways to parasite prevalence than just through the correlation with showiness. The numbers of individuals sampled in a species, the mean number of individuals sampled across species within families, and the number of different species in families each correlated with parasite prevalence (Table VIII). If the number of birds caught in mist nets is related to a species' commonness, then common species have greater parasite loads than rare species. This finding, first, argues against the argument of Cox (1989) in that he suggested that detection of parasites was more likely in rare species because these samples were scanned more carefully. Second, it suggests the possibility that the dependence of the correlation between parasites and showiness on sample size may result from rare species not showing the entire range of variation in parasitemias found in the avifauna as a whole (Pruett-Jones *et al.*, 1990). If, separate from any relationship between parasites and sexual selection, population density affects or correlates with parasite loads, then accurate cross-species comparisons must include all species rather than just rare species.

There was a significant correlation across families between

number of species in the family and parasite prevalence: species in larger families had, on average, greater parasite loads than species in less speciose families (Table VIII). Pruett-Jones *et al.* (1990) and Read (1990) suggest the possibility that it is the process of sexual selection that causes species to be more vulnerable to parasites. If sexual selection otherwise promotes speciation (cf. Lande and Kirkpatrick, 1988), the observed association between parasites and species numbers would be expected.

In a reanalysis of the data originally presented by Hamilton and Zuk (1982), Read and Harvey (1989) show that the correlation between parasites and showiness occurs across species, but not at all higher levels of comparison: within genera, subfamilies, and families. They therefore concluded that the North American data set offered equivocal evidence for the parasite hypothesis and that phylogenetic associations among the avian taxa may be responsible for the correlation. Given the inconsistent correlations, considering all levels of analysis, between parasite burden and male plumage showiness in our data, the results of this study must also be interpreted as equivocal evidence for the Hamilton and Zuk hypothesis. Additionally, male showiness did not correlate with either the variance in parasite load or the opportunity for selection on parasite load arguing against a causal relationship between parasites and sexual selection.

It is important to note, however, that whether the correlation across species between parasite prevalence and male showiness results because of the mechanism proposed by Hamilton and Zuk or because of phylogenic associations among the taxa, this correlation has now been observed in four separate avifaunas encompassing over 10% of the world's bird species (Hamilton and Zuk, 1982; Read, 1987, 1988, 1990; Read and Harvey, 1989; Zuk, in press; this study). Given the diversity of species and families involved in these studies, if the underlying cause of the correlation is phylogeny, we suspect that the details of these relationships, once they are elucidated, will be as interesting as any model of sexual selection based on host–parasite coevolution.

To investigate one possible relationship between phylogeny and the correlation between parasites and sexual selection, we compared presumed phylogenetic order of species and subfamilies with parasite burdens and male showiness (Table X; Fig. 7). Male showiness correlated across species with phylogenetic order regardless of the classification scheme used, strongly suggesting that more recently derived species are also those in which secondary sexual characters are more elaborate. Parasite prevalence, in contrast, correlated with phylogenetic order using one scheme but not another (Table X). Despite the controversy over DNA hybridization as a technique to estimate phylogenies (Sarich *et al.*, 1989), we expect that the classification scheme of Sibley and Ahlquist (1985; Sibley *et al.*, 1988) more accurately reflects phylogeny of the taxa than does that of Peters (1951–1986). Given that parasite prevalences did not correlate with phylogenetic order using the scheme of Sibley and Ahlquist, we conclude that phylogenetic associations as measured through presumed order cannot account for the association between parasites and showiness.

Another obvious possible cause of the association between parasites and showiness is ecology of the species involved, specifically where they forage in relation to the distribution of biting flies (Garvin, 1989). If biting flies are more abundant in the canopy of forests, we expect a correlation between foraging height and parasite burden. As species foraging in the forest canopy are also brighter than those foraging on the ground (Table VIII), brightness should also be associated with parasite load. Our data, however, show that independent of foraging height more heavily parasitized species are also showier.

Some of the results thus offer support for interpreting the correlation between parasites and showiness in light of the Hamilton and Zuk hypothesis, whereas other aspects of the data alternatively suggest that phylogenetic associations, or as yet undetermined factors, may be responsible for this correlation. We view the equivocal nature of interpreting our findings in term of the Hamilton and Zuk hypothesis as less important than the fact that the correlation between showiness and parasite burden has now been demonstrated in another avifauna with an evolutionary history different from the other avifaunas previously studied. The association between parasites and showiness appears to be a general phenomenon, whether the mechanism underlying the association is as yet determined.

6. SUMMARY AND CONCLUSIONS

The hypothesis by Hamilton and Zuk (1982) on parasites and sexual selection predicts that across species there will be a positive correlation between parasite load and plumage showiness in males. In this chapter we present data on blood hematozoa in 799 individuals of 79 species of birds from Papua New Guinea which provide a test of this prediction. Nonpasserines showed significantly lower levels of parasite prevalence than did passerines, differences that are at present not understood. Passerine species with greater prevalence of infection also showed higher intensities of infection. Across species and families of passerines, there were significant correlations between both parasite diversity and parasite prevalence and male showiness but not female showiness. Parasite burdens were not statistically associated with the variance in intensity or the opportunity for selection. Where species foraged in the forest column influenced both parasite loads and showiness, but the significant association between showiness and parasites remained once foraging height was controlled for. Male showiness varied significantly with body size, larger species being showier, but body size did not separately correlate with parasite loads. Male showiness and parasite loads also varied across dietary groups, but within groups showiness was independent of parasite burden. The correlation between showiness and parasites was sensitive to sampling effort across species, as was parasite prevalence generally, suggesting a possible relationship between population density and parasite burdens. Phylogenetic associations among species also appear to interact with parasite burdens, complicating any direct test of the Hamilton and Zuk model using comparative data. Interspecific tests have, to date, raised more questions than provided answers to existing ones and are unlikely, in themselves, to provide a critical test of the parasite hypothesis. The correlation between parasites and plumage showiness appears to be a general pattern, and has now been demonstrated in avifaunas from four continents. Whether this correlation is due to the mechanism suggested by Hamilton and Zuk, phylogenetic associations, or some other cause is, in our opinion, not yet resolved.

ACKNOWLEDGMENTS. Financial support for the research during 1982 and 1983 was provided by National Geographic Society, New York Zoological Society, Chem-tronics Inc. and Dan Brimm, Papua New Guinea Biological Foundation, Frank M. Chapman Memorial Fund, Joseph Henry Fund, Flora and Fauna Preservation Society, Carl Koford Memorial Fund, and Harry Hoogstraal. The work during 1985–1987 was supported by Grant BSR 8416000 from National Science Foundation to Jack Bradbury and S. G. P-J. For assistance and other support we wish to thank our 26 volunteer field assistants and nine Papua New Guinean employees, A. Allison, B. Beehler, T. Pratt, A. Safford, H. Sakulas, the Eltham, Fraser, Harvey-Hall, Violaris, and Winn families, the staffs at Wau Ecology Institute and Museum of Vertebrate Zoology, and New Guinea Goldfields Ltd. Pty. We also thank P. Gullan, P. Sniegowski, T. Schulenberg, D. Stotz, and J. Willis for scoring the bird species for showiness, J. Diamond who scored the species for foraging strata, and D. Clayton, R. Lande, M. Wade, and D. Watt for discussion or comments. The national and Morobe provincial governments in Papua New Guinea allowed us to carry out this study.

7. APPENDIX: PARASITE PREVALENCES AND INTENSITY VALUES FOR SPECIES AND FAMILIES OF PAPUA NEW GUINEA BIRDS[a]

Family/species	Parasite prevalence[b]		Parasite intensity[c]		
	No. individuals examined	No. individuals (%) infected	Mean	Range	SD
NONPASSERINES					
Casuariidae					
Casuarius bennetti	1	0 (0)	0	—	—
Accipitridae					
Accipiter cirrhocephalus	1	0 (0)	0	—	—
Columidae					
Macropygia amboinensis	1	0 (0)	0	—	—
Macropygia nigrirostris	1	1 (100.0)	31	—	—
Gallicolumba rufigula	1	0 (0)	0	—	—
Gallicolumba beccarii	3	0 (0)	0	—	—
Ptilinopus superbus	7	4 (57.1)	26.4	0–8	36.4
Ptilinopus rivoli	3	1 (33.3)	0.3	0–1	0.6
Psittacidae					
Charmosyna papou	4	0 (0)	0	—	—
Micropsitta bruijnii	1	0 (0)	0	—	—
Psittacella brehmii	1	0 (0)	0	—	—
Cuculidae					
Cacomantis variolosus	1	0 (0)	0	—	—
Cacomantis castaneiventris	13	6 (46.1)	39.6	0–128	54.3
Chrysococcyx meyerii	2	1 (50.0)	168.0	0–336	237.6
Alcedinidae					
Halcyon megarhyncha	3	0 (0)	0	—	—

(Continued)

Appendix (*continued*)

Family/species	Parasite prevalence[b]		Parasite intensity[c]		
	No. individuals examined	No. individuals (%) infected	Mean	Range	SD
PASSERINES					
Turdidae					
Zoothera dauma	1	0 (0)	0	—	—
Orthonychidae					
Ptilorrhoa leucosticta	5	2 (40.0)	0	—	—
Sylviidae					
Phylloscopus trivirgatus	4	0 (0)	0	—	—
Maluridae					
Clytomyias insignis	1	0 (0)	0	—	—
Acanthizidae					
Sericornis nouhuysi	14	1 (7.1)	0	—	—
Sericornis perspicillatus	7	2 (28.6)	18.6	0–128	48.3
Sericornis papuensis	1	0 (0)	0	—	—
Rhipiduridae					
Rhipidura brachyrhyncha	1	0 (0)	0	—	—
Rhipidura atra	20	0 (0)	0	—	—
Rhipidura albolimbata	3	1 (33.3)	0.7	0–2	1.2
Rhipidura rufiventris	1	0 (0)	0	—	—
			0		
Myiagridae					
Monarcha axillaris	7	0 (0)	0	—	—
Eopsaltriidae					
Microeca papuana	11	1 (9.1)	0	—	—
Tregellasia leucops	5	1 (20.0)	24.0	0–120	53.7
Eugerygone rubra	2	0 (0)	0	—	—
Poecilodryas albispecularis	11	11 (100.0)	23.7	0–132	36.9
Poecilodryas albonotata	1	1 (100.0)	1	—	—
Amalocichla incerta	1	0 (0)	0	—	—

Peneothello cyanus	8	3 (37.5)	3.5	0–22	7.7
Pachycephalopsis poliosoma	7	1 (14.3)	0.4	0–3	1.1
Pachycephalidae					
Pachycare flavogrisea	2	2 (100.0)	24	0–48	33.9
Rhagologus leucostigma	21	2 (9.5)	0.1	0–1	0.3
Pachycephala soror	25	0 (0)	0	—	—
Pachycephala schlegelii	15	0 (0)	0	—	—
Pachycephala rufinucha	7	0 (0)	0	—	—
Colluricincla megarhyncha	9	1 (11.1)	0	—	—
Pitohui dicrous	2	0 (0)	0	—	—
Pitohui nigrescens	9	1 (11.1)	0	—	—
Climacteridae					
Cormobates placens	10	4 (40.0)	0	—	—
Dicaeidae					
Melanocharis nigra	1	0 (0)	0	—	—
Melanocharis longicauda	3	3 (100.0)	187	43–416	200.5
Melanocharis versteri	13	13 (100.0)	67.2	4–280	85.2
Melanocharis striativentris	26	26 (100.0)	61.7	2–232	58.2
Rhamphocharis crassirostris	7	7 (100.0)	105.9	16–350	119.0
Oreocharis arfaki	4	0 (0)	0	—	—
Zosteropidae					
Zosterops atrifrons	4	0 (0)	0	—	—
Zosterops novaeguineae	7	7 (100.0)	27.9	1–136	50.0
Meliphagidae					
Timeliopsis fulvigula	4	1 (25.0)	0.2	0–1	0.5
Melilestes megarhycha	4	3 (75.0)	4.2	1–10	4.6
Toxorhamphus poliopterus	56	14 (25.0)	8.0	0–216	32.2
Oedistoma iliolophus	3	3 (100.0)	13.0	1–34	18.2
Myzomela rosenbergii	3	3 (100.0)	32.0	3–81	42.7
Meliphaga albonotata	8	1 (12.5)	0.1	0–1	0.4
Lichenostomus subfrenatus	2	1 (50.0)	5.0	0–10	7.1
Ptiloprora plumbea	2	1 (50.0)	18.0	0–36	25.5

(Continued)

Appendix (*continued*)

Family/species	Parasite prevalence[b]		Parasite intensity[c]		
	No. individuals examined	No. individuals (%) infected	Mean	Range	SD
Ptiloprora guisei	11	9 (81.8)	5.0	0–12	5.5
Melidectes belfordi	7	7 (100.0)	27.3	0–20	25.9
Melidectes torquatus	3	3 (100.0)	12.3	1–26	12.7
Melipotes fumigatus	12	11 (91.7)	21.5	0–127	37.2
Estrildidae					
Erythrura trichroa	5	1 (20.0)	1.4	0–7	3.1
Erythrura papuana	20	10 (50.0)	16.2	0–140	34.5
Dicruridae					
Chaetorhynchus papuensis	7	0 (0)	0	—	—
Ptilonorhynchidae					
Ailuroedus melanotis	7	5 (71.4)	0	—	—
Amblyornis macgregoriae	110	90 (81.8)	20.7	0–284	40.8
Paradisaeidae					
Cnemophilus loriae	16	7 (43.8)	0.1	0–2	0.5
Manucodia keraudrenii	2	1 (50.0)	0	—	—
Ptiloris magnificus	3	3 (100.0)	4.0	2–8	3.5
Epimachus albertisi	8	3 (37.5)	0.4	0–2	0.7
Epimachus meyeri	10	9 (90.0)	25.1	0–172	52.2
Astrapia stephaniae	9	9 (100.0)	16.3	1–48	16.1
Lophorina superba	30	24 (80.0)	10.6	0–99	19.7
Parotia lawesii	123	114 (92.7)	7.4	0–130	16.8
Cicinnurus magnificus	9	7 (77.8)	16.9	0–72	22.4
Paradisaea rudolphi	6	5 (83.3)	13.0	0–26	9.9

[a]Taxonomy follows that of Beehler *et al.* (1986).
[b]Prevalence of parasites of any of the types surveyed.
[c]Intensity of *Haemoproteus*/*Plasmodium* protozoa.

REFERENCES

Anderson, R. M., and May, R. M., 1978, Regulation and stability of host–parasite population interactions: I. Regulatory processes, *J. Anim. Ecol.* **47**:219–247.

Anderson, R. M., and May, R. M., 1982, The population dynamics and control of human helminth infections, *Nature* **297**:557–563.

Arnold, S. J., and Wade, M. J., 1984, On the measurement of natural and sexual selection, *Evolution* **38**:709–719.

Beehler, B., 1982, Ecological structuring of forest bird communities in New Guinea, in: *New Guinea Biogeography* (J. L. Gressitt, ed.), W. H. Junk, The Hague, Netherlands, pp. 837–860.

Beehler, B., 1983, *The Behavioral Ecology of Four Birds of Paradise*, unpublished Ph.D. dissertation, Princeton University, Princeton, New Jersey.

Beehler, B., Pratt, T. K., and Zimmerman, D. A., 1986, *Birds of New Guinea*, Princeton University Press, Princeton, New Jersey.

Bennett, P. M., 1986, *Comparative Studies of Morphology, Life History and Ecology among Birds*, unpublished Ph.D. dissertation, University of Sussex, Sussex, England.

Borgia, G., 1979, Sexual selection and the evolution of mating systems, in: *Sexual Selection and Reproductive Competition in Insects* (M. S. Blum and N. A. Blum, eds.), Academic Press, New York, pp. 19–80.

Bradbury, J. W., Vehrencamp, S. L., and Gibson, R., 1985, Leks and the unanimity of female choice, in: *Evolution—Essays in Honour of John Maynard Smith* (J. J. Greenwood and M. Slatkin, eds.), Cambridge University Press, Cambridge, England, pp. 301–314.

Clayton, D. H., Pruett-Jones, S. G., and Lande, R., (in review). A microevolutionary approach to the comparative analysis of parasite-mediated sexual selection, submitted for publication.

Cox, F. E. G., 1989, Parasites and sexual selection, *Nature* **341**:289.

Crofton, H. D., 1971a, A quantitative approach to parasitism, *Parasitol.* **62**:179–193.

Crofton, H. D., 1971b, A model of host–parasite relationships, *Parasitol.* **63**:343–364.

Crow, J. F., 1958, Some possibilities for measuring selection intensities in man, *Hum. Biol.* **30**:1–13.

Diamond, J. M., 1972, *Avifauna of the Eastern Highlands of New Guinea*, Nuttal Ornithological Club, Cambridge, Massachusetts.

Duffy, D. C., 1983, The ecology of tick parasitism on densely nesting Peruvian seabirds, *Ecology* **64**:110–119.

Endler, J. A., and Lyles, A. M., 1989, Bright ideas about parasites, *Trends Ecol. Evol.* **4**:246–248.

Felsenstein, J., 1985, Phylogenies and the comparative method, *Am. Nat.* **125**:1–15.

Garnham, P. C. C., 1966, *Malaria Parasites and Other Haemosporidia*, Blackwell Scientific Publications, Oxford, England.

Garvin, M. C., 1989, *Blood Parasites of Some Passerine Birds in Louisiana and Their Relationship to Sexual Plumage Dimorphism, Plumage Brightness, and Nesting Height*, unpublished M.S. thesis, Louisiana State University, Baton Rouge, Louisiana.

Gressitt, J. L., and Nadkarni, N., 1978, *Guide to Mt. Kaindi*, Wau Ecology Institute, Wau, Papua New Guinea.

Hamilton, W. D., 1982, Pathogens as causes of genetic diversity in their host populations, in: *Population Biology of Infectious Diseases* (R. M. Anderson and R. M. May, eds.), Springer, New York, pp. 269–296.

Hamilton, W. D., and Zuk, M., 1982, Heritable true fitness and bright birds: A role for parasites? *Science* **218**:384–387.

Hamilton, W. D., and Zuk, M., 1989, Reply to Cox, *Nature* **341**:289–290.

Harvey, P. H., Read, A. F., John, J. L., Gregory, R. D., and Keymer, A. E. (in press), An evolutionary perspective: using the comparative method, in: *Parasite-host Associations: Coexistence or Conflict?* (C. A. Toft, A. Aeschlimann, and L. C. Bolis, eds.) Oxford University Press, Oxford.

Hudson, P. J., Dobson, A. P., and Newbord, D., 1985, Cyclic and non-cyclic populations of red grouse: A role for paratisim?, in: *Ecology and Genetics of Host–Parasite Interactions* (D. Rollinson and R. M. Anderson, eds.), Linn. Soc. Sym. Series No. 11, Linn. Soc. Lond., pp. 77–89.

Janetos, A., 1980, Strategies of female choice: A theoretical analysis, *Behav. Ecol. Sociobiol.* **7**:107–112.

Johnson, N. L., and S. Kotz., 1969, *Discrete Distributions*, John Wiley and Sons, New York.

Jones, H. I., 1985, Hematozoa from montane forest birds in Papua New Guinea, *J. Wild. Dis.* **21**:7–10.

Kirkpatrick, M., 1982, Sexual selection and the evolution of female choice, *Evolution* **36**:1–12.

Kirkpatrick, M., 1986, Sexual selection and cycling parasites: A simulation study of Hamilton's hypothesis, *J. Theor. Biol.* **119**:263–271.

Kirkpatrick, M., 1987, Sexual selection by female choice in polygynous animals, *Ann. Rev. Ecol. Syst.* **18**:43–70.

Lande, R., 1981, Models of speciation by sexual selection on polygenic traits, *Proc. Nat. Acad. Sci. USA* **78**:3721–3725.

Lande, R., and Kirkpatrick, M., 1988, Ecological speciation by sexual selection, *J. Theor. Biol.* **133**:85–98.

Margolis, L. G., Esch, W., Holmes, J. C., Kuris, A. M., and Schad, G. A., 1982, The use of ecological terms in parasitology, *J. Parasitol.* **68**:131–133.

May, R. M., 1985, Host–parasite associations: Their population biology and population genetics, in: *Ecology and Genetics of Host–Parasite Interactions* (D. Rollinson and R. M. Anderson, eds.), Linn. Soc. Sym. Series No. 11, Linn. Soc. Lond., pp. 243–262.

May, R. M., and Anderson, R. M., 1983, Parasite–host coevolution, in: *Coevolution* (D. J. Futuyma and M. Slatkin, eds.), Sinauer Associates, Sunderland, Massachusetts, pp. 186–206.

McApline, J. R., Keig, G., and Falls. R., 1983, *Climate of Papua New Guinea*, CSIRO, and Australian National University Press, Canberra, Australia.

Pagel, M. D., and Harvey, P. H., 1988, Recent developments in the analysis of comparative data, *Quat. Rev. Biol.* **63**:413–440.

Peters, J. L., 1951–1986, *Checklist of Birds of the World*, Vols. VII–XV, Museum of Comparative Zoology, Cambridge, Massachusetts.

Pomiankowski, A. N., 1987, The costs of female choice, *J. Theor. Biol.* **128**:195–218.

Pomiankowski, A. N., 1988, The evolution of female mate preferences for male genetic quality, in: *Oxford Surveys in Evolutionary Biology*, Vol. 5, Oxford University Press, Oxford, England, pp. 136–184.

Pratt, T. K., 1983, *Seed Dispersal in a Montane Forest in Papua New Guinea*, unpublished Ph.D. dissertation, Rutgers University, New Brunswick, New Jersey.

Pratt, T. K., and Stiles, E. W., 1985, The influence of fruit size and structure on composition of frugivore assemblages in New Guinea, *Biotropica* **17**:314–321.

Pruett-Jones, M. A., and Pruett-Jones, S. G., 1982, Spacing and distribution of bowers in Macgregor's bowerbird (*Amblyornis macgregoriae*), *Behav. Ecol. Sociobiol.* **11**:25–32.

Pruett-Jones, M., and Pruett-Jones, S., (In Press), Analysis and ecological correlates of tick burdens in a New Guinea avifauna, in: *Bird-Parasite Interactions: Ecology, Evolution, and Behaviour* (J. E. Loye and M. Zuk, eds.), Oxford University Press, Oxford, England (in press).

Pruett-Jones, S. G., and Pruett-Jones, M. A., 1986, Altitudinal distribution and seasonal activity patterns of birds of paradise, *Natl. Geogr. Res.* **2**:87–105.

Pruett-Jones, S. G., and Pruett-Jones, M. A., 1988, The use of court objects by Lawes' Parotia, *Condor* **90**:538–545.

Pruett-Jones, S. G., Pruett-Jones, M. A., and Jones, H. I., 1990, Parasites and sexual selection in birds of paradise, *Amer. Zool.* (30:287–298).

Randolph, S. D., 1975, Patterns of distribution of the tick *Ixodes trianguliceps* Birula on its hosts, *J. Anim. Ecol.* **44**:451–474.

Read, A. F., 1987, Comparative evidence supports the Hamilton and Zuk hypothesis on parasites and sexual selection, *Nature* **327**:68–70.

Read, A. F., 1988, Sexual selection and the role of parasites, *Trends Ecol. and Evol.* **3**:97–102.

Read, A. F., 1990, Parasites and the evolution of host sexual behaviour, in: *Parasitism and Host Behaviour* (C. J. Barnard and J. M. Behnke, eds.), Taylor and Francis, London, pp. 117–157.

Read, A. F., and Harvey, P. H., 1989, Reassessment of evidence for Hamilton and Zuk theory on the evolution of secondary sexual characters, *Nature* **339**:618–620.

Read, A. G., and Weary, D. M., 1990, Sexual selection and the evolution of bird song: A test of the Hamilton–Zuk hypothesis, *Behav. Ecol. Sociobiol.* **26**:47–56.

Ridley, M., 1983, *The Explanation of Organic Diversity: The Comparative Method and Adaptations for Mating*, Oxford University Press, Oxford, England.

Sarich, V. M., Schmid, C. W., and Marks, J., 1989, DNA hybridization as a guide to phylogenies: A critical analysis, *Cladistics* **5**:3–32.

Seger, J., and Hamilton, W. D., 1988, Parasites and sex, in: *The Evolution of Sex* (R. E. Michod and B. R. Levin, eds.), Sinauer Associates, Sunderland, Massachusetts, pp. 176–193.

Sibley, C. G., and Ahlquist, J. E., 1985, The phylogeny and classification of the Australo-Papuan passerine birds, *Emu* **85**:1–14.

Sibley, C. G., Ahlquist, J. E., and Monroe, B. L., Jr., 1988, A classification of the living birds of the world based on DNA–DNA hybridization studies, *Auk* **105**:409–423.

Wade, M. J., 1979, Sexual selection and variance in reproductive success, *Am. Nat.* **114**:742–747.

Wade, M. J., 1987, Measuring sexual selection, in: *Sexual Selection: Testing the Alternatives* (J. W. Bradbury and M. Andersson, eds.), John Wiley & Sons, New York, pp. 197–207.

Wade, M. J., and Arnold, S. J., 1980, The intensity of sexual selection in relation to male sexual behavior, female choice, and sperm precedence, *Anim. Behav.* **28**:446–461.

Ward, P. I., 1988, Sexual dichromatism and parasitism in British and Irish fresh water fish, *Anim. Behav.* **36**:1210–1215.

Williamson, E., and M. H. Bretherton., 1963, *Tables of the Negative Binomial Probability Distribution*, John Wiley & Sons, New York.

Zuk, M., 1989, Validity of sexual selection in birds, *Nature* **340**:104–105.

Zuk, M., (In Press), Plumage brightness is correlated with ectoparasite and hematozoa loads in North American and Neotropical birds, in: *Bird-Parasite Interactions: Ecology, Evolution, and Behaviour* (J. E. Loye and M. Zuk, eds.), Oxford University Press, Oxford, England (in press).

CHAPTER 6

DECEIT OF MATING STATUS IN PASSERINE BIRDS

An Evaluation of the Deception Hypothesis

HANS TEMRIN

1. INTRODUCTION

When several females settle on the breeding territory of a male, reproductive success has been suggested to decrease because more individuals are exploiting a given environment, higher density of individuals may attract more predators, and each female will have reduced male assistance when feeding the young (Orians, 1969). Females should only mate already mated males who are able to offer resources that can compensate for sharing the male and territory with other females. To make accurate comparisons between different mating options, females should be able to assess the mating status of males. However, in polyterritorial bird species, where mated males try to attract additional females in secondary territories, females may be unable to assess their mating status (Haartman, 1969).

Interspecific deceit, especially in predator–prey relationships, is a well studied phenomenon (e.g., Wickler 1968). Deceit or deliberate misleading ought to be commonplace too in intraspecific signalling,

HANS TEMRIN • Division of Ethology, Department of Zoology, University of Stockholm, S–106 91 Stockholm, Sweden.

since it might be a profitable strategy whenever there is any form of assessment, for example in courtship. The deceit should be relatively rare, so that on average it pays the responder to react the way it does, and the responder must at least sometimes be unable to distinguish between fakes and the real thing (Dawkins and Krebs, 1978).

Males could increase their reproductive success by mating with more than one female, although polygyny might be costly because of an increase in the rate of cuckoldry (Alatalo *et al.*, 1984) and a decrease in the number and condition of the young raised by the first female (e.g., Alatalo and Lundberg, 1984, 1986). Females, on the other hand, could maximize their reproductive success by choosing the best male and territory available. Thus, selection has favored males who are successful in attracting mates while females who are able to adequately assess the relative value of the resources offered them have been favored.

It has been suggested that males hide their mating status in order to attract additional females in polyterritorial species like the Pied Flycatcher (*Ficedula hypoleuca*) (Alatalo and Lundberg, 1984) and the Wood Warbler (*Phylloscopus sibilatrix*) (Temrin, 1984), but also in the Great Reed Warbler (*Acrocephalus arundinaceus*) (Catchpole *et al.*, 1985) where polyterritoriality is rare. In general, there is no reason for males to inform females about their mating status during courtship, but the question is whether males take special steps to hide that they are already mated. "The deception hypothesis" (Alatalo *et al.*, 1981) defines the deceit as hiding the true mating status by males. The cost for females of mating indiscriminate of mating status is an average reduction in reproductive success because of reduced male assistance, and, being unable to acquire information about the mating status, would postpone the start of breeding. We then expect a female who chooses an already mated male (a secondary female) to have lower fitness than she would have had if she had chosen one of the alternative unmated males. The loss of fitness is difficult to test directly, so estimations of the relative fledging success of "simultaneously" breeding females of different mating status, measured by the use of linear regression analyses, has been used to test the deception hypothesis (Alatalo *et al.*, 1981; Alatalo and Lundberg, 1984; Stenmark *et al.*, 1988; Temrin, 1989).

2. POLYTERRITORIAL BEHAVIOR

Males in both the Pied Flycatcher and in the Wood Warbler arrive to their breeding areas in Central Sweden in the end of April or beginning of May. The females arrive about one week later. The Pied Flycatcher is a hole-nesting species, while females in the Wood Warbler build a domed nest on the ground. In the Pied Flycatcher and in the Wood Warbler, approximately half of the already-mated males establish a second territory at varying distances from the nest of the first female (45% and 63%, respectively, in Haartman, 1956, and Temrin, 1989). Males start singing in these secondary territories around the commencement of egg laying and some of them succeed in attracting a second female (Table I). In both of these species the first egg laid in the secondary nest is on average one week later than in the primary nest. Monoterritorial mated males in the Wood Warbler establish singing

TABLE I
Some Characteristics of Polygamy in the Pied Flycatcher, the Wood Warbler, and the Great Reed Warbler

Frequency polygynous males (%)	Frequency secondary females (%)	Number of Fledglings			References
		Monogamous ♀♀	1st ♀♀	2nd ♀♀	
Pied Flycatcher:					
13	6.5	—	—	—	Hartman, 1951
3	—	—	—	—	Curio, 1959
18	14	5.1*	4.1	3.3	Alatalo and Lundberg, 1984
—	—	5.5*	5.1	4.7	Stemark *et al.*, 1988
Wood Warbler:					
23	20	4.3	3.7	4.9	Temrin, 1989; Temrin and Jakobsson, 1988
Great Reed Warbler:					
27	25	2.8	3.2	2.4	Urano, 1985
26	23	3.3	2.2	1.1	Catchpole *et al.*, 1985
14	16	2.6	4.0	2.6	Dyrcz, 1986
44	40	2.8	3.6	2.9	S. Bensch and D. Hasselquist (unpublished data of a six-year study)

*Simultaneous females as estimated from linear regression analysis.

posts in outlying parts of their territory (50–75 m from the nest of the first female). The breeding season is also of similar duration in both species: about 30 days as measured from the first to the last commencement of egg laying in the population (Alatalo *et al.*, 1984; Temrin and Jakobsson, 1988).

The behavior of male Great Reed Warblers is different from that of both Pied Flycatchers and Wood Warblers. Polyterritoriality is rare in the Great Reed Warbler and males try to attract additional females in outlying parts of their territories (Dyrcz, 1986). The breeding season is longer than in the two other species: about 60 days between the first and last date of egg laying in both the Japanese study of Urano (1985) and a Swedish study by S. Bensch, D. Hasselquist, and U. Ottosson (unpublished data). See Table I for information about polygyny frequency and fledging success in the three species.

3. TESTING THE DECEPTION HYPOTHESIS

In a situation where a female is unable to immediately assess a male's mating status, females could adopt either a "coy" or a "fast" strategy (Dawkins, 1976). In the case of polyterritoriality and deception of mating status, "coy females" correspond to females who carefully assess the mating status of males at a certain cost, by prolonged courtship or by comparing the behavior of several males in the population. "Fast females," on the other hand, should avoid costful assessments by basing their choice on "easily" available information. We could calculate the outcome of these two strategies by calculating the expected reproductive success of uninformed females (F_u) vs. the expected reproductive success of informed females (F_i). If it should pay females to assess the mating status of males, the reproductive success of females acquiring information about male mating status (F_i) should be higher than the reproductive success of uninformed females (F_u). Following the rationale of the deception hypothesis, we could estimate the reproductive success of an uninformed female as

$$F_u = P_m N_m V_m + P_1 N_1 V_1 + P_2 N_2 V_2$$

where P_m is the probability of mating with a monogamous male, P_1 is the probability of becoming a primary female, P_2 is the probability of becoming a secondary female, N_m, N_1, and N_2 are number of fledged young for monogamously mated, primary and secondary females, and V is the offspring value. [The fitness of the offspring of females of polygynous males in relation to females of monogamous males has been estimated as 0.95 for the offspring of secondary females and 0.99 for the offspring of primary females in the Pied Flycatcher (Alatalo and Lundberg, 1986). This "offspring value" (V) is one of the few attempts to estimate the fitness for the offspring of females of different mating status in passerine birds, and will be used in the following calculations. If we assume that the probability to achieve a certain mating status if uninformed is equal to the frequency of females of different mating status, the reproductive success of uninformed females in the study of Pied Flycatchers by Alatalo and Lundberg (1984) is

$$(0.72 \times 5.1 \times 1) + (0.14 \times 4.1 \times 0.99) + (0.14 \times 3.3 \times 0.95) = 4.7.$$

(See Table II for results from other studies and Appendix I for the calculations.)

The average reproductive success of females acquiring information about male mating status (F_i) could be calculated as:

$$F_1 = \frac{P_m}{P_m + P_1} N_m V_m + \frac{P_1}{P_m + P_1} N_1 V_1 - C_i$$

where C_i is the cost of acquiring information.

The deception hypothesis assumes that the crucial information lacking is the mating status of males. Following the rationale of the deception hypothesis, C_i = number of days to find an unmated male ×

TABLE II
The Reproductive Success of Females Mating Uninformed (F_u) and Females Acquiring Information (F_i) about Male Mating Status, When It Takes 6.6 days and 2 days, Respectively, to Find an Unmated Male

	F_u	F_i (6.6 days)	F_i (2 days)
Pied Flycatcher:			
Alatalo and Lundberg (1984)	4.7	4.6	4.8
Great Reed warbler:			
Catchpole et al. (1985)	2.8	3.1	3.2
Urano (1985)	2.8	2.8	2.9
Dyrcz (1986)	2.7	2.7	2.8
Wood Warbler:			
Temrin (1989)	4.2	3.8	4.0

the decrease in number of fledged young per day to postpone the start of breeding. We do not know the number of days it takes to assess male mating status and to find a suitable male, so instead we could estimate the maximum number of days a female can use to inform herself about male mating status and still do better than uninformed females. The F_u value of 4.7 for the Pied Flycatcher presented above gives females less than four days to find an unmated male if F_i should exceed F_u; i.e., females should inform themselves about mating status if it takes less than four days (Appendix II). In the Great Reed Warbler study of Catchpole *et al.* (1985), a female should try to assess mating status if it takes less than 22 days to find an unmated male (Appendix II). In the Wood Warbler it will not pay females to assess mating status given the fledging success of secondary females observed (Table I and II).

To give a rough idea of the relation between F_u and F_i, we could try to estimate the number of days it take to find an unmated male. Females could either try to assess mating status before they settle with a male or some time after pair formation. Results in both the Pied Flycatcher (Stenmark *et al.*, 1988) and in the Wood Warbler (Temrin, 1989) suggest that there are behavioral differences between unmated and mated males trying to attract females that could make it possible to assess male mating status before pair formation. These studies also show that polygynous males visit their primary females less than 24 hours after the second mating, which could "reveal" males' mating status to secondary females. Alatalo *et al.* (1981) used data from Haartman (1969) to estimate the time it takes to renest in a new nest hole with a new male in the Pied Flycatcher (= 6.6 days) when a female becomes aware of his mating status. Using this figure and data of fledging success from the study of Alatalo and Lundberg (1984) gives the following equation:

$$F_i = \frac{0.72}{0.72+0.14} \times 5.1 \times 1 + \frac{0.14}{0.72+0.14} \times 4.1 \times 0.99 - 6.6 \times 0.058 = 4.5$$

(The decrease in number of fledged young per day for females of monogamous males + primary females is 0.058 in the Pied Flycatcher, as estimated in Alatalo and Lundberg, 1984.) Reproductive success of females acquiring information about mating status in this population of Pied Flycatchers will then be slightly smaller than the reproductive success of uninformed females; i.e., females should not spend time assessing mating status (Table I). In the Pied Flycatcher study of Stenmark *et al.* (1988), the difference in fledging success of females of different mating status is smaller than in Alatalo *et al.* (1981) (Table I), which should favor uninformed females even more.

If the number of days to find unmated males decreases to two days, F_i increases from 4.5 to 4.8 in the Pied Flycatcher and F_i is then slightly larger than F_u (= 4.7), i.e., females should try to assess the mating status of males. The difference is still, however, very small, but it appears that an attempt of the female to assess the true marital status of males is maladaptive if this delays the onset of her reproduction above a relatively short time period (about 1–2 days). Whether females should be "coy" or "fast" depends on the time it takes to assess mating status. In the Wood Warbler it does not pay females to postpone the start of breeding to assess their male's mating status (Table II). In the Great Reed Warbler, there is only a slight reduction in fledging success with time of the season, and the reproductive success of informed females is the same or higher than for uninformed females; i.e., it is not clear from these estimates whether females should spend time assessing mating status or not (Table II).

The probability of finding an already-mated male differs for early and late arriving females. An early female has almost entirely unmated males to choose from, while the frequency of already mated males is high later in the season. The choice is limited for both early and late females; it is difficult for early females to know if the unmated male she chooses will later attract an additional female, while late females have the choice of mating one of the unmated males left over or an already mated male. When the frequency of mated males trying to attract females increases in the population, the time to find an unmated male also increases since it takes time to compare enough males to find an unmated one. Thus, the cost of finding a mate is not likely to be the same for early and late females. However, to simplify the estimates, the same costs have been used over the whole season.

In an earlier study I showed how the frequency of unmated vs. mated male Wood Warblers trying to attract females changes over the season, as well as the frequency of females of different mating status (Temrin, 1989). It turned out that 97% of the females settled with unmated males during the first four days of pair formation in the population, 57% settled with males who stayed monogamous, while 40% settled with males who later attracted a second female. Three percent became secondary females. Of the females who mated later than 12 days after the first females in the population, 73% became secondary females while 27% of the females settled with unmated males who stayed monogamous. The fledging success of females of different mating status also changes during the breeding season (for exact values see the calculations in Appendix I). The reproductive success of uninformed female Wood Warblers who settled early, or late, in the population will exceed that of informed females because of the relatively high fledging success of secondary females (Table II). Thus there is no need even for late female Wood Warblers to assess mating status even if the assessment period is short.

In the Pied Flycatcher, the frequency of unmated vs. mated males trying to attract females also changes during the breeding season (Alatalo and Lundberg, 1984). We do not know the exact figures, but we assume (from data in Alatalo et al., 1984) that early females run no "risk" of mating an already mated male, while about 70% of them will mate unmated males who stay monogamous and about 30% will become primary females of polygynous males. With these figures, early uninformed females will, theoretically, have a reproductive success that exceeds that of informed females (Appendix I), but it is difficult to understand how early females should separate unmated males that will become polygynously mated from those who will stay monogamous. Note that the difference in fledging success between late females of monogamous males and late secondary females is smaller than for early females (Appendix I). The small difference is because of a significant decrease in fledging success with time for females of monogamous males, while the slope is almost zero for secondary females ($b = 0.008$ in Alatalo and Lundbert, 1984). For late females, *Fi* exceeds *Fu* and late females could even use 10 days to find an unmated male and still be favored (see Appendix I). That females are able to change behavior depending on circumstances finds some support in Slagsvold *et al.* (1988), where the premating periods for female Pied Flycatchers who settled with already-mated males tended to be longer than those who mated with unmated males in 1985 1.1 and 0.5 days, respectively, and in 1986 3.4 and 1.6 days).

In the Great Reed Warbler, Urano (1985) has shown how the frequency of females of different mating status changes over the season. Of the females who commenced egg laying the first 10 days in the population, none was a secondary female, about 50% turned out to be primary females, and about 50% were females of monogamous males. Of the females who laid their eggs later in the season (> 20 days after the first ones), roughly 50% were secondary females and 50% females of monogamous males. Using these probability values for the study of Catchpole *et al.* (1985), suggests that it would pay late females to assess mating status since *Fu* is smaller than *Fi* ($2.2 < 3.2$). Even with the slight reduction in fledging success for secondary females in Urano (1985), late females should assess their mating status; i.e., $Fu < Fi$ (see Appendix I).

Whether females should assess mating status or not depends on the relationship between the probability of becoming a secondary female, the fledging success of secondary females, and the costs of assessing male mating status. To clarify this relationship some data on the minimum relative fledging success of secondary females, if deception should be favored, are presented in Table III. When there is a low probability of becoming a secondary female (10%) and the reduction in fledging success is 0.05 young per day when postponing the start of breeding, and when it takes 6.6 days to assess mating status and find an unmated male, the minimum fledging success of secondary females could be as low as 36% of that of primary females and females of monogamous males. If the time to find an unmated male decreases to 2 days, the relative fledging success of secondary females must be at least 84% if uninformed females should be favored.

If the probability of becoming a secondary female increases to 30%, females choosing mates indiscriminately of mating status would only be favored as long as the relative fledging success of secondary females exceeds 82%, when females need 6.6 days to inform themselves about mating status and find an unmated male. If the time to find an unmated male is only 2 days, the fledging success of secondary females must exceed that of primary females plus females of monogamous males (102%) if uninformed females should be favored. With a lower decrease

TABLE III
Minimum Relative Fledging Success of Secondary Females If Deception Should Be Favored under Various Conditions of Daily Decrease in Number of Fledged Young (*N*), Probability of Becoming a Secondary Female (*P*), and Number of Days to Find an Unmated Male (the Fledging Success of Primary Females plus Females of Monogamous Males Is Put as 5 in the Analyses)[a]

Daily decrease in *N* (young/day)	0.05	0.05	0.05	0.05	0.02	0.02	0.02	0.02
P of becoming a secondary female	0.10	0.10	0.30	0.30	0.10	0.10	0.30	0.30
Number of days to find an unmated male	6.6	2	6.6	2	6.6	2	6.6	2
Relative *N* of secondary females for $F_u > F_1$	36%	84%	82%	102%	78%	97%	96%	102%

[a]To interpret these results, read down. For example, for a daily decrease in N of 0.05, a probability of becoming a secondary female of 0.10, and the number of days to find an unmated male of 6.6, an uninformed reproductive strategy will be more successful than an informed strategy when the N of secondary females is > 36% of primary females plus females of monogamous males (1.8/5).
For the calculations see Appendix III (Section 7.3.).

in fledging success per day when postponing the start of breeding (0.02 young per day), fledging success of secondary females must be approximately the same as for primary females plus females of monogamous males, if an "uninformed" strategy should pay.

These examples show that when the probability to become a secondary female is low, secondary females could have a low fledging success and uninformed females would still be favored as long as the decrease in fledging success when postponing the start of breeding is relatively high (0.05 young per day) and it takes at least five to six days to assess mating status and find an unmated male. When the cost of finding an unmated male decreases uninformed females will be favored only as long as there is a slight difference in fledging success for females of different mating status (Table III).

4. FACTORS AFFECTING DECEPTION OF MATING STATUS

In the test of the deception hypothesis above, we followed the rationale of the hypothesis and assumed that polygyny in polyterritorial species is explained by male deception; i.e., that the fledging success we find in the different studies result from females unaware of the mating status of the mates they chose. It is of course possible that females actually include the cost of sharing the male with other females when selecting an already mated male, and that secondary females make a good choice compared with their mating options. Let us try to find out if this is the case, by looking more closely at what has been shown so far in these species where deception has been suggested to explain polygyny.

1. Deception of mating status is primarily concerned with the inability of females when choosing a mate to include the cost of sharing the male with other females (i.e., reduced male assistance). What is the relationship between paternal care and breeding success in these species?
(a) How does paternal care influence breeding success?
(b) How predictable is paternal care?
(c) Could other resources compensate for reduced male assistance?

2. When testing the deception hypothesis it became clear that if secondary female lose in reproductive success the probability to become a polygynously mated female determines whether females should spend time assessing mating status or not. However, becoming a secondary female may not be making a bad choice. How high is the probability of

becoming a secondary female and actually decreasing in fitness compared with the options available?
(a) What is the probability of becoming a secondary female?
(b) What are the mating options?

3. The cost of assessing male mating status may make deception possible (Alatalo and Lundberg, 1984). How high is this cost?
(a) What are the costs of postponing the start of breeding?
(b) Is it possible for females to assess the mating status of males?

4.1. The Relationship between Paternal Care and Breeding Success

4.1.1. How Does Paternal Care Influence Breeding Success for Females of Different Mating Status?

In order to answer this question we must consider how a single unaided parent succeeds in raising the young, if females of different mating status receive different amounts of male assistance, and what the consequences are for females of polygynous males and their young.

In the Pied Flycatcher an experiment where the male or the female in a pair was removed when the young were five days old showed that both unaided males and females increased their feeding rates by 90%, giving similar feeding rates as in broods of pairs (Alatalo *et al.*, 1988). Nestling weight did not differ significantly between the young of pairs and unaided single parents, but there was a tendency for young in nests of unaided parents to be lighter. The nestlings of widowed female in a study of Alatalo *et al.* (1982) were significantly lighter than the young in nests of monogamous pairs. Single unaided parents also raised fewer nestlings than monogamous pairs. Unaided males raised 80% and unaided females 67% as many young as monogamous pairs (Alatalo *et al.* 1988). Widowed females raised 66% as many young as females of monogamous males (Alatalo *et al.*, 1982). These results suggest that male assistance is important for breeding success.

In a recent study of Alatalo and Lundberg (1990), male contribution in feeding was measured as number of visits per hour when the nestlings were 5 and 13 days old. Primary and secondary females received 83 and 46%, respectively, of male assistance compared with females in monogamous pairs. At the end of the nestling period polygamous males made little contribution at the primary nest, while the contribution at secondary nests was minimal just after hatching. Alatalo *et al.* (1982) showed that over the whole nestling period, feeding frequencies did not differ between monogamous, primary, and secondary nests because both primary and secondary females compensated for the reduced assistance. However, females of polygynous males raised significantly fewer young than simultaneous females of monogamous males (Alatalo and Lundberg, 1984; Stenmark *et al.*, 1988). This suggests that females of polygynous males lose in breeding success because of reduced male assistance. It is notable that experimentally induced, unaided parents raised more young (66–80%) than secondary females (64%) in the study of Alatalo and Lundberg (1984).

In the Wood Warbler, males in monogamous pairs do contribute substantially in feeding the young. Over the whole nestling period males feed the young more frequently than do the females (females brood the young the first days after hatching). In primary nests male feeding frequencies are 69% and in secondary nests 58%, of the feeding frequencies of males in monogamous pairs. Females of polygynous males compensate for this reduction in male assistance by increasing their own feeding frequency (Temrin, 1988). These figures are based on studies of few nests, but further studies have shown the same tendency (H. Temrin, unpublished data). However, the reduced male assistance is not reflected in the breeding success for females of polygynous males (Table I).

In the Great Reed Warbler, male contribution in feeding the nestlings was about 50% in nests of monogamous pairs, 40% in primary nests and 4% in secondary and tertiary nests. In the majority of nests of secondary and tertiary females no male feeding was recorded, but of the nine secondary and tertiary nests observed, four males helped their secondary females (with shares of 11 to 22%). In another case the male's share of feeding was 50% (Dyrcz, 1986). It is notable that primary females had the significantly highest fledging success although they receive less male assistance than females in monogamous pairs (Table I).

In all three species, both primary and secondary females have reduced male assistance. This does not seem to influence fledgling success in the Wood Warbler or in the Great Reed Warbler. However, in species where the young are unable to thermoregulate the first days after hatching, reduced male assistance is likely to give the young a disadvantage when the weather conditions are bad.

4.1.2. The Predictability of Paternal Care

There is always a certain unpredictability when females select a mate. A male could be seriously hurt when defending his territory against intruders or be unlucky enough to be taken by a predator. But there are more predictable cases of paternal care. As presented above, females of certain mating status can expect paternal care with different probabilities. Early arriving females can be rather certain that they will settle with unmated males, but uncertain if their mates will later attract a female. Even a primary female of a polygynous male will, on average, have reduced male assistance, and since these females have no choice at the time of pair formation the only way for them to increase their reproductive success is to make it more difficult for the male to attract additional females. The longer she can prevent him from attracting a second female the better. The probability of attracting a second female will decrease, and if he still succeeds, a small or no overlap in time between the primary and secondary broods gives him more time to exclusively feed the young in the primary nest. In the Wood Warbler, females disturb mates who try to attract additional females in the primary territories, while males in secondary territories are rarely disturbed by females (Temrin, 1989). Females also behave aggressively towards other females in both Wood Warblers and Pied Flycatchers (Breiehagen and Slagsvold, 1988; Temrin, 1989). For late arriving females the choice is more accurate. If she succeeds in selecting an unmated male, there is almost no risk that he will become polygynously mated. In the Great Reed Warbler, Dyrcz (1986) observed violent clashes between primary and secondary females, although no special observations were made on female–female aggression.

One factor influencing the amount of male assistance for females of polygynous males is nest predation. If one of the two females of a male Wood Warbler has its nest preyed upon, the other female will have more male assistance than otherwise expected (Temrin and Jakobsson, 1988). But even when nest predation is not involved, secondary females may have exclusive assistance by their males in both Pied Flycatchers and Wood Warblers (Temrin and Jakobsson, 1988; Stenmark *et al.*, 1988). High variability in paternal care was also found by Muldal *et al.*, (1986) in the polygynous Red-winged Blackbird (*Agelaius phoeniceus*). A recent study (Whittingham, 1989) suggests that the amount of male assistance given by a polygynous male in the Red-winged Blackbird is influenced by the ability of a female to meet nestling food demands. Polygynous males also allocated parental effort on the basis of brood size and nestling age.

4.1.3. Could Females Be Compensated for Reduced Male Assistance?

There are two factors that may compensate for reduced male assistance in the three species discussed: easily accessible food, and protection against predation. When females lack male assistance, the accessibility of food is crucial for the prospects of raising the young on its own. In the Pied Flycatcher males defend only a small area around the nest hold (Haartman, 1956), which indicates that the foraging conditions are less likely to be limiting than the quality of the nest hole in this species, but an abundance of suitable food within easy reach should also be important for prospecting females (Askenmo, 1984). However, the single most important criterion for female choice in the Pied Flycatcher may be nest site quality (Alatalo *et al.*, 1986). In Pied Flycatchers breeding in natural cavities, nest predation was found to be by far the most important factor reducing breeding success (Alatalo and Lundberg, 1990). Based on nest hole properties, they were able to predict with 79% accuracy the nest as suffering from predation or being successful. The nest cavity as a protection against predation might be especially important for the Pied Flycatcher where an incubating female runs the risk of being taken by a predator herself (Haartman, 1971; Alatalo and Lundberg, 1990). But other factors of nest site quality might be important too. Even when predation was not involved, unassisted females with nest boxes of high attractiveness had the same fledging success as monogamous and primary females in boxes of medium or low attractiviness (Askenmo, 1984). Another important factor for nest site selection is how well protected the birds will be from adverse weather (Askenmo, 1984).

In both Wood Warblers and Great Reed Warblers nest predation was found to be the most important factor affecting breeding success (Urano, 1985; Dyrcz, 1986; Temrin and Jakobsson, 1988). In the Great Reed Warbler polygynous males could offer their females relatively abundant food supplies since nests of polygynous males were often situation near old trees and abundant shrub vegetation, which served as foraging grounds for the parents feeding their offspring (Dyrcz, 1986). It seems that in both the Pied Flycatcher and the Great Reed Warbler there is a variation of quality of resources that makes it possible for capable males to compensate secondary females for reduced assistance.

4.2. The Probability of Becoming a Secondary Female and Losing in Fitness

4.2.1. What Is the Probability of Becoming a Secondary Female?

In the Pied Flycatcher, the polygyny frequencies vary considerably between different studies. When nest box densities are close to the natural situation, the frequency of polygynous males is low (Curio, 1959) while an excess of nest boxes gives polygyny frequencies of 13 to 39% (Winkel and Winkel, 1984; Askenmo, 1977). In the extensive studies of Pied Flycatchers breeding in nest boxes by Haartman (1951), 13% of the females were mated with polygynous males, which gives a probability of 6.5% of becoming a secondary female. Haartman (1969) states that this low frequency may almost seem negligible, but it is not accidental since a majority of the males tried to obtain a second territory. Alatalo and Lundberg (1984) have shown how nest box densities influence the frequency of polygyny. When boxes are in "shortage" 6% of the females are secondary and when in "excess" 11 to 39% are secondary. In an area where Pied Flycatchers bred in natural cavities, 21 nests were found, and three of the females were secondary (Alatalo and Lundberg, 1984).

Is being a secondary female making a bad choice? Occasionally males may have two females within the same territory, and hence not all secondary females are unaware of their secondary status (Alatalo and Lundberg, 1984). As shown in Alatalo *et al.* (1986) mate choice is not random in the Pied Flycatcher; females use several cues of which nest site quality is the most important. Some of the females selecting males on these criteria may make good choices. Under natural conditions, nest predation is common in the Pied Flycatcher (Nilsson, 1984), which will substantially increase the frequency of females obtaining exclusive male assistance (Temrin and Jakobsson, 1988). Given these conditions, the frequency of secondary females is not the same as the frequency of females making a bad choice.

In the Wood Warbler almost 20% of the females are secondary. One-third of the secondary females were attracted to monoterritorial males and should be aware of their mating status at the time of pair formation. A high proportion of primary females had their nests preyed upon, giving almost half of the secondary females exclusive assistance by their males (Temrin and Jakobsson, 1988). In a study of the Great Reed Warbler (Dyrcz, 1986), there were 21 polygynous males of which four had "disrupted territories." And even in one of those four cases, the secondary female built the nest close to that of the primary female. Small territories were quite often found and Dyrcz suggests that deception in these cases seems very improbable and that the attractiveness of a territory could have been so strong that females choose to take on a secondary status. Hence, in both the Wood Warbler and the Great Reed Warbler, the frequency of secondary females does not seem to be correlated with the proportion of females making a bad choice.

4.2.2. Mating Options

There are two major objections that can be raised against comparing the fledging success of secondary females with that of "simultaneous" females of monogamous males as estimated from linear regression analysis (Davies, 1989). Firstly, secondary females may do worse than a recently settled female of a monogamous male simply because the best option available is much worse than the previous best option, and secondly, females of different status may have different reproductive capabilities.

In the Pied Flycatcher estimates from linear regression analysis suggest that secondary females suffer in fledging success, but we do not know much about their options. In a paper primarily concerned with mate choice in the Pied Flycatcher (Alatalo *et al.*, 1984) it is possible to look more closely at the actual options for females selecting already mated males. Alatalo *et al.* state that 15% (5 of 34) of the males stayed unmated and that unmated males were always in excess of unmated females. In three cases of females selecting already mated males, there were 3, 3, and 1 unmated male(s) in the population. One male who remained unmated that season appeared in all three cases. The other four unmated males, who were at hand when females chose already-mated males, settled in the area one or a few days before males became polygynously mated. Five males stayed unmated through the breeding season. Four of those were the last to arrive in the area, and they arrived when 28 of 30 males were mated. The three polygynous males were the first, the second, and the fifth male to arrive to the study area. Askenmo (1984) suggested that the early arriving males are able to occupy second nest holes before the last males to arrive have choosen their primary sites, and they may be more attractive as mates than unmated males. That secondary females are able to select good quality nest boxes is supported by the results of Askenmo (1984), who found that secondary boxes advertised without success by monogamous males tended to have lower attractiveness scores than boxes accepted by secondary females.

In the Wood Warbler secondary females seem to have the choice between unmated males on territories where Wood Warblers rarely breed or already mated males trying to attract females to outlying parts of their territory or in secondary territories at some distance from the first (Temrin, 1989). Their options thus seem limited. However, this could not explain all causes of polygyny, since early secondary females have several unmated males at hand that later attract females.

In the Great Reed Warbler the territories of unmated males were often situated unfavorably as regards accessibility to foraging grounds, and it was obvious to Dyrcz (1986) that in all such territories there were no good nest sites as few and scarce reeds would not hide the nest well. In a study of the Great Reed Warbler by Urano (1985) the bachelor males were the last to settle in the area, on average one month later than monogamous males.

4.3. The Cost of Assessing Male Mating Status

4.3.1. What Are the Costs of Postponing the Start of Breeding?

A decrease in fledging success in a population with time of breeding season has been used to explain why females do not have time to adequately assess the quality of males or territories (e.g., Alatalo *et al.*, 1981; Catchpole *et al.*, 1985; Whooten *et al.*, 1986). I will not go into the problem of why there is a decrease in fledging success over time in some species, but merely point out that whether a decrease in fledging success influences the behavior of a single female depends on the causes of the decrease. The decision of an individual female should not be influenced by a decrease in fledging success over time if the cause is that the condition or quality of females decrease with time of the season. If the cause is a decrease in the quality of males and territories or abundance of food, female opportunities will be influenced.

In the Pied Flycatcher there is no decrease in fledging success over time for secondary females ($b = 0.008$, $N = 70$; Alatalo and Lundberg, 1990). Thus, with the reduction in fledging success found in their study, it should always pay secondary females to leave their mates as soon as learning their mating status and as long as unmated males are available. However she may be eager to start breeding simply because both she and her offspring must have time to prepare themselves for migration. If she is an early female, there is also the possibility of renesting in cases of nest predation.

In both the Wood Warbler and the Great Reed Warbler there is no significant decrease in fledging success over time if nests preyed upon are included (Temrin and Jakobsson, 1988; Bensch *et al.*, unpublished data). In the Wood Warbler the family remains together two to three weeks after the young have left the nest, and during this time parents feed the young (Aschenbrenner, 1966; H. Temrin, personal observation). The young then have about two months to prepare themselves for migration.

4.3.2. Is It Possible for Females to Assess the Mating Status of Polyterritorial Males?

In both the Pied Flycatcher and the Wood Warbler, males in secondary territories return to their primary females frequently during the intense singing periods, making it possible for a human observer, at least, to separate most already-mated males from unmated males (Temrin *et al.*, 1984; Stenmark *et al.*, 1988; Temrin, 1989). There remains a possibility that mated males trying to attract females are able to change behavior when females are around, and that they then stay in their secondary territories or leave them less frequently. However, in both species, males seem to reveal their mating status less than 24 hours after attracting a second female when they visit their primary females (Stenmark *et al.*, 1988; Temrin, 1989).

Polyterritoriality is rare in the Great Reed Warbler. In the study of Dyrcz (1986) four cases of disrupted territories were found with an average distance of 112m. In the 13 compact territories of polygynous males the average distance between the first and second nest was about 28m, and in some cases did not exceed 10m. In the study of Catchpole *et al.* (1985) one case of successive polyterritoriality was found, but Catchpole *et al.* suggest that there is a reduction of visibility in reed beds which makes it possible to deceive secondary females of their mating status. However, the female Great Reed Warbler has a distinct call which facilitates the location of the nest site. This call is heard when their males try to attract females, and should be easy to locate for prospecting females (S. Bensch and D. Hasselquist, unpublished data).

5. CONCLUSIONS

There seems to be a more pronounced effect of reduced male assistance in the Pied Flycatcher than in Wood Warblers and Great Reed Warblers. Both primary females and secondary females of polygynous males in the Pied Flycatcher have a relatively low fledging success, while primary females in the Great Reed Warbler and secondary females in the Wood Warbler have a relatively high fledging success, although both categories of females, on average, suffer in male assistance in relation to females of monogamous males. That the nestlings are unable to thermoregulate the first days after hatching could give females with reduced assistance difficulties in raising the young on their own. In the Wood Warbler the young are fed by their parents for approximately two weeks after leaving the nests, which might strengthen any possible disadvantage in receiving reduced male assistance.

The conspicuous behavior of the first female in the Great Reed Warbler and the short distances between the primary and secondary nests makes deception unlikely as a general explanation of polygyny in this species. Even if the reed beds in which this warbler nests are dense the sound of the two females of a polygynous male should be possible to detect. There are however rare cases of polyterritoriality where it is possible that females are unable to assess the mating status of males.

Polyterritorial behavior may make it more difficult to assess mating status than in cases of monoterritoriality, and will be unfavorable for both females of a polygynous male; primary females will be unable to drive prospecting secondary females away and secondary females will be unable to include the cost of sharing the male when comparing different mating options. There might be advantages with polyterritoriality that could compensate for this; after finding a nest, predators have been found to increase the searching effort in the surroundings (e.g, Tinbergen *et al.*, 1967). With several nests in a territory this might confer a disadvantage for each female. Long distances between nests could also decrease the aggressive interactions between the two females of a polygynous male. However, so far no evidence that this is the case is available in polyterritorial species.

In the Wood Warbler there is no reduction in fledging success for secondary females and these females seem to make accurate choices of mates. We cannot rule out the possibility that some secondary females lose in fitness because of inability to compensate for reduced male assistance, resulting in lower survival rates and lower future fecundity for both female parents and their offspring (see Gustafsson and Sutherland, 1988). However, this explanation could only be valid in some cases, since almost half of the secondary females have exclusive assistance by their males because the primary nests had been preyed upon.

There are at least two alternative, but not mutually exclusive, explanations to reduced fledging success for secondary females in these studies of the Pied Flycatcher. The first is that the phenomenon results from the artificial situation with nest boxes: as a result of nest boxes being provided in excess a higher proportion of already mated males are able to take up secondary nest sites, and unmated young males are able to offer females high quality nest boxes. Late in the season, when suitable nest sites are scarce under natural conditions, a female offered a nest site over a certain threshold quality should make a quick choice since it is unlikely that she will encounter another nest cavity of similar quality; coyness could result in another female getting hold of the resource. Since nest predation is the single most important factor influencing breeding success under natural conditions, the effect of a high quality nest site is likely to compensate for a reduction in male assistance. In a nest box area, such a mechanism will not be adequate since high quality nest boxes are in excess and the single most important factor influencing breeding success is male assistance.

Secondly, secondary females of polyterritorial males do worse, on average, than primary females and females of monogamous males because of inability to distinguish unmated males from mated males trying to attract females. The probability of becoming a secondary female and actually losing in fitness is so low under natural conditions that it will not pay females to postpone the start of breeding in order to assess the male's mating status. The estimates of fledging success of "simultaneous" females has not take into account the mating options for late females. For late females the probability of encountering an already mated male trying to attract females is relatively high, and it might pay them to be more choosy than early females. However, as the frequency of mated males trying to become polygynously mated increases in the population, so does the cost of assessing mating status and of finding an unmated male.

A crucial factor affecting which behavior females should adopt is the cost of assessing mating status. This study shows that when the probability of becoming a secondary females is low, the number of days to find an unmated male and the decrease in fledging success when postponing the start of breeding substantially affected the minimum fledging success secondary females could have, if uninformed females should still be favored. Unfortunately, behavioral studies of female choice have not been made in the three species discussed, which make it even more difficult for us to estimate the number of days needed for a female to make an accurate choice of mate (but see Slagsvold *et al.*, 1988).

The reliability of estimates of the success for a "fast" strategy vs. a "coy" strategy are only as reliable as the data used. Data from naturally breeding Pied Flycatchers would increase our understanding of how competition for high quality nest cavities, and also nest predation, influence the relative breeding success of secondary females. By removing primary females in the Pied Flycatcher so that polygynous males could give secondary females exclusive assistance, we may be able to determine whether reduced male assistance is the only factor responsible for the relatively low fledging success of secondary females.

We also need to define the costs of postponing the start of breeding and study the mating options for females chosing already mated males. Behavioral studies of female mate choice will increase our knowledge of whether females are "coy" or "fast," and also tell us if early and late females differ in their behavior when selecting a mate.

6. APPENDIXES

6.1. Calculations of Reproductive Success for Uninformed Females (F_u) vs. Informed Females (F_i)[a]

For Wood Warblers (Temrin, 1989):

F_u = (0.60 × 4.3 × 1) + (0.20 × 3.7 × 0.99) + (0.20 × 4.9 × 0.95) = 4.24
F_i (6.6) = (0.60/ (0.60 + 0.20) × 4.3 × 1) + (0.20/ (0.60 + 0.20) × 3.7 × 0.99) − (6.6 × 0.049[b]) = 3.82
F_i (2) = 4.05 − (2 × 0.049) = 4.04

F_u (1–4) = (0.57 × 6.2 × 1) + (0.40 × 6.0 × 0.99) + (0.03 × 7 × 0.95) = 6.11
F_i (1–4) = (0.57/ (0.57 + 0.40) × 6.2 × 1) + (0.40/ (0.57 + 0.40) × 6.0 × 0.99) − (6.6 × 0.049) = 5.77
F_i (1–4) = 6.09 − (2 × 0.049) = 5.99

F_u (>12) = (0.27 × 5.5 × 1) + 0 + (0.73 × 5.8 × 0.95) = 5.51
F_i (>12) = (0.27/0.27 × 5.5 × 1) + 0 − (6.6 × 0.049) = 5.18
F_i (>12) = 5.5 − (2 × 0.049) = 5.40

For Pied Flycatchers (Alatalo and Lundberg, 1984):

F_u (early) = (0.70 × 5.7 × 1) + (0.30 × 4.5 × 0.99) = 5.33
F_i (early) = 5.33 − (6.6 × 0.058) = 4.95
F_i (early) = 5.33 − (2 × 0.058) = 5.21

F_u (late) = (0.50 × 4.6 × 1) + 0 + (0.50 × 3.5 × 0.95) = 3.96
F_i (late) = 4.6 − (6.6 × 0.058) = 4.22
F_i (late) = 4.6 − (2 × 0.058) = 4.48
F_i (late) = 4.6 − (10 × 0.058) = 4.02

For Great Reed Warblers:

Dyrcz (1986):

F_u = (0.68 × 2.6 × 1) + (0.16 × 4.0 × 0.99) + (0.16 × 2.1 × 0.95) = 2.72
F_i (6.6) = (0.68/ (0.68 + 0.16) × 2.6 × 1) + (0.16/ (0.68 + 0.16) × 4.0 × 0.99) − (6.6 × 0.023[c]) = 2.71
F_i (2) = 2.86 − (2 × 0.023) = 2.81

Catchpole *et al.* (1985):

F_u = (0.54 × 3.3 × 1) + (0.23 × 3.2 × 0.99) + (0.23 × 1.1 × 0.95) = 2.75
F_i (6.6) = (0.54/0.77 × 3.3 × 1) + (0.23/0.77 × 3.2 × 0.99) − (6.6 × 0.023) = 3.11
F_i (2) = 3.26 − (2 × 0.023) = 3.21

F_u (early) = (0.50 × 3.3 × 1) + (0.50 × 3.2 × 0.99) = 3.23
F_i (early) (early)= (0.50 × 3.3 × 1) + (0.50 × 3.2 × 0.99) − (6.6 × 0.023) = 3.08
F_i (early) = 3.23 − (2 × 0.023) = 3.18

F_u (late) $= (0.50 \times 3.3 \times 1) + (0.50 \times 1.1 \times 0.95) = 2.17$
F_i (late) $= 3.3 - (6.6 \times 0.023) = 3.15$
F_i (late) $= 3.3 - (2 \times 0.023) = 3.25$

Urano (1985):

$F_u = (0.50 \times 2.8$ RMU $1) + (0.25 \times 3.2 \times 0.99) + (0.25 \times 2.4 \times 0.95) = 2.76$
$F_i = (0.50/0.75 \times 2.8 \times 1) + (0.25/0.75 \times 3.2 \times 0.99) - (6.6 \times 0.023) = 2.77$
$F_i = 2.92 - (2 \times 0.023) = 2.88$

F_u (early) $= (0.50 \times 2.8 \times 1) + (0.50 \times 3.2 \times 0.99) = 2.98$
F_i (late) $= 2.98 - (6.6 \times 0.023) = 2.82$
F_i (late) $= 2.98 - (2 \times 0.023) = 2.93$

F_u (late) $= (0.50 \times 2.8 \times 1) + 0.5 \times 2.4 \times 0.95) = 2.54$
F_i (late) $= 2.8 - (6.6 \times 0.023) = 2.65$
F_i (late) $= 2.8 - (2$ RMU $0.023) = 2.75$

[a]See text for formulae for F_u and F_i. Days to assess mating status and find an unmated male are given in parentheses.

[b]When nests preyed upon are excluded there is a tendency for fledging success in primary females and females of monogamous males to decrease 0.049 young per day, as estimated from data in Temrin (1989).

[c]No data showing a decrease in fledging success over time have been presented in the three studies of the Great Reed Warbler. In a five-year study in Sweden, fledging success decreased 0.023 young per day for females of monogamous males (S. Bensch, D. Hasselquist, and U. Ottosson, unpublished data).

6.2. Calculations of the Maximum Number of Days a Female Could Use to Inform Herself about Mating Status and Find an Unmated Male, and Still Do Better Than Uninformed Females

The Pied Flycatcher study of Alatalo and Lundberg (1984):

$F_u = 4.7$
$F_i = (0.72/0.86 \times 5.1 \times 1) + (0.14/0.86 \times 4.1 \times 0.99) - (0.058 \times 3.97) = 4.7$

The Great Reed Warbler study of Catchpole *et al.* (1985):

$F_u = 2.75$
$F_i = (0.54/0.77 \times 3.3 \times 1) + (0.23/0.77 \times 3.2 \times 0.99) - (0.023 \times 22.2) = 2.75$

6.3. Examples of the Relationship between the Probability of Becoming a Secondary Female, the Fledging Success (N_2) of Secondary Females, and the Costs of Assessing the Mating Status of Males[a]

$F_u = (0.90 \times 5 \times 1) + (0.10 \times N_2 \times 0.95)$

$F_i = 5 - (6.6 \times 0.05) = 4.67$; $F_u = F_i$ when $N_2 = 1.79$
$F_i = 5 - (2 \times 0.05) = 4.90$; $F_u = F_i$ when $N_2 = 4.21$

F_u = (0.70 × 5 × 1) + (0.30 × N_2 × 0.95)
F_i = 5 − (6.6 × 0.05) = 4.67; $F_u = F_i$ when N_2 = 4.11
F_i = 5 − (2 × 0.05) = 4.90; $F_u = F_i$ when N_2 = 5.09

F_u = (0.70 × 5 × 1) + (0.30 × N_2 × 0.95)
F_i = 5 − (6.6 × 0.02) = 4.87; $F_u = F_i$ when N_2 = 4.81
F_i = 5 − (2 × 0.02) = 4.96; $F_u = F_i$ when N_2 = 5.09

[a]See text for formulae and definition of coefficients, except here $F_i = N_m$ (or N_1) − C_i for simplicity reasons.

ACKNOWLEDGMENTS. I thank Magnus Enquist, Olof Leimar, and Christer Wiklund for constructive comments on the manuscript.

REFERENCES

Alatalo, R. V., and Lundberg, A., 1984, Polyterritorial polygyny in the Pied Flycatcher *Ficedula hypoleuca*—evidence for the deception hypothesis, *Ann. Zool. Fennici* **21**:217–228.

Alatalo, R. V., and Lundberg, A., 1986, The sexy son hypothesis: Data from the Pied Flycatcher *Ficedula hypoleuca*, *Anim. Behav.* **34**:1454–1462.

Alatalo, R. V., and Lundberg, A., 1990, Polyterritorial polygyny in the Pied Flycatcher, in: *Advances in the Study of Behaviour*, (P. J. B. Slater, ed.), Academic Press, New York (in press).

Alatalo, R. V., Carlson, A., Lundberg, A., and Ulfstrand, S., 1981, The conflict between male polygamy and female monogamy: The case of the Pied Flycatcher *Ficedula hypoleuca*, *Am. Nat.* **117**:738–753.

Alatalo, R. V., Lundberg, A., and Ståhlbrandt, D., 1982, Why do Pied Flycatcher females mate with already-mated males? *Anim. Behav.* **30**:585–593.

Alatalo, R. V., Lundberg, A., and Ståhlbrandt, K., 1984, Female mate choice in the Pied Flycatcher, *Behav. Ecol. Sociobiol.* **14**:253–261.

Alatalo, R. V., Lundberg, A., and Glynn, C., 1986, Female Pied Flycatchers choose territory quality and not male characteristics, *Nature* **323**:152–153.

Alatalo, R. V., Gottlander, K., and Lundberg, A., 1988, Conflict or cooperation between parents in feeding nestlings in the Pied Flycatcher, *Ornis Scand.* **19**:31–34.

Askenmo, C., 1977, Some Aspects of the Reproduction Strategy of the Pied Flycatcher *Ficedula hypoleuca* (Pallas), Ph.D. thesis, University of Gothenburg, Gothenburg, Sweden.

Askenmo, C., 1984, Polygyny and nest site selection in the Pied Flycatcher, *Anim. Behav.* **32**:972–980.

Aschenbrenner, L., 1966, *Der Waldlaubsänger* (*Phylloscopus sibilatrix*), A. Ziemsen Verlag, Wittenberg Lutherstadt.

Breiehagen, T., and Slagsvold, T., 1988, Male Polyterritoriality and female—female aggression in the Pied Flycatcher *Ficedula hypoleuca*, *Anim. Behav.* **36**:604–605.

Catchpole, C., Leisler, B., and Winkler, H., 1985, Polygyny in the Great Reed Warbler *Acrocephalus arundinaceus*: A possible case of deception, *Behav. Ecol. Sociobiol.* **16**:285–291.

Curio, E., 1959, Verhaltenstudien am Trauerschnäpper, *Z. Tierpsychol.* Beiheft **3**:1–118.

Davies, N. B., 1989, Sexual conflict and the polygamy threshold, *Anim. Behav.* **38**:226–234.

Dawkins, R., 1976, *The Selfish Gene*, Oxford University Press, Oxford, England, 224 pp.

Dawkins, R., and Krebs, J. R., 1978, Animal signals: Information or manipulation? in: *Behavioural Ecology* (J. R. Krebs, and N. B. Davies, eds.), Blackwell Scientific Publications, Oxford, England, 282–309.

Dyrcz, A., 1986, Factors affecting facultative polygyny and breeding results in the Great Reed Warbler *Acrocephalus arundinaceus, J. Orn.* **127**:447–461.

Gustafsson, L., and Sutherland, W. J., 1988, The costs of reproduction in the Collared Flycatcher *Ficedula albicollis, Nature* **335**:813–815.

Haartman, L. von, 1951, Successive polygamy, *Behaviour* **3**:256–274.

Haartman, L. von, 1956, Territory in the Pied Flycatcher *Muscicapa hypoleuca, Ibis* **98**460–475.

Haartman, L. von, 1969, Nest-site and evolution of polygamy in European passerine birds, *Ornis Fenn.* **46**:1–12.

Haartman, L. von, 1971, Population Dynamics, in: *Avian Biology*, Vol. 1 (D. S. Farner and J. R. King, eds.), Academic Press, New York, pp. 391–459.

Muldal, A. M., Moffatt, J. D., and Robertson, R. J., 1986, Parental care of nestlings by male Red-winted Blackbirds, *Behav. Ecol. Sociobiol.* **19**:105–114.

Nilsson, S. G., 1984, Clutch size and breeding success of the Pied Flycatcher *Ficedula hypoleuca* in natural tree holes, *Ibis* **126**:407–410.

Orians, G. H., 1969, On the evolution of mating systems in birds and mammals, *Am. Nat.* **103**:589–603.

Slagsvold, T., Lifjeld, J. T., Stenmark, G., and Breiehagen, T., 1988, On the cost of searching for a mate in female Pied Flycatchers *Ficedula hypoleuca, Anim. Behav.* **36**:433–442.

Stenmark, G., Slagsvold, T., and Lifjeld, J. T., 1988, Polygyny in the Pied Flycatcher *Ficedula hypoleuca*: A test of the deception hypothesis, *Anim. Behav.* **36**:1646–1657.

Temrin, H., 1984, Why are some Wood Warbler (*Phylloscipus sibilatrix* males polyterritorial? *Ann. Zool. Fennici* **21**:243–247.

Temrin, H., 1988, *Polyterritorial Behaviour and Polygyny in the Wood Warbler Phylloscopus sibilatrix*, Ph.D. thesis, University of Stockholm, Stockholm, Sweden.

Temrin, H., 1989, Female pairing options in polyterritorial Wood Warblers *Phylloscopus sibilatrix, Anim. Behav.* **37**:579–586.

Temrin, H., and Jakobsson, S., 1989, Female reproductive success and nest predation in polyterritorial Wood Warblers (*Phylloscopus sibilatrix*), *Behav. Ecol. Sociobiol.* **23**:225–231.

Temrin, H., Mallner, Y., and Windén, M., 1984, Observations on polyterritoriality and singing behaviour in the Wood Warbler *Phylloscopus sibilatrix, Ornis scand.* **15**:67–72.

Tinbergen, N., Impekoven, M., and Frank, D., 1967, An experiment on spacing-out as a defense against predation, *Behaviour* **28**:307–321.

Urano, E., 1985, Polygyny and the breeding success of the Great Reed Warbler *Acrocephalus arundinaceus, Res. Popul. Ecol.* **27**:393–412.

Whittingham, L. A., 1989, An experimental study of paternal behavior in Red-winged Blackbirds, *Behav. Ecol. Sociobiol.* **25**:73–80.

Whootton, J. T., Bollinger, E. K., and Hibbard, C. J., 1986, Mating systems in homoge-

noeous habitas: The effects of female uncertainty, knowledge costs, and random settlement, *Am. Nat.* **128**:499–512.
Wickler, W., 1968, *Mimicry in Plants and Animals*, McGraw Hill, New York.
Winkel, W., and Winkel, D., 1984, Polygynie des Trauerschnäppers *Ficedula hypoleuca* am Westrand seines Areals in Mitteleuropa, *J. Ornithol.* **125**:1–14.

CHAPTER 7

AGE-SPECIFIC FORAGING PROFICIENCY IN BIRDS

JOSEPH M. WUNDERLE, JR.

1. INTRODUCTION

The fact that most birds exhibit high mortality during the juvenile period (Lack, 1954) suggests that this is a critical time in their life history. Much juvenile mortality may be attributed to a lack of proficiency in some aspects of behavior. For instance, juveniles must acquire skills in predator avoidance, foraging, and social interaction, particularly with competitors (Brown, 1975; Galef, 1976; Slater, 1983). Furthermore, they must learn to efficiently allocate time and energy to these activities so as to balance conflicting demands in reducing their vulnerability to both predation and starvation (Weathers and Sullivan, 1989). Thus the rate at which these behaviors develop is critical to the survival of juveniles and has implications for a variety of avian life history traits, particularly the extent of parental care (Ashmole and Tovar, 1968) and delayed maturation (Lack, 1968).

This chapter focuses on the factors which influence juvenile foraging skills, particularly the constraints which prevent juveniles from rapidly acquiring adult levels of foraging proficiency. Periods of high mortality, such as the juvenile period, provide strong opportunities for

---

JOSEPH M. WUNDERLE, JR. • Institute of Tropical Forestry, Southern Forest Experiment Station, United States Department of Agriculture Forest Service, Rio Piedras, Puerto Rico 00928-2500; and Department of Biology, University of Puerto Rico, Cayey, Puerto Rico 00633.

the action of natural selection (Arnold and Wade, 1984). Yet, despite potentially high selection pressures on the rate of behavioral development, juvenile proficiency often lags behind that of adults for a considerable period of time. This suggests that the rate of behavioral development is constrained by one or more factors. The factors constraining development are of particular interest for understanding the evolution of behavioral or morphological traits (Maynard Smith *et al.*, 1985; Irwin, 1988). It has been argued that more attention should be paid to the development of foraging skills, since this may be when selection acts more strongly on foraging performance (Zach and Smith, 1981).

Whereas the factors influencing behavioral ontogeny can conveniently be divided into internal and external factors, understanding their interaction and influence on behavior can be difficult (e.g., Hailman, 1967; S. M. Smith, 1983). For example, during the ontogeny of foraging behavior, improvements in foraging skills might be attributed to maturation of the nervous and muscular systems as well as learning (e.g., Cruze, 1935). Learning can improve a predator's ability to locate and identify appropriate prey (e.g., Kamil and Yoerg, 1982; Shettleworth, 1984), and also contribute to improved neuromuscular coordination necessary for prey capture or handling. These factors all depend on at least some level of interaction with the environment. For instance, the physical environment, such as light levels or weather conditions, can influence foraging proficiency of the young (e.g., Carl, 1987). The biotic environment, such as the distribution and abundance of prey, can also affect the efficiency with which the young forage (e.g., Sutherland *et al.*, 1986). Social factors, such as observational learning and local enhancement from experienced adults, further contribute to the improvement of juvenile foraging skills (e.g., Palameta and Lefebvre, 1985) in contrast to aggression from dominant adults which can hinder juvenile foraging proficiency (e.g., Goss-Custard *et al.*, 1987a, b, c). The identification of these factors, their mode of interaction, and assessment of their relative importance during the development of behavior is the goal of those studying behavioral ontogeny, and is the approach taken in this chapter.

This review does not attempt to examine the ontogeny of foraging through all phases of development, but rather is limited to the period when the young are capable of feeding independently and are free from all parental care. At this time the juveniles are free-flying and have obtained most of the morphological and physiological traits of the parents. During this stage most morphological growth is complete and the juveniles have usually obtained a mass equivalent to that of a mature adult, yet their feeding skills are not equivalent to those of other adults. The weakness of this approach is that it ignores the early stages of behavioral development upon which a juvenile bird's later foraging proficiency might depend. That early developmental stages are important was emphasized by Hailman (1982) who notes that the current behavioral phenotype of an organism is dependent upon previous behavioral phenotypes as well as their interaction with the genotype and environment. Although not as thoroughly examined, reference will be made therefore to appropriate studies on earlier stages of development. Throughout this review, I use the term juvenile to indicate young individuals who have not reached sexual maturity. Thus, juvenile passerines are individuals in their first year of life whereas juvenile nonpasserines may be in their first, second, or third year (e.g., gulls).

If juveniles are less proficient foragers than adults they may compensate in a variety of ways—the most obvious is by foraging longer. Juveniles might also forage in different habitats or patches, use different search patterns, consume different sizes or types of food, or steal from more adept foragers. The compensation options available to juvenile foragers are of considerable interest in light of recent theoretical work on the alternative strategies applied to individual differences in behavior (e.g., Maynard Smith, 1982; Partridge and Green, 1985).

One valuable theoretical approach to foraging behavior is to assume that foragers make a series of hierarchical decisions (Gass and Montogomerie, 1981; Orians, 1981). This hierarchical decisionmaking process is initiated when a forager selects a general habitat in which to feed. Next, it selects a patch in which to feed. Once in a patch the forager must choose a particular search pattern, followed by a decision as to which food items to consume. In some instances, a decision is required as to which capture technique is to be used to subdue the prey. Finally, for certain food types, the forager must choose a method for handling the food as it is consumed. This hierarchical approach to studying foraging behavior has proven useful and is fundamental to the organization of optimal foraging theory (see reviews in Krebs *et al.*, 1983; Krebs and McCleery, 1984; Pyke, 1984). Age-related differences can occur in each of these foraging stages due to a young bird's lack of proficiency in a specific stage and/or because the young forager compensates for its lack of skills at another stage.

Thus this review focuses on the mechanics of juvenile foraging proficiency in an effort to:

1. Identify the specific stage or stages during foraging which account for the reduced foraging proficiency of juveniles relative to adults during a typical foraging sequence (i.e., the progression involving foraging site, search patterns, diet choice, prey capture, and prey handling).
2. Determine the factor or factors which constrain juvenile forag-

ing proficiency (i.e., maturation, learning, adult dominance, and predation).

3. Determine how juveniles compensate for their lack of foraging proficiency (i.e., more time feeding, searching in different habitats, selecting different prey, relying on observational learning, and piracy).

2. THE NATURE OF THE LITERATURE

This review is based on a relatively specialized literature that is subject to certain characteristics and limitations. The review is of studies in which the foraging behavior of self-feeding young (juveniles) is compared to that of older individuals (adults), as summarized in Section 6. The predominant interest of this literature, at least in its earliest years, has been to document age-related differences in foraging proficiency in support of Lack's (1968) delayed breeding hypothesis (e.g., Orians, 1969; Recher and Recher, 1969). Thus much of the initial focus was to simply demonstrate that juveniles were more inept foragers than adults, rather than uncover how, where, and why juveniles were less proficient. A few studies on the ontogeny of foraging behavior have also contributed to this subject by identifying factors controlling foraging development (e.g., Norton-Griffiths, 1967, 1968, 1969). Only recently have workers tried to identify how juvenile foragers differ from adults, determine the consequences for juveniles, and uncover how juveniles compensate for their less efficient foraging abilities (e.g., Goss-Custard *et al.*, 1982a, b, c; Goss-Custard and Durell, 1983, 1987a, b, c).

Age-related differences have usually been quantified in three different ways: foraging success (number of prey items captured per capture attempt); foraging rate (number of prey items captured per unit time); and interfood time interval (time interval between capture of one prey item and the next). Each of these measures incorporates different components of the foraging sequence. Foraging success measures the ability to capture prey once it has been located; it includes the ability to recognize and select capturable prey as well as capture ability, but it does not include searching or handling abilities. Foraging rate and interfood time interval both quantify search, recognition, selection, capture, and handling abilities together. Unfortunately, these measures do not allow quantification of proficiency at each individual stage of a foraging sequence (i.e., search only, recognition only, etc.). Therefore, some workers have measured time used in each stage (e.g., search time, handling time, etc.) to quantify age differences (Sutherland *et al.*, 1986).

However, most studies have not identified either at what stage of the foraging sequence age-related foraging differences are greatest or the relative rates of development of foraging abilities.

Approximately 90% of the field studies on age-related foraging proficiency have not involved color-banded individuals so that the actual rate of development of foraging proficiency cannot be determined for most species. Although many workers have suggested that juvenile foraging improves with time, their observations are based on unbanded birds at different times of the year. Such apparent improvement might simply result from the interim mortality of less efficient individuals (cf. Orians, 1969). This explanation could also apply to those species in which different age classes can be distinguished on the basis of plumage (e.g., gulls) and in which workers have documented improved proficiency with advancing age class of unbanded birds. Thus in the absence of marked individuals it is difficult to determine at what age juvenile birds obtain adult levels of foraging proficiency and the rate of its development.

Although studies of age-related foraging proficiency have been made with a variety of species, they do not accurately represent the diversity of avian taxa and the full spectrum of foraging methods (Section 6). For instance, aquatic species are well represented with studies of 26 species of Charadriiformes, six species of Ciconiiformes, three species of Pelecaniformes, two species of Gruiformes, and one species of Falconiformes. Most of the aquatic species fed primarily or partially on fish (62% of aquatic species), and their most frequent hunting technique, plunge diving, was studied in pelicans, Ospreys (*Pandion haliaetus*), gulls, and terns (16 studies). Gulls have been the most frequently studied group, in which 23 studies have documented age-related differences in a variety of foraging situations. The predominance of certain taxa and foraging modes in studies of age-related differences in foraging proficiency is undoubtedly due to the ease of observing foraging sequences and the presence of age-related plumage differences. For example, Falconiformes are poorly represented in published studies, most likely because of the difficulty of observing foraging sequences. Also, studies of age-related foraging differences in Passeriformes are underrepresented. Furthermore, many avian orders simply have not been studied for age-related foraging proficiency. Therefore in light of the skewed representation of avian taxa and foraging modes, caution must be exercised in developing generalities regarding age differences in avian foraging proficiency. Nonetheless, the available studies demonstrating that juveniles are less proficient foragers than adults, regardless of diet, supports the view that this is a general phenomenon.

The rates at which juveniles obtain adult levels of proficiency, however, varies widely.

3. FORAGING COMPONENTS

3.1. Foraging Site

Habitat selection in birds has been relatively well studied (for a recent review see Cody, 1985). While relatively few studies have focused on the ontogeny of habitat preferences, their results demonstrate that the development of habitat preferences involves a complex interaction of a diversity of internal and external factors, such as genetic predispositions, sensory physiology, imprinting, associative or trial-and-error learning, and social interactions (reviewed in Klopfer and Ganzhorn, 1985). Juveniles are often less specific in their habitat preferences than older birds (Bairlein, 1981) suggesting that experience (Partridge, 1979) or other factors influence their later choices. The development of foraging site preferences specifically has been found to involve a complex interaction of innate bias, exploration, and learning (Greenberg, 1984, 1987).

The selection of a habitat is one of the first decisions faced by a forager followed by the selection of a patch or patches in which to feed. However, what constitutes a habitat or a patch will vary from species to species, and the effective size of a habitat or patch will vary with the size and mobility of a forager. Therefore the problem of defining either a habitat or a patch is a problem of ecological grain, which often cannot be easily divided into a simple dichotomy (Emlen, 1973). Many of the workers studying age-related differences in foraging seem to have used the term "habitat" or "patch" indiscriminately. Because of the difficulty of distinguishing between habitat and patch selection by foragers, I will focus this discussion primarily on patch selection with occassional reference to habitat selection studies. Although adults and juveniles might frequently use different habitats, the factors responsible for age-related differences in habitat use are frequently the same as those accounting for age-related differences in patch use.

Most studies of patch selection assume that a forager's decisions are based on the energetic return realized from feeding in a patch per unit effort (e.g., Royama, 1970; MacArthur, 1972; Pyke *et al.*, 1977; Krebs and McCleary, 1984). Although, other factors, such as nutrition, competition, and predation may interfere with this scheme (e.g., Werner, 1976; Clark, 1980), the majority of studies indicate that adult foragers attempt to feed in the most profitable patches available (e.g., Royama, 1970; Smith and Dawkins, 1971; Smith and Sweatman, 1974; Wakeley, 1978; Pulliam, 1980). This might also be true for juveniles, although a variety of factors might limit their use of the most profitable patches.

Studies on species as varied as pelicans, cormorants, oystercatchers, gulls, herons, ospreys, kites, parrots, titmice, and warblers have found age-related differences in habitat or patch use (e.g., Moyle, 1966; Ficken and Ficken, 1967; Howe, 1974; Davis, 1975; Morrison *et al.*, 1978; Goss-Custard *et al.*, 1982b; Burger and Gochfeld, 1983; Brandt, 1984; Goss-Custard and Durell, 1983; Bourne, 1985; Draulins and Van Vessem, 1985; Magrath and Lill, 1985; Maccarone, 1987; Goss-Custard and Durell, 1987a, b, c; Gustafsson, 1988; East, 1988; Edwards, 1988, 1989b; Hesp and Barnard, 1989). These studies have suggested several explanations which might account for age-related differences in foraging sites, such as: (1) competition in which dominant adults displace juveniles from certain foraging sites; (2) inability of juveniles to accurately evaluate patch differences; (3) diet differences arising from lack of juvenile proficiency for searching, capturing or handling prey; and (4) different nutritional requirements. Explanations 1 and 2 represent developmental constraints in which juveniles have not yet obtained adult levels of proficiency and generally decrease survival of juvenile birds. Explanations 3 and 4 generally have a positive effect on juvenile survival in that they compensate for a lack of juvenile skills or provide certain nutrients. It is conceivable that more than one factor might be responsible for a species' age differences.

Competition in which dominant adults displace subordinate juveniles from the most profitable patches is the most frequently cited reason for age-related differences in foraging site use. For example, Glaucous-winged Gulls (*Larus glaucescens*) in immature plumage were forced by aggressive adults to feed in suboptimal habitats while foraging along a salmon stream (Moyle, 1966). During a two year study, Maccarone (1987) found that adult and juvenile European Starlings (*Sturnus vulgaris*) used available foraging substrates differently. Adults were more numerous than expected in managed grassland both years, and they were more abundant than expected at human-assisted food sources in one year. Adult aggression often prevented juveniles from using these habitats. Juveniles were more numerous in unmanaged grassland and in soybean and wheat fields both years, and in agricultural fields other than soybean and wheat in one year. Maccarone suggested that juveniles might reduce competition with adults on managed grassland by using agricultural fields and unmanaged grassland, in which adults were less common. Another example occurs in Coal Tits (*Parus ater*) on the Baltic island of Gotland where juveniles are more generalized foragers than adults as illustrated by their tendency to forage on both the outer and inner parts of trees, whereas adults use the central part. Gustafsson (1988) attributed this difference to adult competition for food and noted that as juveniles became older they shifted their foraging to the central parts of the tree. A similar difference between adult and juvenile Willow Tits (*P. atricapillus*) was found by Ekman and Askenmo (1984) who attributed it to dominant adults trying to avoid predators rather than a response to food supplies. Ekman (1987) notes that these two explanations are not mutually exclusive, because foraging in a safe locale will reduce vigilance thereby allowing more time for feeding. In another case, Magrath and Lill (1985) noted that since juvenile Crimson Rosellas (*Platycercus elegans*) were behaviorally subordinate to adults at artificial feeding stations, juveniles may be excluded from the best habitat and thus have to spend more time feeding than adults.

Frequently, it is both the subordinate status of young birds and their lack of foraging proficiency which results in differential patch use. Unfortunately, it is often difficult to disentangle the separate effects of foraging skill and interference on the intake rate of young birds (Draulines and Van Vessem, 1985; East, 1988). However, this was achieved by Goss-Custard and his associates in a detailed series of studies on the interaction of aggression and juvenile foraging proficiency in different patches or habitats by Eurasian Oystercatchers (*Haematopus ostralegus*) (Goss-Custard *et al.*, 1982a, b; Goss-Custard and Durell, 1983; Goss-Custard and Durell, 1987a, b, c). Their work with color-banded birds nicely demonstrated that adults (> 4 years of age) displaced immatures (2–4 years of age) from the preferred mussel beds. When adults were present, the immatures moved to low ranked beds and had a tendency to change mussel beds more frequently than adults. However, once the adults left in the spring to breed, immatures foraged in the high ranked mussel beds, until the adults returned and displaced them again in the autumn. As the birds matured, the mean rank of the bed on which they fed in winter increased, while their status and competitive ability increased. Most adults seen at least five times in one year occurred on only one habitat and ate only one kind of prey. In contrast, immature birds were often seen on two or three habitats, and consumed two prey species more often than one. At high water, immatures occurred more frequently than adults in fields and foraged on more easily obtainable prey. Immature oystercatchers foraging in fields during mild winters have greater mortality because of accidents and a wider array of parasites, while during cold winters the fields are unavailable to them. All these factors contributed to the higher mortality rates of the younger birds (Heppleston, 1971; Goss-Custard *et al.*, 1982c).

It is apparent from the studies on age-related differences in patch or habitat use, where competition is believed to be important, that adult aggression may force juveniles into suboptimal habitats and/or require juveniles to feed in a wider range of foraging sites, both of which could require juveniles to feed on different prey types. By forcing juveniles into suboptimal patches, adults may retard the rate at which juveniles obtain foraging proficiency on the most profitable prey types. This may be the case in immature Eurasian Oystercatchers which begin to eat mussels only when they are mature enough to compete with adults on the mussel beds (Goss-Custard and Durell, 1983).

Under certain conditions, juveniles may compensate for their lack of skill in finding, capturing, or handling prey by feeding in sites where their lack of skill is less likely to hinder them, while older birds forage in sites requiring more skill. For instance, young gulls often congregate on dumps, presumably because food is either easier to find or capture than when using more traditional sites, which may require learning and practice (Schreiber, 1968; Spaans, 1971; Cooke and Ross, 1972). The higher proportion of juvenile European Starlings feeding in cherry orchards has been attributed to their preference for cherries as a result of their inability to capture the larger insects preferred by adults feeding elsewhere (Stevens, 1985). Juvenile Olivaceous Cormorants (*Phalacrocorax olivaceus*) had higher prey capture rates in an area where juveniles outnumbered adults than in an area where adults were more numerous, suggesting that juveniles hunted in areas where they were most successful (Morrison *et al.*, 1978). Burger and Gochfeld (1983) found that young Laughing Gulls (*Larus atricilla*) concentrated their feeding in areas where the foraging task was easiest and where their feeding success approached that of adults. In habitats where food was readily available (e.g., garbage and fish offal) the differences between age classes in foraging success was negligible. Thus it appears that young Laughing Gulls compensate for their less developed foraging skills by concentrating their foraging in specific areas. In Brown Pelicans (*Pelecanus occidentalis*), prey density had the largest effect on patch use by both adults and juveniles (Brandt, 1984). However, diving proficiency (success rate) had the least influence on adult patch use, since adult diving proficiency varied little between patches. In contrast, juvenile diving proficiency was quite variable between patches and thus together with prey density had a strong effect upon juvenile patch use.

Another example of age-related differences in patch use resulting from compensatory site selection by unskilled juvenile foragers has been described by Hesp and Barnard (1989). They found that young Common Black-headed Gulls (*Larus ridibundus*) were less effective at stealing earthworms from Northern Lapwings (*Vanellus vanellus*) than were adults, and predicted that juveniles should opt for alternatives to kleptoparasitism when their efficiency at the latter is low and if alternatives offer potentially greater intake rates. As predicted, they found a higher proportion of juveniles on plowed or stubble fields and in flocks following plows than in flocks ("patches") of lapwings. Also, adults were more abundant in high density lapwing flocks than low density flocks, where juveniles were occasionally found. Presumably because juveniles lacked many of the skills of adults in kleptoparasitism, they were proportionately less abundant in lapwing flocks and more abundant in plowed/stubble fields. This is consistent with the hypothesis that juveniles are selecting the most profitable patches in relation to their foraging proficiency.

The last factor which can explain age-related differences in patch use results from the inability of juveniles to discriminate between the most profitable patches utilized by adults. In this instance, juveniles cannot compensate for their lack of foraging abilities because they are unable to differentiate between patches. For example, Edwards (1988) showed that adult Ospreys foraging during the postfledgling period exhibited a statistical preference for the littoral habitat which had the greatest fish abundance, rather than the nearby limnetic habitat. In contrast, juveniles foraging at the same time did not display any habitat preference, but rather used littoral and limnetic habitats in relation to their availability, suggesting to Edwards (1989b) that juveniles had not yet obtained the ability to recognize habitat differences in prey abundance. Unfortunately, such demonstrations are unusual since very few studies have sampled the prey density or abundance in different patches while monitoring foraging by different age classes, and thus the current literature is inadequate to assess how often juveniles fail to accurately select the most profitable patches.

Although age-related differences in patch use by Brown Pelicans were found by Brandt (1984), none could be attributed to the inability of juveniles to select profitable patches. Both adults and juveniles were strongly influenced by the foraging efforts of the other; adults tended to join juveniles foraging in a patch and vice versa. Interestingly, this attraction to other foragers (local enhancement) augmented the foraging success only of juveniles. In this case, local enhancement which contributed to juvenile diving success, may have been the factor allowing juveniles to achieve "adult" levels of patch selection proficiency. Porter and Sealy (1982) found that juvenile Glaucous-winged and California Gulls (*Larus californicus*) did not initiate feeding flocks, were less adept at finding food, and had a lower feeding success rate. One of the ways in which juveniles compensated for these inadequacies was by joining actively feeding groups of birds. In addition, juveniles were more likely than adults to approach gull models floating in a bay. Thus local enhancement may be one way in which juveniles compensate for their inability to select profitable patches.

It is important to note that age-related differences in habitat or patch use have not always been found, even when the birds showed age-related differences in foraging proficiency (Puttick, 1979; Gochfeld and Burger, 1984; Greig-Smith, 1985; Burger and Gochfeld, 1986; Bildstein, 1987). These studies suggest that specific, but as yet unidentified, conditions are necessary to permit differential habitat use by adults and juveniles.

Finally, although we can frequently document which habitats or patches are used by adult and juvenile foragers, no field workers have yet documented the search time necessary for a bird to locate these habitats or patches. Burger and Gochfeld (1983) noted this problem and recommended that "habitat search time" be added to Schoener's (1971) concept of "search time" to more accurately reflect the time a bird searches for a given food item. If and when "habitat search time" is measured in the field (possibly with radio telemetry?) it would seem likely that juveniles will be less proficient than adults.

3.2. Search Methods and Patterns

Both the pattern and method of searching by foragers in response to variation in resource density can affect net energy intake (e.g., Tinbergen *et al.*, 1967; Croze, 1970; Cody, 1971; J. N. M. Smith, 1974a, b; Charnov *et al.*, 1976). It should not be surprising that adults and juveniles differ in their search methods. Unfortunately quantification of seraching methods in the field is very difficult and only a few field studies have documented age-related differences. A common searching difference detected between adults and juveniles is in the speed with which foragers move through a patch. For example, adult American Robins (*Turdus migratorius*) tend to search more slowly (steps/min and steps/bout) and cover less area than do juveniles, but as a consequence are more successful than juveniles (Gochfeld and Burger, 1984). Similarly, adult (> 2 years) Herring Gulls (*Larus argentatus*) tend to walk more slowly than the less successful first and second year gulls foraging in a dump (Greig *et al.* 1983). Verbeek (1977a) also noted that adults remove more objects in their search for food in dumps and commented that juvenile gulls may find it easier to steal because they have not yet learned to dig by removing objects. However, in some instances juveniles appear to put more time and effort into the searching, such as in Northwestern Crows (*Corvus caurinus*) in which juveniles spend more time searching for clams in mud banks than do adults (Richardson and Verbeek, 1987).

One of the few studies to quantify differences in search patterns is that of Wunderle and Lodge (1988), which compared the foraging abilities of naive, hand-reared juvenile Bananaquits (*Coereba flaveola*) to those of captive wild adults foraging. The artificial patch contained a random array of either open or closed flowers. Foragers removed the tops from covered flowers in order to feed, thereby marking all visited flowers. Open flowers provided no clues as to previous visitation. Adults searched the patch more thoroughly and made fewer flower revisits during each foraging bout than did juveniles. Flower revisitations, within a foraging bout, by juveniles were more than expected by change while those made by adults were fewer than expected by chance. This was true for both open and covered flower types. The increase in foraging proficiency with age involved several different characteristics. On open flowers, adults and juveniles differed in the frequency of their patch departures and returns, and in their pattern of turning in the patch, thus resulting in a more complete search of the patch by adults. On covered flowers, however, the two age classes differed primarily in their pattern of turning, and adults made a more complete search of the patch. Both age classes used the presence or absence of flower covers as a clue to previous visitation. It is likely that adults use their memory to reduce their chance of revisiting flowers during a foraging sequence. The importance of memory is suggested by the departures of adults from depleted areas of the patch and their movement to unvisited areas when they foraged on open flowers (i.e., flowers with no clues indicating previous visitation). It is likely that a particular pattern of the random array of flowers in one section of the patch is remembered by the departing adult and thus avoided on its return. With experience, juveniles may associate memory and avoidance of previously visited flowers with increased energy intake per foraging effort. Thus young Bananaquits may improve their searching abilities by learning to use their memory by a trial-and-error process to reduce flower revisitations.

Recent studies on search strategies and spatial learning have indicated that different species may show either a preference for locations where they have just found food (win-stay) or avoidance of those locations (win-shift). A major factor influencing win-shift or win-stay learning may be whether the resource is depleted on a single visit (Olton *et al.*, 1981). For example, if a food resource is easily depleted during a single visit, a forager would do best to shift to another locality rather than revisit the original site. This win-shift pattern is common among adult nectarivores, which rarely revisit flowers during a foraging bout (Gill and Wolf, 1977; Kamil, 1978). However, if the food resource is not exhausted during a visit it would be advantageous for the forager to return to the same location to feed. The win-stay pattern has been found to occur in titmice (*Parus* sp., Smith and Sweatman, 1974) and adult Red-winged Blackbirds (*Agelaius phoeniceus*, Beauchamp *et al.*, 1987).

Laboratory studies have demonstrated an age-related effect in controlling the relative balance of win-stay and win-shift responses, which can influence the foraging proficiency of juveniles. For example, Beauchamp *et al.* (1987) conducted experiments with Red-winged Blackbirds in a three-choice maze, where individual foragers were allowed to either consume all the food from the first visited site (depletion condition) or only some of it (non-depletion condition). After a bird's first choice, it was subsequently re-tested to determine if it returned to the previously visited site (win-stay) or whether it chose an alternate site (win-shift). Under the depletion condition, both adults and juveniles were more likely to show the win-shift response. However, under the non-depletion condition adults showed a win-stay response in contrast to juveniles which responded randomly. Thus in this case juvenile blackbirds must learn the win-stay response to efficiently feed in non-depleting patches.

In contrast, juveniles of the nectarivorous Bananaquit must learn the win-shift response to prevent revisitation to recently emptied flowers (Wunderle and Soto-Martinez, 1987). Laboratory experiments with artificial flowers demonstrated that wild-caught adults were quick to learn win-shift tasks, but had difficulty learning to return to flowers in a position just visited (win-stay). However, naive juveniles learned win-shift and win-stay problems with equal facility. This juvenile learning flexibility (i.e., no predetermined win-shift response) may account for their relatively high level of flower revisitations, which reduces their foraging efficiency (Wunderle and O'Brien, 1985; Wunderle and Lodge, 1988).

In summary, only a few lab studies have examined adult-juvenile differences in searching behavior in detail, while some field studies suggest that differences exist. Obviously, age-related differences in search behavior should be studied for more species before conclusions can be made regarding the factors controlling and constraining the development of this behavior. However, the existing evidence suggests that some of the important factors in the development of searching behavior are learning via association and trial-and-error.

3.3. Food Recognition and Selection

The development of food recognition by young birds involves an intricate interplay of learning with genetic predispositions. During normal development young birds initially show curiosity and exploratory behavior by pecking at both appropriate and inappropriate items, but with time concentrate their pecking on edible items (e.g., Hogan-Warburg and Hogan, 1981; Davies and Green, 1976; Moreno 1984). Trial-and-error plays an important role in the development of food recognition with a complex interaction between avoidance and acceptance learning, as elegantly demonstrated by experimentation with domestic chicks (e.g., Hogan, 1973a, b, 1977a, b; Hale and Green 1979, 1988).

Mueller (1974) summarized the theories concerning prey recognition in birds and mentioned three hypotheses: (1) juvenile learning is from their parents; (2) juvenile learning is from personal experience; (3) juvenile learning is mostly independent of personal experience and is thus primarily genetically based. Each of these hypotheses is likely to be correct depending upon the species (S. M. Smith, 1983). For example, learning from the parents and personal experience is probably important for many precocial species (e.g., Turner, 1964). While for other species such as American Kestrels (*Falco sparverius*) and Loggerhead Shrikes (*Lanius ludovicianus*), parental teaching is likely to be unimportant (Mueller, 1974; S. M. Smith, 1972, 1973). S. M. Smith (1983) notes that for a variety of species there is strong evidence for a genetic basis in the development of food recognition, which may allow for its rapid development.

After the young have achieved independence from their parents, age-related differences may occur in prey recognition and selection producing differences in diet. These two stages of foraging can differ with age because: (1) juveniles are less proficient than adults at prey recognition and selection; (2) juveniles are less skilled at another foraging stage (i.e., capture or handling) and thus compensate by selecting a diet different than that utilized by adults; (3) juveniles have different nutrient requirements than adults (e.g., Savory, 1977; Stokkan and Steen, 1980); (4) adult dominance influences juvenile diet selection by preventing juveniles from selecting the most profitable prey types (e.g., Goss-Custard and Durell, 1983; Amat and Aguilera, 1990). This discussion will focus primarily on the age-related proficiency differences (1 and 2 above) in prey recognition and selection when adults and juveniles are foraging in the same locality. Cases of age-related diet differences arising from differences in foraging site were discussed previously.

Juveniles may be less proficient at recognizing suitable prey, as a result of their inexperience. For example, Mueller and Berger (1970) found that traps baited with sparrows captured more adult Sharp-shinned Hawks (*Accipiter striatus*) than juveniles, while traps baited with pigeons took more juveniles than adults. Mueller and Berger concluded that the adult Sharp-shinned Hawks were less likely to attack inappropriately large prey than were juveniles. Juvenile Royal Terns (*Sterna maxima*) were more likely to be caught on fishooks than were adults whose ability to recognize capturable prey was an important factor in their fishing success (Buckley and Buckley, 1974). Although naive Pinyon Jays (*Gymnorhinus cyanocephalus*) showed an immediate preference for pinyon seeds over other objects, their ability to distinguish between good and bad seeds (endosperm absent) improved with learning (Ligon and Martin, 1974). Naive Red-winged Blackbirds attacked aposematically colored stinkbugs at the same rate as they attacked mealworms (Alcock, 1973). However, as the young blackbirds obtained experience with the distasteful stinkbugs they were less likely to attack them. This, of course, is an important assumption for the evolution of aposematic coloration and mimicry. Numerous studies have demonstrated that young birds learn to avoid unpalatable prey species, although some species appear to have inborn recognition of dangerous prey (for a review see S. M. Smith, 1983).

Inappropriate objects might also be attacked by young foragers, as demonstrated by juvenile Gray Herons (*Ardea cinerea*) which frequently seized inanimate objects (such as floating feathers and sticks), items adults rarely seized (Cook, 1978). Juvenile Green-backed Herons (*Butorides striatus*) used more inappropriate objects (plant material) while bait fishing than did adults which relied more on animal baits that were more effective at attracting fish (Higuchi, 1986). For kleptoparasites, recognition of an appropriate "victim" from which to steal probably requires experience which juveniles may lack. For example, adult Laughing Gulls preferentially attacked juvenile pelicans, presumably because they were easier to exploit than adult pelicans, while yearling gulls showed no preference (Carroll and Cramer, 1985).

To accurately recognize prey, juvenile birds may need to develop search images (Dawkins, 1971), which may initially be imperfectly retained. This may occur in Black-winged Stilts (*Himantopus himantopus*) studied by Espin *et al.* (1983) who suggest that inaccurate prey recognition might contribute to the significantly higher number of intention movements made by juvenile stilts. In Royal Terns, juveniles repeatedly circled back and forth over an area, making many intention movements to dive and often abandoning dives close to the surface, behaviors rarely observed in adults (Buckley and Buckley, 1974). The higher incidence of intention movements by juveniles, suggesting imperfect search image formation, has also been found in several gull species (Porter and Sealy, 1982; MacLean, 1986).

Juveniles might compensate for their inability to accurately recognize suitable prey items by observing other foragers (Galef, 1976). This may take the form of social facilitation, in which one of the behaviors already in an animal's repertoire is released by the presence of another doing that behavior (Thorpe, 1963). Or, as discussed earlier, there may be local enhancement in which an individual is attracted to a novel environmental stimulus by the presence of another forager and then achieves a food-finding ability by individual learning (Thorpe, 1963). A naive observer copying an experienced individual's actions regardless of its outcome is "blind" imitation or "mindless" copying (Gould, 1982). Each of these phenomena could cause a naive forager to have a higher probability or rate of learning after witnessing an experienced forager. Furthermore, studies have demonstrated that naive birds will frequently adopt the diet of the experienced model (e.g., Klopfer, 1959; Murton, 1971; Partridge, 1976; Palameta and Lefebvre, 1985). In this way juveniles might quickly adopt the diets of adults (Richardson and Verbeek, 1987). Even interactions between naive siblings can result in diet similarities (Edwards, 1989b).

For a predator to select an optimal diet, it should rank prey types according to their profitability (energy yield per unit handling time) and their overall abundance. When the most profitable prey types are abundant the predator should specialize; as the abundance of the most profitable prey declines, the predator should generalize (for review see Pyke *et al.*, 1977). A variety of adult animals seem to fit the optimal diet model (for review see Krebs *et al.*, 1983) which depends on a predator's ability to access prey profitability and abundance.

It is presently difficult to determine when and how quickly juveniles can select an optimal diet. Some studies show that recently independent juveniles can select the most profitable prey types, within the limits imposed by their foraging inefficiency (Richardson and Verbeek, 1987; Sullivan, 1988b), and that juveniles can sometimes evaluate prey densities as well as can adults (Brandt, 1984). These results suggest that some juveniles, possibly at an early age, might have the ability to select an optimal diet. However, an inability to recognize appropriate prey, as discussed earlier, and a failure to respond to fluctuations in prey abundance (Edwards, 1989b) suggest that for other species the ability to select an optimal diet might require considerable time for development. Obviously more work is needed to determine when and how optimal diet selection arises in young birds.

Ineffective capture skills may cause juveniles to select different types or sizes of prey than those taken by adults. For example, inexperienced juveniles may feed on easily captured or found food items while more experienced adults capture more difficult prey. Stevens (1985) suggests that this may be the case in European Starlings, in which juvenile pecking success was 5.97% for large insects (mostly leatherjackets) compared to 48.1% success for adults, while the difference was considerably less when feeding on small insects. As expected, adults took large prey such as leatherjackets, which represented 82.2% of their diet (by dry weight) versus 34.9% taken by juveniles. However, juveniles pecked more at cherries on the ground and at insects in cow dung, which are easily accessible in comparison with leatherjackets. Stevens related this to the previous studies which indicated an extraordinary high proportion of juveniles in cherry orchards. Feare (1980) earlier suggested that different nutritional requirements could explain the preference of juveniles for cherries. Stevens (1985) argues that this nutrition hypothesis is unlikely because growing juveniles have high protein needs and do their best to gather animal food: juveniles did more searching for grubs than did adults and fed for longer time periods in a variety of locations where insects were easier to capture. Thus Stevens concluded that juvenile European Starlings are unable to gather enough animal food and therefore tried to compensate by eating the more accessible and energy-rich cherries. Similarly Breitwisch *et al.* (1987) have suggested that juvenile Northern Mockingbirds (*Mimus polyglottos*) may consume a considerable amount of fruit, because of the ease with which it may be obtained. It would be valuable to know if such a pattern exists in other opportunistically frugivorous species and if a disproportionate amount of frugivory results from inept juvenile foragers.

Walton (1979) suggested that a lack of juvenile capture skills might explain age-related diet differences in Meadow Pipits (*Anthus pratensis*) during June when juveniles consumed more Coleoptera and Hymenoptera than did adults, but far fewer Dipetera (14% versus 47% for adults). The proportion of Coleoptera taken by juveniles in June and July (36% and 38%) was considerably higher than that taken by adults (19% and 22%). Walton noted that it is unlikely that juveniles were able to capture the more abundant but faster moving Diptera, while Coleoptera provided a scarcer, but more easily caught food supply. In an another case, juvenile Ospreys shifted their selection among size classes of fish over time (Edwards, 1989b). Initially, the juveniles selected the smallest size classes and ignored the largest. However, over time the juvenile Ospreys began to take more large fish. Edwards suggested that the initial absence of large fish in the diet was due to the inability of juveniles to capture large fish.

Another factor influencing juvenile diet selection is their limited handling abilities (see also Section 3.5). For example, because of their limited handling skills, small prey may be more profitable (energy gained/handling time) for juveniles whereas adult skills allow proficient handling of larger prey with profitabilities equivalent or larger than those of smaller prey (Sullivan, 1988b). Age-related differences in handling ability may explain many of the instances where juveniles feed on smaller prey items than those taken by adults. This tendency is widespread across a diversity of species, such as Great Blue Heron (*Ardea herodias;* Quinney and Smith, 1980), Gray Heron (Cook, 1978; Draulans, 1987), Cattle Egret (*Bubulcus ibis;* Siegfried, 1972), Ringed Plover (*Charadrius hiaticula;* Pienkowski, 1983), Eurasian Oystercatcher (Goss-Custard and Durell, 1983), Curlew Sandpiper (*Calidris ferruginea;* Puttick, 1978), Snail Kite (*Rostrhamus sociabilis;* Bourne, 1985), Osprey (Edward, 1989b), American Robin (Gochfeld and Burger, 1984), Northern Mockingbird (Breitwisch *et al.*, 1987), Northwestern Crow (Richardson and Verbeek, 1987), Yellow-eyed Junco (*Junco phaeonotus;* Sullivan, 1988b), Common Chaffinch (*Fringilla coelebs;* Kear, 1962), European Greenfinch (*Carduelis chloris;* Kear, 1962; Newton, 1967), and Eurasian Linnet (*Carduelis cannabina;* Newton, 1967). This widespread and possibly general phenomenon deserves more attention.

Age-related differences in diet could arise as a result of short-term persistence of diet specialization (Bryan and Larkin, 1972). This is suggested in the work of Edwards (1989b) who examined the ontogeny of diet selection in color-banded Ospreys. In the first year of the study, all young differed from adults with respect to diet selection, but in the second year 9 of 14 young differed from adults. Individual differences in the diet selection of juveniles were present both years, with preference for each fish species in each year shown by at least one juvenile. Also, in both years some young exhibited no preferences for any fish species, but rather captured fish in relation to their availability. Interestingly, the diet preferences of individual juveniles remained constant throughout the postfledgling period while at the same time adults readily changed their diet in relation to fish abundance. The individual preference patterns of juveniles were related to whichever fish species were captured first. This suggests that the first few prey captured play an important role in initiating the development of a preference. Edwards suggests that this might lead to an early search image which could benefit juveniles by decreasing time and energy expended sampling for other prey. How long an individual juvenile's diet preference remains is unknown, but it would seem unlikely to persist for long once the birds arrive on their wintering grounds. Similar persistence in diet resulting from the earliest independent foraging experiences of juveniles has been documented in other species (Rabinowitch, 1968, 1969).

If diet selection involves the integration of information from all the foraging stages, then a lack of juvenile proficiency at any stage could have important consequences for the choice of diets by juveniles. It is likely that optimal diet selection in juveniles develops as rapidly as proficiency in other foraging activities (Richardson and Verbeek, 1987).

3.4. Prey Capture

The development of prey capture techniques, commonly proceeds from simple pecking to more complex capture movements often requiring considerable coordination and skill. For example, young Reed Warblers (*Acrocephalus scirpaceus*) initially captured insects with a stand and catch technique which declined in use with age, while leap catching and flycatching appeared later in development (Davies and Green, 1976). Northern Wheatears (*Oenanthe oenanthe*) initially used ground gleaning and stand gleaning motions to capture insects and only after the fifth day following fledging did they attempt the more complex perch-to-ground-sallying and aerial hawking movements, which are the common adult techniques (Moreno, 1984). Similarly, newly independent juvenile Northern Mockingbirds foraged primarily on the ground while adults used more aerial attacks to capture their insect prey; however juveniles increased their use of aerial techniques as the summer proceeded (Breitwisch *et al.*, 1987). Thus the capture techniques dependent upon flight performance are usually the last to develop. However, sometimes juveniles are capable of proficient prey capture by flight but use more energy conservative capture techniques than those used by adults (Davies and Green, 1976). For example, adult Reed Warblers may use more energy-demanding techniques (i.e., flycatching) in order to obtain sufficient food for their young as well as themselves.

Much of the initial improvement in capture success by young birds may be attributed largely to maturation (Cruze, 1935). However, once juveniles begin to feed independently, it is uncertain how important maturation is to the development of their capture proficiency. One case in which maturation has been correlated with improved capture skills in independent juveniles occurs in European Starlings (Stevens, 1985). The gradual improvements of juvenile capture skill is correlated with bill growth. This of course, does not eliminate the importance of learning from practice, which is likely to contribute to capture proficiency, sometimes long after most maturational changes have ceased (e.g., Richardson and Verbeek, 1987). Most workers have cited learning and practice as a cause, if not the major cause, of improved juvenile capture skill.

Generally, the greater the skill needed to capture prey, the less successful are juveniles in comparison with adults, and presumably a longer period of development is required. For instance, young Laughing Gulls were as successful as adults when snatching offal in the air, less successful when aerial dipping, and much less successful when plunge-diving for fish (Burger and Gochfeld, 1983). Not surprisingly, young Laughing Gulls most frequently used the method requiring the least skill and which provided them with highest success rate (aerial snatching of offal). In contrast, adult gulls most frequently used plunge-diving techniques for fish, presumably the method requiring the most skill. Similarly, young gulls of 15 different species avoided the most difficult foraging tasks such as plunge diving (Burger, 1987). When Sandwich Terns (*Sterna sandvicensis*) plunge dove from low heights into shallow water, no difference was found between adult juvenile capture success; however, when diving from greater heights into deeper water, with fish further below the surface, adult capture success was higher than that of juveniles (Dunn, 1972). In European Starlings, the difference in pecking success is large for difficult capture methods, such as grubbing, whereas little difference was found between age classes when catching prey on cow dung, a technique requiring relatively little skill (Stevens, 1985).

A commonly studied capture technique in which experience has been attributed to improved capture success is the plunge-diving method of Brown Pelicans, terns, gulls, and Ospreys (e.g., Orians, 1969;

Dunn, 1972; Buckley and Buckley, 1974; Burger and Gochfeld, 1983; Edwards, 1989a,b). Using this method a forager must contend not only with the evasive tactics of the prey, but with surface glare, refraction, and wind. Carl (1987) studied this technique in four age classes (yearlings; juveniles, 12–21 months; subadults 22–39 months; and adults) of Brown Pelicans, and found that both diving success and technique varied with age class. He showed that diving success among older birds improved with greater dive heights and steeper dive angles. He attributed the positive association between age and diving skills to several factors involving light physics, weather, neurological control, and learning. He noted that yearling pelicans may lack experience with changing marine conditions and perhaps lack the neurological coordination for high dives or steep angles. This lack of yearling diving proficiency restricts their choice of prey to those located near the surface and restricts the time of foraging. In contrast, older pelicans are able to exploit both shallow and deeper prey under a variety of conditions.

Experience with a capture technique may not always be the major factor responsible for improved juvenile capture success. For example, Coblentz (1986) describes a "midflight pause" in adult Brown Pelicans, which he suggests might allow a brief evaluation of the forager's probability of success with resumption of flight if success appears unlikely. Juvenile pelicans never showed a midflight pause prior to plunge-diving. This difference may have allowed adult pelicans to better discriminate against plunge-dives in which capture was unlikely, and thus contribute to the higher success of adults. Similarly, Buckley and Buckley (1974) concluded that the ability of adult Royal Terns to recognize capturable prey was a major factor in their fishing success. In this species both adults and juveniles had similar diving success (i.e., number of captures/dive), but adults increased their catch per unit time by increasing their diving rate to 1.6 that of juveniles. These two studies indicate that care must be exercised in interpreting capture success data because two different factors could be involved. Age-related capture differences may arise because of differences in ability to recognize capturable prey before a capture attempt, or because of actual differences in capture ability.

Under certain conditions, young foragers may be able to compensate for their lack of capture skills by obtaining food from other individuals. Begging is the obvious strategy for inept altricial young during the early period of parental care, but as they mature self-feeding becomes a more effective strategy, particularly as the parents become more reluctant to feed the young (Davies, 1976). Young Northern Wheatears switched between begging and self-feeding strategies on a daily basis as they approached independence (Moreno, 1984). Early in the morning, when prey were inactive, the young wheatears chased their parents intensively for food, but switched to self-feeding around 0800 when rising temperatures led to an increase in insect activity and capture rates. However after 1000, capture success decreased, presumably due to faster moving insects evading capture, and the young returned to chasing their parents. Here Moreno demonstrated a clear relationship between the chasing and begging intensity of the young wheatears and their self-feeding success rates. These observations indicate that the young of altricial species may be able to choose the most profitable strategy according to their own ability to capture prey, at least during the end of the parental care period.

Another way in which juveniles might compensate for their lack of capture skills is by piracy or kleptoparasitism. Optimal foraging theory predicts that a forager will rob food only if the net energy gain from robbing exceeds or matches the net energy gained from feeding by other methods (Charnov, 1976; Dunbrack, 1979; but see Kushlan, 1978). Juveniles will at times resort to robbing, often more frequently than adults, as demonstrated in several studies. Verbeek (1977a,b) found that juvenile Herring Gulls feeding on refuse and on starfish were less skillful than adults and used chasing and stealing as an important method for obtaining food. Similarly, Greig *et al.* (1983) found that juvenile Herring Gulls made more attempts than adults to obtain food from other gulls, by displacement attacks and by intrusions onto the feeding patches of others, rather than searching the refuse itself. Juveniles were considerably less skillful than adults both in finding and obtaining food from refuse, whereas adults obtained almost all their food by searching the refuse pile itself. In another example, juvenile common Black-headed Gulls were less successful at robbing food from Northern Lapwings than were adults, and presumably as a result of their reduced success rate they attacked more often (Hesp and Barnard, 1989). However, when other feeding opportunities (e.g., feeding in plowed fields) arose, juveniles stopped robbing and adopted other foraging methods while adults continued to rob lapwings. In a similar manner, juvenile Bald Eagles (*Haliaeetus leucocephalus*) made more attempts at piracy from other eagles than did adults, possibly because juveniles are less proficient than adults at obtaining food independently (Fischer, 1985). Thus in these instances, juveniles probably rob to compensate for their inability to obtain other foods.

Other studies on several species of gulls have found that adults, rather than juveniles, were the most frequent pirates on conspecifics or other host species (Burger and Gochfeld, 1979, 1981). In their study of piracy by four gull species, Burger and Gochfeld (1981) found that adults were victims less often than juveniles, and once chased, juveniles were more likely to lose their food than adults. This suggests that under the circumstances of their studies piracy by juveniles might be very costly, probably more so than for adults. Not surprisingly, juveniles appeared to be foraging in a different manner than adults—going after smaller food pieces (easier for them to handle), avoiding dense feeding flocks (where they are most likely to be attacked), and attempting piracy from other young. Such feeding behaviors may have provided the juveniles with alternative food sources as opposed to the relatively more costly piracy behavior.

Observations on piracy by juveniles are consistent with optimal foraging predictions suggesting that juveniles may compensate for their relatively inept foraging abilities by robbing, when the rewards from robbing are greater than other food sources, or switch to other food sources when the costs of robbing are too high. Unfortunately none of the current literature on age-related differences in piracy provides a test of the theory, which requires measurements of the net energy gained from robbing versus other food sources. To rigorously test the theory, studies using time and energy budget analysis are needed to determine the relative costs and benefits of prey robbing versus other feeding methods for juveniles and adults. Examples of the necessary approach are found in the studies of adult egret robbing behavior by Kushlan (1978) and Dunbrack (1979).

To be a successful pirate requires a variety of skills: the ability to identify potential victims, judging when to pursue a victim, the ability to follow the victim closely in flight, agility in flight, the ability to snatch food from the victim or to catch food dropped by the escaping victim, and knowing when to give-up an unsuccessful pursuit (Burger and Gochfeld, 1981). Most of these skills are likely to require experience and hence it is nor surprising that juveniles are frequently less proficient at robbing than adults (Burger and Gochfeld, 1979, 1981; Gochfeld and Burger, 1981; Hesp and Barnard, 1989), but not always (Verbeek, 1977a, b; Greig *et al.*, 1983). As mentioned previously, juveniles are sometime less proficient at identifying the most vulnerable victims (Carroll and Cramer, 1985). Judging when to pursue a victim appeared to be a difficulty for juvenile Common Black-headed Gulls when attacking Northern Lapwings for earthworms (Hesp and Barnard, 1989). In this case, juvenile gulls were less successful because their attacks were more likely to be mistimed and detected by the victim. Juveniles were also less selective in initiating long distance attacks and choosing vantage points within flocks. The method of approach largely explains the difference in success between adult and juvenile Ring-billed Gulls (*Larus delawarensis*) pirating food from European Starlings (Burger and Gochfeld, 1979). Juvenile Ring-billed Gulls frequently walked toward Starlings, giving them time to fly off with food, whereas adult gulls were more likely to fly and hence startle the victim into dropping food. Adult Magnificent Frigatebirds (*Fregata magnificens*) were more likely to enter foraging gull flocks than were juveniles and consequently adults were more successful robbers (Gochfeld and Burger, 1981). In some gull species, adults were more likely to abandon unsuccessful piracy attempts earlier in the act than were juveniles (Verbeek, 1977b; Burger and Gochfeld, 1981). It thus appears that experience, at all stages of the robbing process, is crucial for obtaining proficiency.

Social interactions between adults and juveniles may reduce juvenile capture success. For instance, once juveniles have obtained food by robbing or capture they are often more likely to be victims and lose more food when pursued by adults (Burger and Gochfeld, 1981; Gochfeld and Burger, 1981; Carroll and Cramer, 1985; but see Schnell *et al.*, 1983). Under certain circumstances, it may be advantageous for juveniles to forage together in flocks in the absence of adult interference, as suggested for Royal Terns, White Ibis (*Eudocimus albus*), and Crimson Rosellas (Buckley and Buckley, 1974; Bildstein, 1983; Magrath and Lill, 1985). For example, juvenile Common Black-headed Gulls feeding in flocks without adults foraged more efficiently than juveniles in flocks with adults (Ulfstrand, 1979). Even when in mixed-age flocks, juvenile White Ibis were more likely to have juveniles as nearest neighbors than adults, although this did not appear to influence their foraging proficiency (Bildstein, 1983). Sometimes within mixed-age flocks, juveniles tend to hunt on the flock periphery, presumably as a result of adult dominance (e.g., Bildstein, 1983; Gochfeld and Burger, 1984). If this is so, it suggests that a peripheral location is disadvantageous, possibly because of lower food densities or because higher levels of vigilance are required (Bildstein, 1984). Thus the tendency for juveniles to be subordinate to adults may also contribute to their lower capture proficiency.

Foraging in a group often confers a variety of benefits, including improved foraging success (for a review see Morse, 1980). However, juveniles often lack the skills necessary to take advantage of the benefits of group foraging. For example, in Great Blue Herons, adult foraging success was greater in flocks of more than five individuals, while juvenile success was not related to flock size (Quinney and Smith, 1980). In Black-necked Stilts (*Himantopus mexicanus*), individuals feeding in groups obtained more food than those feeding nearby, and those that fed near dense flocks obtained more food items than juveniles that fed solitarily (Burger, 1980). Burger suggests that perhaps the group movement stirred up insects, making food generally more available to members of the flock. Juvenile stilts frequently fed solitarily, while adults rarely fed outside the groups, suggesting that adults learn that group foraging is more efficient. However, in other circumstances juveniles can benefit by foraging in a group. The benefits of group foraging by juveniles are apparent in Yellow-eyed Juncos in which Sullivan (1988a) showed that independent juveniles foraging alone scanned more than juncos in flocks, and consequently had lower pecking rates than those foraging in flocks. She concluded that the juvenile's increased pecking rates when foraging in flocks may be attributable, in part, to their decreased scanning rate, but local enhancement, social facilitation, and copying are probably also involved. In Ospreys, capture success improves more rapidly among siblings that forage together than in siblings that forage solitarily (Edwards, 1989a). Furthermore, sibling Ospreys had similar capture techniques suggesting that socially facilitated learning was involved. These studies indicate that under some circumstances juvenile foraging proficiency may be aided by group foraging, while under other circumstances juvenile foraging proficiency does not appear to improve while foraging in a group.

In summary, prey capture proficiency improves with maturation, experience, and coordination. Complex capture techniques are likely to require considerable time for development, mostly because the predator requires time to obtain experience. Inept juvenile foragers may seek alternative capture methods or diets to compensate for their lack of capture skills. Certain social interactions with other foragers have the potential to accelerate the development of juvenile capture proficiency, but adult piracy and dominance may inhibit juvenile proficiency.

3.5. Food Handling Time and Techniques

Handling time is the time required for a forager to eat a captured food item, and is a widely used measure of time and energy investment by foragers (Krebs *et al.*, 1983). Handling time can represent a substantial portion of the foraging time for many species such as raptors, gulls, parrots, oystercatchers, shrikes, titmice, finches, and corvids. Among adults, handling time is known to vary with size of prey, bill size, hunger, competition, previous experience, the need for vigilance, and the risk associated with dropping items (Krebs, 1980; Barnard and Stephens, 1981; Sherry and McDade 1982; Paszkowski and Moremond,

1984; Greig-Smith, 1984). Considering the variety of factors which can influence food handling times it should not be surprising that adults and juveniles often differ in their food handling abilities.

Juveniles of a variety of species and diet types (piscivores, insectivores, mollusc eaters, scavengers, carnivores, seed eaters) have been found to be less adept at handling food items than adults. For instance, juveniles have a tendency to drop food items more frequently than adults (Buckley and Buckley, 1974; Barash *et al.*, 1975; Davies and Green, 1976; Burger, 1981; Greig-Smith, 1985; MacLean, 1986), and juveniles frequently have longer handling times than adults (Davies and Green, 1976; Ingolfsson and Estrella, 1978; Gochfeld and Burger, 1981; Bourne, 1985; Sutherland *et al.*, 1986; Richardson and Verbeek, 1987; Enoksson, 1988; but see Greig-Smith, 1985). In some species, adults and juveniles have equivalent handling times with small prey items, but adults are noticeably faster than juveniles when handling larger prey items (Quinney and Smith, 1980; Sullivan, 1988b). Furthermore, the development of food-handling proficiency can be a slow process and consequently is often one of the last feeding skills to develop in certain birds (e.g., rosellas, Cannon, 1979; three gull species, MacLean, 1986).

Improvement in juvenile handling time and ability can largely be attributed to learning and to maturation. For example, Davies and Green (1976) demonstrated that naive hand-reared Reed Warblers were less adept at handling flies (longer times and more dropping) than were experienced warblers; yet naive and experienced warblers were equally capable of catching flies. Although the naive warblers did improve their handling skills with time, they never obtained the proficiency of the experienced birds. Davies and Green suggested that a sensitive period in the early stages of growth, when the bill was flexible, allowed rapid development of handling skills in contrast to later life when the bill was less flexible. The combination of bill maturation (i.e., growth in length) and experience were also cited as the reason for improved handling skills of juvenile European Nuthatches (*Sitta europaea*) in the field (Enoksson, 1988).

An instance of the importance of food handling experience during a sensitive period has been documented in shrikes. For example, in young Loggerhead Shrikes learning how to impale or wedge prey is by imprinting (S. M. Smith, 1972). These birds use a type of behavior designated as dabbing in which they take an item of food, turn sideways and place it on their perch. Dabbing begins at about 21 days after hatching and after approximately 36 hours of this behavior the young shrikes add pulling components—young turn sideways, reachout, then pull the food item along the perch toward them (dragging behavior). Initially there is no orientation to this behavior, but when the food item accidently catches on something, the young shrike concentrates its dragging efforts along that part of the branch in which the food item was snagged. Thus the young shrikes quickly learn to direct their dragging to appropriate places such as forks or thorns. However, this learning must occur within a critical period; those young shrikes having experience between 20 and 40 days posthatching all learned to impale normally, but those raised in cages with dowel perches until 75 days posthatching did not learn how to impale,, even when provided with the opportunity to observe experienced shrikes impale food items nearby. Also, young shrikes seem to prefer the kind of impaling device with which they were raised, even when it is less efficient (Wemmer, 1969).

The relative importance of self-learning in the development of prey impaling and wedging varies among different species of shrikes as shown in a comparative study of European shrikes (Lorenz and von St. Paul, 1968). For example, personal experience appeared to be unimportant to Red-backed (*Lanius collurio*) and Woodchat Shrikes (*L. senator*) before being able to direct dragging behavior to thorns (impaling), but both required experience before learning to direct dragging to branch forks (wedging). The opposite pattern was found in Northern Shrikes (*L. excubitor*) in which inexperienced young directed dragging to forks, but they required experience to impale items on thorns. Lorenz and von St. Paul argued that these patterns were consistent with the natural history of the three species.

Possibly one of the most complex forms of food handling behavior is the shell-dropping behavior of gulls, rooks, crows, and Lammergeyers (*Gypaetus barbatus*) used to break open hard-shelled prey to consume the edible inner parts. The complexity of this behavior results from the need for many different decisions if the forager is to obtain the maximum rate of net energy intake for its efforts. Each of the following must be selected by the forager: the size and type of food item for dropping, a suitable substrate on which to drop the item, an appropriate height from which to drop the item, and the number of times to drop an item before giving up if it does not break (Barash *et al.*, 1975; Siegfried, 1977; Zach, 1979; Kent, 1981; Maron, 1982; Richardson and Verbeek, 1987). Therefore, considering the complexity of this behavior it should not be surprising that it takes considerable time for young birds to obtain the proficiency of adults.

Juveniles appear to be less skilled than adults at almost all stages of the shell-dropping process. For example, first year Herring Gulls and Glaucous-winged Gulls differed from adults (> 2 years) by dropping more food items on inappropriate substrates (soft sand or water) and by dropping items from a greater range of heights which is disadvantageous because items dropped at great heights are more likely to be stolen and those dropped from low heights are unlikely to break (Barash *et al.*, 1975; Ingolfsson and Estrella, 1978). Furthermore, yearling Glaucous-winged Gulls were more likely than adults to "accidently" drop clams while flying. Yearling Northwestern Crows took more time to drop clams than adults and required more drops to break them open (Richardson and Verbeek, 1987).

Learning in various forms appears to play a prominent role in the improvement of juvenile shell-dropping behavior. Initially, observational learning and imitation from the parents might be important as the young birds follow their parents after fledging (Richardson and Verbeek, 1987). Yearling gulls sometimes attempt to rob prey from older birds engaged in shell-cracking (Verbeek, 1977b), which might aid in associating the shell-breaking pattern with hard surfaces (Ingolfsson and Estrella, 1978). However, after the period of parental association, trial-and-error learning is likely to account for most of the refinement of a youngster's shell-dropping skills. Consistent with the likelihood of a trial-and-error learning process are the accounts of yearling birds dropping shells only to retrieve them again in midair, dropping items without investigating the results, dropping items while standing on the ground, dropping two items at once, and making flights similar to drop flights, but without actually releasing an object (Ingolfsson and Estrella, 1978; Richardson and Verbeek, 1987). In contrast to the importance of learning, maturation appears to be an unlikely explanation for the lack of juvenile proficiency in shell-dropping behavior—juveniles at fledging have a mass similar to adults and their flight capabilities appear to be similar within a few months of fledging (Barash *et al.*, 1975; Richardson and Verbeek, 1987).

Another suggested factor contributing to the lack of juvenile proficiency in shell-dropping behavior is related to potential problems arising from overspecialization as a juvenile (Olive, 1982). Richardson and Verbeek (1987) have elaborated on this idea by suggesting that a lack of juvenile proficiency might arise from the need to retain behavioral plasticity until reaching maturity. They state that proficiency at a skill like shell-dropping obtained during an early age might be maladaptive if it inhibits the development of proficiency in some other specialized skill later in life. Unfortunately, they provide no examples to support this idea, and thus it presently appears to be untested.

It appears that most *simple* food handling skills, such as using the feet to hold food items by titmice (Vince, 1960) and opening seed by finches (Kear, 1962; Newton, 1967), can develop independently of interaction with other individuals, including the parents. However, as the food handling process becomes more complex it seems likely that learning from parents or other individuals is important, as suggested previously for shell-dropping behavior. The classic example of the importance of learning a food-handling skill from the parents is that of the Eurasian Oystercatcher in which the young develop the same mussel feeding techniques of their parents (Norton-Griffiths, 1967, 1969). Young oystercatchers whose parents stab mussels develop the stabbing technique while those whose parents hammer develop the hammering technique. Norton-Griffiths showed that the young learn the method of opening mussels from their parents by switching the eggs of hammers and stabbers. Although oystercatchers acquire the favored mussel opening technique of their parents, they will at times use other techniques (Goss-Custard and Sutherland, 1984). How the young actually learn the necessary skills is unknown, but young using either of the two mussel feeding techniques remain with their parents 18–26 weeks whereas those oystercatcher young feeding on other prey stay with their parents for only 6–7 weeks. It thus appears that this specialized handling technique has the cost of requiring a longer period of parental care. Another instance where parental interaction might prove important is in the rapid development of food handling skills in fish-eating birds such as the Common Tern (*Sterna hirundo*) studied by Gochfeld (1980). He suggested that the consistency with which parents present fish to newly hatched young might enhance the development of the chick's ability to handle and manipulate food items. This untested idea would appear amenable to experimental testing.

That the limited food handling abilities of juveniles can influence their diet choice was nicely demonstrated by Sullivan (1988b) in a detailed field study of age-specific food handling times and prey profitability in Yellow-eyed Juncos. Sullivan presented four different age classes of juncos (recently independent juveniles, young experienced juveniles, old experienced juveniles, and adults) with large and small mealworms and measured handling times. She found that handling times were significantly related to both mealworm size and age class. The large mealworms took significantly longer to handle than the small mealworms for all age classes of juncos. Recently independent juveniles took longer to handle the large mealworms than either age class of experienced juveniles. Adult juncos took significantly less time to handle the large mealworms than did experienced juveniles. She next used the handling times to calculate the profitability (joules/second) when foraging on small and large mealworms within each junco age class.

Her profitability calculations indicated that for recently independent juveniles, small mealworms were much more profitable than large mealworms and, not surprisingly, these birds selected significantly more small mealworms than large mealworms. However, for the other three age classes, small and large mealworms were equally profitable and these birds displayed no preferences in mealworm size. Thus the diet of adults and juveniles may differ because of differences in the relative profitability of prey for each age class, arising primarily from differences in prey handling ability.

Sullivan's (1988b) profitability analysis is likely to be important for understanding why juveniles tend to select similar prey items than adults in a wide variety of species (see Section 3.3). For example, Great Blue Herons fit the pattern of juvenile selection of smaller prey than adults, and furthermore juveniles and adults have equivalent handling times for small prey, but adults are quicker with larger prey (Quinney and Smith, 1980). In this case too, it is likely that the smaller prey are more profitable for juveniles because of their inept handling of larger prey, which are preferred by the adults. It would be valuable to determine the profitability of different sizes and types of prey items for adults and juveniles of those species in which juveniles tend to prefer smaller prey items or prey types than adults.

Handling skills may influence habitat or patch selection if prey sizes and types are not evenly distributed between localities. This occurs in adult oystercatchers in which the two prey handling types ("hammers" and "stabbers") feed in the mussel beds appropriate to their respective handling skills (Norton-Griffiths, 1967). Handling differences might account for habitat or patch differences between adults and juveniles, particularly if inept handling skills restrict juveniles to a particular prey type, and if that prey type is limited to certain patches or habitats. This hypothesis has not yet been fully confirmed in the field. Bourne's (1985) field study of Snail Kites would seem to have many of the necessary requisites: juveniles have longer handling times, juveniles select smaller snails, and juveniles select different habitats than adults. However, Bourne concluded that there was no support for the hypothesis that kite diet differences are caused by handling time differences, although he provided no alternative explanation.

Finally, the lack of food-handling proficiency of young birds can have major implications for survival. The lack of food-handling skills can increase the likelihood that food items will be stolen, as a result of longer handling times (Gochfeld and Burger, 1981; Barnard and Thompson, 1985) or due to inept handling (Barash *et al.* 1975), thereby resulting in more time and energy expended in foraging. Moreover, inept and careless food-handling will require greater time and energy expenditure for foraging by juveniles, thus exposing them to greater risks of predation or making it difficult to acquire sufficient energy (Sullivan, 1989). Thus if food-handling behavior is one of the last foraging skills to develop it could serve as a major constraint on proficiency for a variety of species requiring complex food-handling abilities.

4. DISCUSSION

So far we have seen that juveniles may be less proficient than adults in selecting a foraging site, in search behavior, in food recognition and diet selection, in capture methods, and/or in handling techniques. However, few studies have examined the relative contribution of these components to age-related foraging differences. One notable exception is the work of Sutherland *et al.* (1986) which compares the relative importance of encounter rate with handling time in contributing to adult/juvenile differences in foraging rates by Common Moorhens (*Gallinula chloropus*). Juvenile moorhens take longer to pick up and swallow items and are less efficient at finding food while searching. To examine the relative importance of these two foraging components, Sutherland *et al.* used Holling's (1959) equation for the number of prey found (N_a) in time (T) at a given encounter rate (λ) and handling time (H_t):

$$N_a = \frac{\lambda \bullet T}{1 + \lambda \bullet H_t}$$

By altering handling time or encounter rate, this expression can be used to assess the relative contribution of handling time and encounter rate to adult/juvenile foraging differences. For instance, juvenile moorhens found 31.7 items in 120 s. If juveniles had the same handling times as adults they would locate 35.6 prey in 120 s, producing an improvement of 11%. If they had instead the same encounter rate as adults they would find 49.7 per 120 s, for an improvement of 36%. Therefore in this case the age difference in encounter rate was more important than the difference in handling time in accounting for the difference in feeding rate.

The real contribution of the Sutherland *et al.* (1986) analysis, however, is in reemphasizing the importance of food density to the relative contribution of the two foraging components to adult/juvenile foraging differences. They note that the encounter rate is a combination of food density and a forager's ability to find it. Hence, doubling the food density will double the encounter rate. At very high food densities, and consequently high encounter rates, the feeding rate will be limited by handling time. In other words, at high food densities, the feeding rate will be the reciprocal of handling time and the discrepancy in encounter rate will be irrelevant. In contrast, at very low food densities the handling time will be trivial and the encounter rate will be the important determinant of feeding rate.

Consistent with these predictions are the results of Draulans (1987) which demonstrated that yearling Grey Herons were less successful (measured as success and frequency of strikes) than adults at low prey densities, but that yearlings obtained a foraging performance similar to adults at high prey densities in fish farm ponds. Draulan found no age-related difference in handling times and attributed the juveniles' relative lack of success at low prey densities to their lower encounter rates which included prey recognition and capture skills. Thus at high prey densities, age differences in foraging may be trivial in those species in which adults and juveniles do not differ in handling times. More importantly, this finding emphasizes that not all age-related proficiency differences are important, at least under certain conditions.

Most of the field studies of age-specific foraging differences have been conducted in the post-breeding period (fall or winter in the temperature zone) when prey densities are likely to be relatively low. At this time, age-related foraging differences are likely to be evident for species in which younger individuals are relatively inept at encountering prey, due to a lack of proficiency in selection of foraging site, search pattern, prey recognition, and/or capture. However, few studies (but see Orians, 1969) of age-related foraging differences have been conducted during the breeding season when food densities might be expected to be highest, and consequently inept prey-encounter abilities are less of a hindrance to the proficiency of younger foragers. Therefore it is conceivable for some species (those with no or minor age differences in handling), that age-related differences become trivial during the breeding season when high prey densities make the inept encounter abilities of yearlings, second-year, and third-year birds irrelevant. Studies are needed to test this possibility particularly because the inept foraging ability of younger individuals is commonly cited as a reason for delayed breeding in many birds (reviewed in Ryder, 1980; Burger, 1988).

To understand the mechanisms involved in the development of foraging proficiency it is valuable to examine the sequence in which the various aspects of foraging behavior are improved (e.g., Davies and

Green, 1976). MacLean (1986) has described this sequence in three gull species (all unbanded) and found that the rate at which juveniles obtained proficiency in the different foraging stages varied between species. He found in Bonaparte's Gull (*Larus philadelphia*) that search and pursuit improved simultaneously and handling skills improved later. In the Ring-billed Gull, searching ability improved first, followed by pursuit and capture abilities, and handling improved last. In contrast, pursuit improved first in Herring Gulls followed by handling and searching abilities. What accounts for these species differences was left unexplained, but future studies are likely to uncover more species differences in the sequence in which the different aspects of foraging behavior are improved. Studies of additional species may allow us to use the comparative method to identify the factors controlling the sequence in which foraging proficiency is obtained.

As demonstrated throughout this chapter, lack of experience is the factor most likely to hinder juvenile foraging proficiency, and its effect is greatest for the methods requiring the most skill for mastery. For piscivores, raptors, and kleptoparasites the greatest skills are probably required for recognizing capturable prey, the actual capture technique, and sometimes handling. Foragers picking prey items from a surface or probing below the surface must be most adept at prey recognition, although handling skills may become paramount if the food is encased in a protective covering. In contrast, searching skills are likely to be most important for nectarivores. Therefore, the foraging stage or stages requiring proficiency is dependent upon the prey type. As a forager's diet becomes more variable, it is likely to require higher levels of skill at more stages of the foraging sequence. This of course, requires experience, which takes time and hence is costly to develop. Thus the necessity of mastering many different skills requiring experience limits how broad or general an individual forager's diet may become.

A stable food supply will encourage individual specialization on particular foraging skills, which can enhance foraging proficiency (Partridge and Green, 1985, 1987). For example, recent experiments with Jackdaws (*Corvus monedula*) demonstrated that specialization improves foraging efficiency (Partridge and Green, 1987). In this study, hand-reared birds were trained to be either "specialist" feeders on one of three artificial laboratory feeding tasks or "generalist" feeders on all three. The results indicated that the specialist birds were usually more efficient at obtaining food from their tasks than were the generalists. This suggests that specialization enhances foraging proficiency and, in addition, diet specialization may improve digestive efficiency (Partridge and Green, 1985). These advantages may encourage short-term diet specialization, even in juveniles (e.g., Edwards, 1989b), or lead to individual diet specialization (Partridge and Green, 1985).

If the idea that specialization enhances foraging proficiency is a general phenomenon, then it suggests that specialist juveniles may obtain adult levels of proficiency at faster rates than generalist juveniles. A rigorous test of this hypothesis is not presently possible because of the limited number of studies with marked birds, necessary for an accurate determination of juvenile development rates. However, some of the present studies are consistent with the hypothesis. For instance, the highly specialized handling skills used by Eurasian Oystercatchers to open muscles may be obtained by juveniles in only a few months (Goss-Custard and Durell, 1987a; *contra* Norton-Griffiths, 1969). In contrast, most gull species have a generalized diet and in some species juveniles do not obtain adult levels of proficiency until after three years (MacLean, 1986; Burger, 1987). A valuable test of the hypothesis, might be made by comparing the rate at which young gulls (i.e., extreme generalists) obtain adult levels of proficiency while plunge-diving with the rate at which proficiency is obtained by specialist plunge-divers (Brown Pelicans, Ospreys, terns).

As a consequence of their inept foraging abilities, juveniles are likely to spend more time foraging than adults (Schreiber, 1968; Spaans, 1971; Davis, 1975; Verbeek, 1977b; Cook, 1978; Morrison *et al.*, 1978; Puttick, 1979; Burger and Gochfeld, 1979; Burger, 1980; Ryan and Dinsmore, 1980; Porter and Sealy, 1982; Schnell *et al.*, 1983; Magrath and Lill, 1985; MacLean, 1986; Sutherland *et al.*, 1986; East, 1988; Sullivan, 1988a). Accordingly, adults often spend more time resting than juveniles, presumably as a result of their better foraging skills (Buckley and Buckley, 1974; Schnell *et al.*, 1983; Magrath and Lill, 1985; MacLean, 1986; Sutherland *et al.*, 1986; Sullivan, 1988a). However, sometimes juveniles loaf more, possibly watching for adult feeding flocks (Porter and Sealy, 1982). Such observations suggest that juveniles may have difficulty in maintaining a positive energy budget, which is consistent with the findings that starvation is a major cause of juvenile mortality in a variety of species (e.g., Lack, 1966; Southern 1970; Hirons *et al.*, 1979; Perrins, 1980; Sullivan, 1989; Catterall *et al.*, 1989). In addition, as juveniles forage more in an effort to maintain a positive energy budget they become more prone to predation. This may occur because juveniles will tolerate greater risks in obtaining food (e.g., Caraco *et al.*, 1982; Schneider, 1984), decrease their vigilance (e.g., Lendrem, 1983; East, 1988; Sullivan, 1988a), and accelerate their return to feeding areas after the passage of a predator (e.g., Hegner, 1985; de Laet, 1985; East, 1988). The time period during which juvenile foraging inefficiency accounts for much of their increased mortality is virtually unknown for most species (however, see Sullivan, 1989), but is valuable for understanding avian survivorship and population regulation.

5. SUMMARY

We know that juveniles can be less proficient than adults in their selection of foraging sites, search patterns, recognition and selection of food, and in capture and handling techniques. In all species which have been examined, age-related differences in foraging proficiency have been found in at least one aspect of foraging behavior. We frequently do not know the relative importance of these differences. Unfortunately, for only a very few species do we know the rate at which foraging proficiency improves, the factors influencing this rate, and whether predictable species differences occur in foraging development rates. Experience, necessary for both cognitive and motor skills, is the factor most likely to delay the juvenile's acquisition of adult levels of foraging proficiency. Sometimes, juveniles compensate for their inept foraging skills and differ from adults by foraging in different sites, selecting different sizes and types of food, using different capture methods, resorting to piracy, and/or joining feeding groups. While these age-related foraging differences have been reasonably well documented, we generally do not know the conditions in terms of cost of benefits which encourage or prevent these juvenile foraging differences. Because juveniles are usually subordinate to adults they are sometimes displaced from the prime foraging sites, prevented from taking the most profitable prey, and/or more likely to be victims of piracy. Therefore adult dominance may interact with the juvenile's inept foraging skills, and contribute to the inability to obtain an optimal diet or maintain a positive energy budget. Untangling the effects of adult dominance and the juvenile's lack of foraging skills is a difficult task for field workers concerned with identifying the factors contributing to high juvenile mortality rates and has been accomplished in only a few studies with marked individuals. Finally, experimentation and laboratory studies have yet to fully contribute to solving problems associated with the acquisition of foraging proficiency by juvenile birds.

1. APPENDIX: Summary of Species in Which Age-Related Differences in Foraging Proficiency Have Been Documented*

Brown Pelican (*Pelacanus occidentalis*)

Orians (1969). Plunge-diving. PR, PS, PC. Adults had higher capture success in dry season (60% vs. 49%) but not in wet season (57% vs. 61%).

Schnell *et al.* (1982). Plunge-diving. PR, PS, PC. Adults had higher success (84% vs. 75%).

Brandt (1984). Plunge-diving. FS, PR, PS, PC. No difference in foraging site use, adults had higher capture success (83% vs. 49%).

Carl (1987). PR, PS, PC. Adults had higher capture success (14% vs. 8%).

Olivaceous Cormorant (*Phalacrocorax olivaceus*)

Morrison *et al.* (1978). Surface diving, FS, PR, PS, PC. Adults had higher success rates in two habitats (18% vs. 10% and 19% vs. 12%).

Magnificent Frigatebird (*Fregata magnificens*)

Gochfeld and Burger (1981). Piracy of gulls, PR, PS, PC, PH. Adults had higher capture success (13% vs. 4%). Adults had shorter handling time (1.9 vs. 6.0 s).

Great Blue Heron (*Ardea herodias*)

Quinney and Smith (1980). Striking fish, PR, PS, PC. Adults took larger prey and had higher capture success (62–77% vs. 14–33%).

DesGranges (1981). Striking fish, PR, PS, PC. Adults had higher capture success (80% vs. 52%).

Grey Heron (*Ardea cinerea*)

Cook (1978). Striking fish, FS, PS, PC. Adults foraged in different sites and had higher capture success (50% vs. 29%).

Draulans and Van Vessem (1985). Striking fish, FS. Adults foraged in different sites than juveniles.

Draulans (1987). Striking fish, PS, PC, PH. Adults took larger prey and capture success differences were related to prey density; no age differences in prey handling.

Little Blue Heron (*Egretta caerulea*)

Recher and Recher (1969). Striking fish, FS, PR, PS, PC. No differences in foraging site but juveniles had more misses per strike and hence less g/min intake.

Cattle Egret (*Bubulcus ibis*)

Siegfried (1972). Insect gleaning, PR, PS, PC. In this diet study, adults took larger prey and had more food in their guts than juveniles.

Green-backed Heron (*Butorides striatus*)

Higuchi (1986). Bait fishing. Juveniles used inappropriate bait.

White Ibis (*Eudocimus albus*)

Bildstein (1983). Probing for crabs, FS, PR, PS, PC. Age differences in foraging site were found and yearlings had a capture rate 40% of the adult rate.

Bildstein (1984). Probing for crabs, PR, PS, PC. Second year birds had a capture rate 60% of the adult rate.

Osprey (*Pandion haliatus*)

Edwards (1989a). Plunge-diving, PR, PS, PC. Sibs had greater capture success than singletons, but singletons eventually caught up; sibs had similar capture techniques.

Edwards (1989b). Plunge-diving, FS, PR, PS. Juveniles differed from adults in foraging sites and diet; juveniles selected their diet on the basis of their first capture.

Snail Kite (*Rostrhamus sociabilis*)

Bourne (1985). Snail capture, FS, PH. Juveniles differed from adults in foraging site, took smaller snails, and had longer handling times regardless of snail size.

Northern Harrier (*Circus cyaneus*)

Bildstein (1987). Aerial attack and pounce on terrestrial prey, S, PC. No age difference in foraging site or capture technique. Juveniles differed from adults in temporal pattern of foraging, foraging height and speeds, and had lower capture success.

Sharp-shinned Hawk (*Accipiter striatus*)

Mueller and Berger (1970). Aerial attack, PR, PS. Juveniles took inappropriate prey.

European Sparrowhawk (*Accipiter nisus*)

Barnard (1979). Aerial attack, PC. A juvenile used innappropriate technique and made more prey passes than an adult.

Common Moorhen (*Gallinula chloropus*)

Sutherland *et al.* (1986). Surface picking, PR, PS, PC, PH. Adults had a higher capture rate (89 items/min vs. 40 items/min) and shorter handling times than juveniles.

American Coot (*Fulica americana*)

Ryan and Dinsmore (1980). Surface picking. Time budget analysis indicates that yearlings and second year birds spend more time feeding than third year birds.

Northern Lapwing (*Vanellus vanellus*)

Barnard and Thompson (1985). Surface picking, PR, PS, PC. Juveniles differed from adults in having lower capture rates and longer handling times. Juveniles were more likely to be victims of gull attacks.

Masked Lapwing (*Vanellus miles*)

Burger and Gochfeld (1985b). Surface picking, S, PR, PS, PC, PH. Interfood time interval was less for adults and juveniles pecked more without success.

Ringed Plover (*Charadrius hiaticula*)

Pienkowski (1983). Surface picking, S, PS, PC, PH. Juveniles differed from adults by having shorter waiting and giving-up times, took smaller, less variable prey, and lower pecking rates. Handling ability improved with age.

Eurasian (Northern Pied) Oystercatcher (*Haematopus ostralegus*)

Norton Griffiths (1967, 1968, 1969). Mussel probing, FS, PS, PH. Juveniles required 3 years to acquire adults levels of handling proficiency; juveniles learn mussel opening method of parent, either stabbing or hammering, which influences foraging site selection.

Goss-Custard *et al.* (1982), Goss-Custard and Durell (1983, 1987a, b, c). Mussel and other invertebrate probing, FS, PS, PH. Age difference found in foraging site use, diet selection, and handling ability. Juveniles were displaced by adults on prime mussel beds, so juveniles took different prey in different patches and were more likely to lose food to older birds.

Black-necked Stilt (*Himantopus mexicanus*)

Burger (1980). Surface/subsurface pecking, PR, PS, PC, PH. Interfood intervals were shorter for adults than juveniles, but only at mid-day; juveniles spent more time and energy foraging than adults at certain times of day.

Black-winged Stilt (*Himantopus himantopus*)

Espin *et al.* (1983). Surface/subsurface pecking, PS, PR, PC, PH. Adults have a higher capture succes rate and move more slowly than juveniles, but do not differ in handling times.

American Avocet (*Recurvirostra americana*)

Burger and Gochfeld (1986). Pecking and Scything, FS, PR, PS, PC. Juveniles were most similar in capture times to adults in habitats where adults were most efficient. Capture time (time to obtain 10 prey items) was shorter for adults under all conditions.

Ruddy Turnstone (*Arenaria interpres*)

Groves (1978). Surface pecking, FS, PR, PS, PC. Adults had higher foraging rates and success in two sites than juveniles in both sites; adults had higher foraging rates (.22/s vs. .19/s and .38/s vs. .25/s) and success rates (.19/s vs. .16/s and .26/s vs. .16/s) than juveniles. Adults were dominant to juveniles.

Curlew Sandpiper (*Calidris ferruginea*)

Puttick (1978). Substrate probing, PR, PS. Stomach analysis indicates that juveniles took smaller prey items than adults.

Puttick (1979). Substrate probing, FS, PR, PS, PC. No difference in foraging site; adults had higher foraging rate (43 vs. 28 probes/min) and higher success rate (10 vs. 4 prey items/min) than juveniles.

Gulls—15 species

Burger (1987). Variety of foraging methods, PR, PS, PC, PH. Interfood interval used for comparison. Adult interfood interval was shorter in all but 5 feeding situations. Variance in interval was explained by age, species, food type, method, and habitat.

Laughing Gull (*Larus atricilla*)

Burger (1981). Dump foraging, PR, PS, PC, PH. No age differences in success rate due to attacks by other gulls. Juveniles were more likely to drop food items.

Burger and Gochfeld (1981). Piracy, PR, PS, PC, PH. Adults had higher success as pirates; adults less likely to be victims; adults more likely to pirate large items and keep them; juveniles more likely to steal from other juveniles; juveniles more likely to drop items and took longer to handle food than adults.

Burger and Gochfeld (1983), Burger (1987). Different foraging methods (aerial snatching, aerial dipping, plunge-diving), FS, PR, PS, PC, PH. Adults differed in success in relation to habitat; juveniles concentrated in habitats where their interfood interval was closest to adults. Adults had shorter interfood intervals and higher success rates than did juveniles.

Carroll and Cramer (1985). Piracy from pelicans, PR, PS. Adult gulls preferred juvenile pelicans as victims; juvenile gulls selected their victims at random.

Burger and Gochfeld (1985). Dump foraging, S, PR, PS, PC, PH. Adult interfood interval is shorter than juveniles when not overwhelmed by pirates. When overwhelmed by pirates no differences occurred.

Common Black-headed Gull (*Larus ridibundus*)

Ulfstrand (1979). Surface picking for worms, S, PR, PS, PC, PH. Adult success rate (items/min) was higher than juvenile rate; juvenile success rate was higher in pure juvenile flocks than in mixed adult/juvenile flocks due to adult interference.

Burger (1987). Different foraging methods. Interfood time interval was shorter for adults except when aerial dipping for invertebrates on a bay.

Hesp and Barnard (1989). Piracy on lapwings, FS, PS, PC. Adults often in lapwing flocks while juvenile sometimes foraged in plowed fields; juveniles tended to select their victims at the wrong time; juveniles mistimed their attack and thus had lower success rates than adults (21.7–22.3% vs. 39.4–53.1%).

Amat and Aguilera (1990). Piracy on egrets, stilts, and godwits, PS. Juveniles more likely to attack stilts while adults preferentially attack godwits. Adults force juveniles to use suboptimal victims (stilts).

Bonaparte's Gull (*Larus philadelphia*)

MacLean (1986). Plunge-diving, S, PR, PS, PC, PH. Juveniles had higher search times, lower capture efficiency, longer interfood intervals, and were more likely to drop items than adults. He documented juvenile improvements over time.

Burger (1987). Different foraging methods. Interfood interval was shorter for adults than juveniles, except when picking up insects from a mudflat.

Ring-billed Gull (*Larus delawarensis*)

Burger and Gochfeld (1979, 1981). Piracy on starlings, PS, PC. Adults were more successful than juveniles or subadults (66% success vs. 33% vs. 36%). Adult success was due to use of an effective method (flight), while juveniles were more likely to walk.

MacLean (1986). Plunge-diving, S, PS, PC, PH. Juvenile search time was longer; adults were more successful at prey capture, had shorter interfood interval, but juveniles improved with time; juveniles were more likely to drop items. He showed that juveniles improved with time.

Burger (1987). Different foraging methods. Interfood interval shorter for adults in all foraging situations.

Burger and Gochfeld (1985). Dump foraging. Interfood interval shorter for adults under all conditions.

California Gull (*Larus californicus*)

Porter and Sealy (1982), Plunge-diving and dipping, PR, PS, PC. Adult success rate was higher than rate for subadult or juvenile (95% vs. 76% vs. 26%). Juveniles used local enhancement to find food.

Burger (1987). Different foraging methods. Adult interfood interval was shorter than interval for juveniles under all conditions.

Herring Gull (*Larus argentatus*)

Verbeek (1977a). Dump digging, S. Adults moved more items when searching for food and thus encountered more food items/min than juveniles.

Verbeek (1977b). Plunge-diving for starfish, PS, PC. Adults had higher capture efficiency than juveniles (64% vs. 16%).

Ingolfsson and Estrella (1978). Shell-dropping, PH. Juveniles were less likely to crack shells on the first drop (43% vs. 65%) and less likely to be successful after several attempts.

Burger (1981). Dump foraging, PR, PS, PC, PH. Foraging efficiency (items/attempt) varied with age (adults, 58% vs. subadults, 28% vs. juveniles 35%). Juveniles were more likely to drop items and were more likely to attempt piracy than adults.

Burger and Gochfeld (1981). Piracy on gulls. Adults were pirates more often than juveniles; adults were successful pirates more often; adults were victims less often; once chased juvenile were more likely to drop food.

Greig *et al.* (1983). Dump foraging, S, PR, PS, Adults turned over objects more quickly than juveniles; juveniles walked more per food item (133 paces/item vs. 41 paces/item) than adults. Juveniles were more likely to use displacement attacks for food.

Burger and Gochfeld (1985). Dump foraging, S PR, PS, PC, PH. Adult interfood interval was shorter than interval for juveniles, but only when not overwhelmed by pirates. If overwhelmed by pirates no age differences were found.

MacLean (1986). Plunge-diving PR, PS, PC, PH. Juveniles showed more intention movements; interfood interval shorter for adults; capture efficiency was lower for juvenile, but improved with time; juveniles were more likely to drop food items.

Burger (1987). Different foraging methods. Adults had shorter interfood intervals except when picking worms from a surface.

Glaucous-winged Gull (*Larus glaucescens*)

Barash *et al.* (1975). Clam dropping, PH. Juveniles were more likely to drop clams in flight (29%) than adults (7%) and they were more likely to to drop items on the wrong substrate and from the wrong height.

Searcy (1978). Plunge-diving, PR, PS, PC. Adults were more efficient than subadults or juveniles (69% vs. 50% vs. 44%).

Burger (1987). Different foraging methods. Adult interfood interval was shorter than the interval for juveniles under all conditions.

Royal Tern (*Sterna maxima*)

Buckley and Buckley (1974). Plunge-diving, S, PR, PS, PC, Ph. Adults covered the feeding area more quickly, were more likely to recognize capturable prey, and thus caught more prey per unit time than did juveniles. Juveniles were more likely to drop prey items. Capture success rate did not differ with age.

Sanwich Tern (*Sterna sandvicensis*)

Dunn (1972). Plunge diving, PR, PS, PC, PH. Adults had a higher capture success than juveniles (17% vs. 13%) and higher capture rate (.24 captures/min vs. .16 captures/min). There was no difference in the rate of diving and juveniles were more likely to drop prey.

Crimson Rosella (*Platycercus elegans*)

Magrath and Lill (1985). Insect and plant gleaning. FS, PS. Juveniles fed in different sites and took more insects than adults which fed on fern sori. Time budget analysis indicated that juveniles spent more time feeding.

Northwestern Crow (*Corvus caurinus*)

Richardson and Verbeek (1987). Clam dropping, S, PS, PH. Juveniles spent more time searching than adults (64 s vs. 35 s), took smaller clams, required more drops to break clams (2 drops vs. 1.7 drops/clam), and thus obtained ⅓ the net energy intake of adults.

Rook (*Corvus frugilegus*)

East (1988). Seed-eating, FS, PR, PS, PC, PH. Juveniles foraged in different sites and were initially slower at consuming grain (65 vs. 133 s) than adults, but eventually obtained the same consumption rate.

European (Wood) Nuthatch (*Sitta europaea*)

Enoksson (1988). Seed-eating, PH. In summer juveniles took fewer seeds per visit than adults (1.4 vs. 2.2), but no difference in October.

Reed Warbler (*Acrocephalus scirpaceus*)

Davies and Green (1976). Insect gleaning and flycatching, PR, PS, PC, PH. Juveniles initially pecked at all objects, but quickly became selective; capture success improved with age; naive birds were more likely to drop prey, but handling time and ability improved with age.

Spotted Flycatcher (*Muscicapa striata*)

Davies (1976). Insect gleaning and flycatching, PR, PS, PC. Juvenile capture success improved with time and hence they relied less on begging from parents.

Northern Wheatear (*Oenanthe oenanthe*)

Moreno (1984). Insect gleaning and flycatching, PR, PS, PC. Naive juveniles initially pecked at inedible objects, and their capture success improved with age.

Eastern Bluebird (*Sialia sialis*)

Goldman (1975). Perch-to-ground sallying and flycatching, PR, PS, PC. Adults had higher capture success than juveniles (67% vs. 32%).

American Robin (*Turdus migratorius*)

Gochfeld and Burger (1984). Soil probing for invertebrates, S, PR, PS, PC. Juveniles took more steps per bout than adults, took smaller prey items, and juvenile capture success was 75% that of adults. Juveniles took 136% as long to capture each item, and required 161% as many steps as adults, obtained 25% less food per unit time.

Northern Mockingbird (*Mimus polyglottos*)

Breitwisch *et al.* (1987). Ground gleaning and perch-to-ground sallying, PR, PS, PC. Adults took larger prey and had higher capture success (70% vs. 25%) than juveniles. Adults used more aerial attacks, but juveniles increased aerial attacks over time.

European Starling (*Sturnus vulgaris*)

Stevens (1985). Ground probing and pecking, S, PR, PC. Juveniles more likely to feed on fruit, possibly due to ease of capture. Adults had higher capture success when feeding on invertebrates (177 mg prey/100 pecks vs. 35 mg).

Silvereye (*Zosterops lateralis*)

Jansen (1990) Insect gleaning and probing, PR, PS, PC. Older birds had higher success rate than first-year birds for all capture techniques and substrates except dead leaves and bark. Second-year birds had higher success rate than first-year for gleans and in needles.

Bananaquit (*Coereba flaveola*)

Wunderle and Soto (1987). Nectar feeding, S, PR. Juveniles were more likely to re-visit a previously visited flower than were adults.

Wunderle and Lodge (1988). S, PR. Juveniles made more flower revisitations than expected by chance alone; adults made fewer flower revisitations than expected by chance. Movement patterns of adults were different than those of juveniles.

Yellow-eyed Junco (*Junco phaeonotus*)

Sullivan (1988a). Picking insects from the ground, PS, PR, PC, PH. Juveniles initially had higher pecks per capture than adults, but reached adult levels by independence. Time budget indicated that juveniles spent more time searching for, pecking at, and handling food items than adults.

Sullivan (1988b). Picking insects from the ground, PS, PH. Juveniles selected smaller mealworms than adults, and took less time to handle small prey than large. In contrast, adults took less time than juveniles to handle large prey items. Age-related diet differences were due to the relative profitability of each prey for each age class.

Weathers and Sullivan (1989). Picking insects from the ground. Time and energy analysis indicates that adults obtain nearly three times as much energy per hour foraging as recently independent juveniles (14.5 vs. 5.3 kJ/h).

Red-winged Blackbird (*Agelaius phoeniceus*)

Beauchamp *et al.* (1987). Search in three choice maze, S. Adults were more likely to return to the site where they were initially rewarded with food and in which food supply was not depleted. Juveniles returned at random.

Eurasian Bullfinch (*Pyrrhula pyrrhula*)

Greig-Smith (1985). Fruit and seed eating, FS, PS, PH. No age differences in foraging site, but adults ate more ash seeds and juveniles more likely to drop food items.

*Given for each reference is foraging mode and any differences for the following foraging components: foraging site (FS), search (S), prey recognition (PR), prey selection (PS), prey capture (PC), and prey handling (PH).

ACKNOWLEDGMENTS. This manuscript benefitted from the constructive comments of Thomas C. Edwards, Jr., Dennis M. Power, Kimberly A. Sullivan, and Marcia Wilson. Scott Lanyon assisted by making the library facilities of the Field Museum of Natural History in Chicago available to me. Numerous articles were obtained via interlibrary loan with the help of JoAnne Feheley, librarian at the Institute of Tropical Forestry.

REFERENCES

Alcock, J., 1973, The feeding response of hand-reared Red-winged Blackbirds (*Agelaius phoeniceus*) to a stinkbug (*Euschistus conspersus*), *Am. Midl. Nat.* **89**:307–313.

Amat, J. A., and Agiulera, E., 1990, Tactics of Black-headed Gulls robbing egrets and waders, *Anim. Behav.* **39**:70–77.

Arnold, S. J., and Wade, M. J., 1984, On the measurement of natural and sexual selection: Theory, *Evolution* **38**:709–719.

Ashmole, N. P., and Tover, S. H., 1968, Prolonged parental care in Royal Terns and other birds, *Auk* **85**:90–100.

Bairlein, F., 1981, Ökosystemanalyse der Rastplätze von Zugvögeln: Beschreibung und Deutung der Verteilungsmuster von ziehenden Kleinvögeln in verschiedenen Biotopen der Stationen des "Mettnau-Reit-Illmitz-Programmes," *Ökologie der Vögel* **3**:7–137.

Barash, D. P., Donovan, P., and Myrick, R., 1975, Clam dropping behavior of the Glaucous-winged Gull (*Larus glaucescens*), *Wilson Bull.* **87**:60–64.

Barnard, C. J., 1979, Interactions between House Sparrows and Sparrowhawks, *Brit. Birds* **72**:569–573.

Barnard, C. J., and Stephens, H., 1981, Prey size selection by lapwings in lapwing/gull associations, *Behaviour* **77**:1–22.

Barnard, C. J., and Thompson, D. B. A., 1985, *Gulls and Plovers: The Ecology and Behaviour of Mixed-Species Feeding Groups*, Croom Helm, Beckenham, England.

Beauchamp, G., Cyr, A., and Houle, C., 1987, Choice behaviour of Red-winged Blackbirds (*Agelaius phoeniceus*) searching for food: The role of certain variables in stay and shift strategies, *Behav. Processes* **15**:259–268.

Bildstein, K. L., 1983, Age-related differences in the flocking and foraging behavior of White Ibises in a South Carolina salt marsh, *Colonial Waterbirds* **6**:45–53.

Bildstein, K. L., 1984, Age-related differences in the foraging behavior of White Ibises, *Colonial Waterbirds* **7**:146–148.

Bildstein, K. L., 1987, Behavioral ecology of Red-tailed Hawks (*Buteo jamaicensis*), Roughlegged Hawks (*Buteo lagopus*), Northern Harriers (*Circus cyaneus*), and American Kestrels (*Falco sparverius*), in south central Ohio, *Ohio Biol. Surv. Biol. Notes* **18**:1–53.

Bourne, G. R., 1985, The role of profitability in Snail Kite foraging, *J. Anim. Ecol.* **54**:697–709.

Brandt, C. A., 1984, Age and hunting success in the Brown Pelican: Influences of skill and patch choice on foraging efficiency, *Oecologia* **62**:132–137.

Breitwisch, R., Diaz, M., and Lee, R., 1987, Foraging efficiencies and techniques of juvenile and adult Northern Mockingbirds (*Mimus polyglottos*), *Behaviour* **101**:225–235.

Brown, J. L., 1975, *The Evolution of Behavior*, W. W. Norton, New York.

Bryan, J. E., and P. A. Larkin, 1972, Food specialization by individual trout, *J. Fish. Res. Board Canada* **29**:1615–1624.

Buckley, F. G., and Buckley, P. A., 1974, Comparative feeding ecology of wintering adult and juvenile Royal Terns (Aves: Laridae, Sterninae), *Ecology* **55**:1053–1063.

Burger, J., 1980, Age differences in foraging Black-necked Stilts in Texas, *Auk* **97**:633–636.

Burger, J., 1981, Feeding competition between Laughing Gulls and Herring Gulls at a sanitary landfill, *Condor* **83**:328–335.

Burger, J., 1987, Foraging efficiency in gulls: A congeneric comparison of age differences in efficiency and age of maturity, *Studies in Avian Biology* **10**:83–90.

Burger, J., 1988, Effects of age on foraging in birds, in: *Proc. 19th International Ornithological Congress* (H. Ouellet, ed.), Univ. Ottawa Press, Ottawa, pp. 1127–1140.

Burger, J., and Gochfeld, M., 1979, Age differences in Ring-billed Gull kleptoparasitism on Starlings, *Auk* **96**:806–808.

Burger, J., and Gochfeld, M., 1981, Age-related differences in piracy behaviour of four species of gulls, Larus, *Behaviour* **77**:242–267.

Burger, J., and Gochfeld, M., 1983, Feeding behavior in Laughing Gulls: Compensatory site selection by young, *Condor* **85**:467–473.

Burger, J., and Gochfeld, M., 1985a, The effects of relative numbers on aggressive interactions and foraging efficiency in gulls: The cost of being outnumbered, *Bird Behaviour* **5**:81–89.

Burger, J., and Gochfeld, M., 1985b, Age-related differences in the Masked Lapwing *Vanellus miles* in Australia, *Emu* **85**:47–48.

Burger, J., and Gochfeld, M., 1986, Age differences in foraging efficiency of American Avocets *Recurvirostra americana*, *Bird Behavior* **6**:66–71.

Cannon, C. E., 1979, Observations on the behavioural development of young rosellas, *Sunbird* **10**:25–32.

Caraco, T., Martindale, S., and Pulliam, H. R., 1982, Avian time budgets and distance from cover, *Auk* **97**:872–875.

Carl, R. A., 1987, Age-class variation in foraging techniques by Brown Pelicans, *Condor* **89**:525–533.

Carroll, S. P., and Cramer, K. L., 1985, Age differences in kleptoparasitism by Laughing Gulls (*Larus atricilla*) on adult and juvenile Brown Pelicans (*Pelicanus occidentalis*), *Anim. Behv.* **33**:201–205.

Catterall, C. P., Kikkawa, J., and Gray, J., 1989, Interrelated age-dependent patterns of ecology and behaviour in a Silvereye population, *J. Anim. Ecol.* **58**:557–570.

Charnov, E. L., 1976, Optimal foraging: Attack strategy of a mantid, *Am. Nat.* **110**:141–151.

Charnov, E. L., Orians, G. H., Hyatt, K., 1976, Ecological implications of resource depression, *Am. Nat.* **110**:247–259.

Clark, D. A., 1980, Age- and sex-dependent foraging strategies of a small mammalian omnivore, *J. Anim. Ecol.* **49**:549–563.

Coblentz, B. E., 1986, A possible reason for age-differential foraging success in Brown Pelicans, *J. Field Ornithol.* **57**:63–64.
Cody, M. L., 1971, Finch flocks in the Mojave Desert, *Theor. Pop. Biol.* **2**:142–158.
Cody, M. L., 1985, *Habitat Selection in Birds*, Academic Press, New York.
Cook, D. C., 1978, Foraging behaviour and food of Grey Herons (*Ardea cinerea*) on the Ythan Estuary, *Bird Study* **25**:17–22.
Cooke, F., and Ross, R. K., 1972, Diurnal and seasonal activities of a post-breeding population of gulls in southeastern Ontario, *Wilson Bull.* **84**:164–172.
Croze, H., 1970, Searching image in Carrion Crows, *Z. Tierpsychol.* **5**:1–85.
Cruze, W. W., 1935, Maturation and learning in chicks, *J. Comp. Psychol.* **19**:371–409.
Davies, N. B., 1976, Parental care and the transition to independent feeding in the young Spotted Flycatcher (*Muscicapa striata*), *Behaviour* **59**:280–295.
Davies, N. B., and Green, R. E., 1976, The development and ecological significance of feeding techniques in the Reed Warbler (*Acrocephalus scirpaceus*), *Anim. Behav.* **24**:213–229.
Davis, J. W. F., 1975, Specialization in feeding location by Herring Gulls, *J. Anim. Ecol.* **44**:795–804.
Dawkins, M., 1971, Perceptual changes in chicks: Another look at the "search image" concept, *Anim. Behav.* **19**:566–574.
DesGranges, J., 1981, Observations sur l'alimentation du grand héron *Ardea herodias* au Québec (Canada), *Alauda* **49**:25–34.
Draulans, D., 1987, The effect of prey density on foraging behaviour and success of adult and first-year Grey Herons (*Ardea cinerea*), *J. Anim. Ecol.* **56**:479–493.
Draulans, D., and Van Vessem, J., 1985, Age-related differences in the use of time and space by radio-tagged Grey Herons (*Ardea cinerea*) in winter, *J. Anim. Ecol.* **54**:771–780.
Dunbrack, R. L., 1979, A re-examination of robbing behavior in foraging egrets, *Ecology* **60**:644–645.
Dunn, E. K., 1972, Effect of age on the fishing ability of Sandwich Terns *Sterna sandvicensis*, *Ibis* **114**:360–366.
East, M., 1988, Crop selection, feeding skills and risks taken by adult and juvenile Rooks *Corvus frugilegus*, *Ibis* **130**:294–299.
Edwards, T. C., 1988, Temporal variation in prey preference patterns of adult Ospreys, *Auk* **105**:244–251.
Edwards, T. C., Jr., 1989a, Similarity in the development of foraging mechanics among sibling Ospreys, *Condor* **91**:30–36.
Edwards, T. C., Jr., 1989b, The ontogeny of diet selection in fledgling Ospreys, *Ecology* **70**:881–896.
Emlen, J. M., 1973, *Ecology: An Evolutionary Approach*, Addison-Wesley, Reading, Massachusetts.
Ekman, J., 1987, Dominance, exposure and time use in Willow Tit flocks: The cost of subordination, *Anim. Behav.* **35**:445–452.
Ekman, J., and Askenmo, C., 1984, Social rank and habitat use in Willow Tit groups, *Anim. Behav.* **32**:508–514.
Enoksson, B., 1988, Age- and sex-related differences in dominance and foraging behaviour of nuthatches *Sitta europea*, *Anim. Behav.* **36**:231–238.
Espin, P. M. J., Mather, R. M., and Adams, J., 1983, Age and foraging success in Black-winged Stilts *Himantopus himantopus*, *Ardea* **71**:225–228.
Feare, C. J., 1980, The economics of Starling damage, in: *Bird Problems in Agriculture* (E. N. Wright, I. R. Inglis, and C. J. Feare, eds.), BCPC Publications, London.

Ficken, M. S., and Ficken, R. W., 1967, Age-specific differences in the breeding behavior of the American Redstart, *Wilson Bull.* **79**:188–199.

Fischer, D. L., 1985, Piracy behavior of wintering Bald Eagles, *Condor* **87**:246–251.

Galef, B. G., Jr., 1976, Social transmission of acquired behavior: A discovery of tradition and social learning in vertebrates, *Advances in the Study of Behavior* **6**:77–100.

Gass, C. L., and Montgomerie, R. D., 1981, Hummingbird foraging behavior, decision-making and energy regulation, in: *Foraging Behavior: Ecological, Ethological, and Psychological Approaches* (A. C. Kamil and T. D. Sargent, eds.), Garland STPM Press, New York, pp. 159–194.

Gill, F. B., and Wolf, L. L., 1977, Nonrandom foraging by sunbirds in a patchy environment, *Ecology* **58**:1284–1296.

Gochfeld, M., 1980, Learning to eat by young Common Terns: Consistency of presentation as an early cue, *Colonial Waterbirds* **3**:108–118.

Gochfeld, M., and Burger, J., 1981, Age-related differences in piracy of frigatebirds from Laughing Gulls, *Condor* **83**:79–82.

Gochfeld, M., and Burger, J., 1984, Age differences in foraging behavior of the American Robin, *Behaviour* **88**:227–239.

Goldman, P., 1975, Hunting behavior of Eastern Bluebirds, *Auk* **92**:798–801.

Goss-Custard, J. D., and Durell, S. E. A. leV. dit, 1983, Individual and age differences in the feeding ecology of Oystercatchers *Haematopus ostralegus* wintering on the Exe Estuary, Devon, *Ibis* **125**:155–171.

Goss-Custard, J. D., and Durell, S. E. A. leV. dit, 1987a, Age-related effects in oystercatchers *Haematopus ostralegus*, feeding on mussels, *Mytilus edulis*. I. Foraging efficiency and interference, *J. Animl. Ecol.* **56**:521–536.

Goss-Custard, J. D., and Durell, S. E. A. leV. dit, 1987b, Age-related effects in oystercatchers *Haematopus ostralegus*, feeding on mussels, *Mytilus edulis*. II. Aggression, *J. Anim. Ecol.* **56**:537–548.

Goss-Custard, J. D., and Durell, S. E. A. leV. dit, 1987c, Age-related effects in oystercatchers *Haematopus ostralegus*, feeding on mussels, *Mytilus edulis*. III. The effect of interference on overall intake rate, *J. Anim. Ecol.* **56**:549–558.

Goss-Custard, J. D., Durell, S. E. A. leV. dit, and Eng, B. J., 1982a, Individual differences in aggressiveness and food stealing among wintering oystercatchers, *Haematopus ostralegus* L., *Anim. Behav.* **30**:917–928.

Goss-Custard, J. D., Durell, S. E. A. leV. dit, McGrorty, S., and Reading, C. J. 1982b, Use of mussel *Mytilus edulis* beds by Oystercatchers *Haematopus ostralegus* according to age and population size, *J. Anim. Ecol.* **51**:543–554.

Goss-Custard, J. D., Durell, S. E. A. lev. dit, Sitters, H. P., and Swinfen, R., 1982c, Age structure and survival of a wintering population of oystercatchers, *Bird Study* **29**:83–98.

Goss-Custard, J. D., and Sutherland, W. J., 1984, Feeding specialization in oystercatchers *Haematopus ostralegus*, *Anim. Behav.* **32**:299–301.

Gould, J., 1982, *Ethology*, W. W. Norton, New York.

Greenberg, R., 1984, The role of neophobia in the foraging site selection of a tropical migrant birds: An experimental study, *Proc. Natl. Acad. Sci.* **81**:3778–3780.

Greenberg, R., 1987, Development of dead leaf foraging in a tropical migrant warbler, *Ecology* **68**:130–141.

Greig, S. A., Coulson, J. C., and Monoghan, P., 1983, Age-related differences in foraging success in the Herring Gull (*Larus argentatus*), *Anim. Behav.* **31**:1237–1243.

Greig-Smith, P. W., 1984, Food-handling by Bullfinches in relation to the risks associated with dropping seeds, Anim. Behav. **32**:929–931.

Greig-Smith, P. W., 1985, Winter survival, home ranges and feeding of first-year and adult

Bullfinches, in: *Behavioral Ecology: the Ecological Consequences of Adaptive Behaviour* (R. M. Sibley and R. H. Smith, eds.), Blackwell Scientific Publ., Oxford, pp. 387–392.

Groves, S., 1978, Age-related differences in Ruddy Turnstone foraging and aggressive behavior, *Auk* **95:**95–103.

Gustafsson, L., 1988, Foraging behaviour of individual Coal Tits, *Parus ater,* in relation to their age, sex, and morphology, *Anim. Behav.* **36:**696–704.

Hailman, J. P., 1967, The ontogeny of an instinct, *Behaviour* (Suppl.) **15:**1–159.

Hailman, J. P., 1982, Ontogeny: Toward a general theoretical framework for ethology, in: *Perspectives in Ethology: Ontogeny,* Vol. 5 (P. P. G. Bateson and P. H. Klopfer, eds.), Plenum Press, New York, pp. 133–189.

Hale, C., and Green, L., 1979, Effect of initial-pecking consequences on subsequent pecking in young chicks, *J. Comp. Physiol. Psychol.* **93:**730–735.

Hale, C., and Green, L., 1988, Effects of ingestional experiences on the acquisition of appropriate food selection by young chicks, *Anim. Behav.* **36:**211–224.

Hegner, R. H., 1985, Dominance and anti-predator behaviour in Blue Tits, *Parus caeruleus, Anim. Behav.* **33:**762–768.

Heppleston, P. B., 1971, The feeding ecology of oystercatchers (*Haematopus ostralegus* L.) in winter in Northern Scotland, *J. Anim. Ecol.* **40:**651–672.

Hesp, L. S., and Barnard, C. J., 1989, Gulls and plover: Age-related differences in kleptoparasitism among Black-headed Gulls (*Larus ridibundus*), *Behav. Ecol. Sociobiol.* **24:**297–304.

Higuchi, H., 1986, Bait-fishing by the Green-backed Heron *Ardeola striata* in Japan, *Ibis* **128:**285–290.

Hirons, G., Hardy, A., and Stanley, P., 1979, Starvation in young Tawny Owls, *Bird Study* **26:**59–63.

Hogan, J. A., 1973a, How young chicks learn to recognize food, in: *Constraints on Learning* (R. A. Hinde and J. Stevenson-Hinde, eds.), Academic Press, London, pp. 119–139.

Hogan, J. A., 1973b, Development of food recognition in young chicks: I. Maturation and nutrition, *J. Comp. Physiol. Psychol.* **83:**355–366.

Hogan, J. A., 1977a, The development of food recognition in young chicks: IV. Associative and nonassociative effects of experience. *J. Comp. Physiol. Psychol.* **91:**839–850.

Hogan, J. A., 1977b, The ontogeny of food preferences in chicks and other animals, in: *Learning Mechanisms in Food Selection* (L. J. Barker, M. Best, and M. Domjan, eds.), Baylor University Press, Waco, Texas, pp. 71–97.

Hogan-Warburg, A. J., and Hogan, J. A., 1981, Feeding strategies in the development of food recognition in young chicks, *Anim. Behav.* **29:**143–154.

Holling, C. S., 1959, Some characteristics of simple types of predation and parasitism, *Can. Ent.* **91:**385–398.

Howe, H. F., 1974, A case of age-related territory usurption in the American Redstart, *Jack-pine Warbler* **52:**92.

Ingolfsson, A., and Estrella, B. T., 1978, The development of shell-cracking behavior in Herring Gulls, *Auk* **95:**577–579.

Irwin, R. E., 1988, The evolutionary importance of behavioural development: The ontogeny and phylogeny of bird song, *Anim. Behav.* **36:**814–824.

Jansen, A., 1990, Acquisition of foraging skills by Heron Island Silvereyes *Zosterops lateralis chorocephala, Ibis* **132:**95–101.

Kamil, A. C., 1978, Systematic foraging by a nectar-feeding bird, the Amakihi (*Loxops virens*), *J. Comp. Physiol. Psychol.* **92:**388–396.

Kamil, A. C., and Yoerg, S. I., 1982, Learning and foraging behavior, in: *Perspective in*

Ethology: Ontogeny, Vol. 5 (P. P. G. Bateson and P. H. Klopfer, eds.), Plenum Press, New York, pp. 325–364.

Kent, B. W., 1981, Prey dropping by Herring Gulls on soft sediments, *Auk* **98**:350–354.

Kear, J., 1962, Food selection in finches with special reference to interspecific differences, *Proc. Zool. Soc. London* **138**:163–204.

Klopfer, P. H., 1959, Social interactions in discrimination learning with specific reference to feeding behaviour in birds, *Behaviour* **14**:282–299.

Klopfer, P. H., and Ganzhorn, J. U., 1985, Habitat selection: Behavioral aspects, in: *Habitat Selection in Birds* (M. L. Cody, ed.), Academic Press, New York, pp. 436–453.

Krebs, J. R., 1980, Optimal foraging, predation risk and territory defence, *Ardea* **68**:83–90.

Krebs, J. R., and McCleery, R. H., 1984, Optimization in behavioural ecology, in: *Behavioural Ecology: An Evolutionary Approach* (J. R. Krebs and N. B. Davies, eds.), Sinauer Associates, Sunderland, Massachusetts, pp. 91–121.

Krebs, J. R., Stephens, D. W., and Sutherland, W. J., 1983, Perspectives in optimal foraging, in: *Perspectives in Ornithology* (A. H. Brush and G. A. Clark, Jr., eds.), Cambridge Univ. Press, New York, pp. 165–216.

Kushlan, J. A., 1978, Nonrigorous foraging by robbing egrets, *Ecology* **59**:649–653.

Lack, D., 1954, Natural Regulation of Animal Numbers, Clarendon, Oxford.

Lack, D., 1966, *Population Studies of Birds*, Clarendon, Oxford.

Lack, D., 1968, Ecological Adaptation for Breeding in Birds, Methuen, London.

Lact De, J. F., 1985, Dominance and anti-predator behaviour of Great Tits *Parus major*: A field study, *Ibis* **127**:372–377.

Lendrem, D. W., 1983, Predation risk and vigilance in the Blue Tit *Parus caeruleus*, *Behav. Ecol. Sociobiol.* **14**:9–13.

Ligon, J. D., and Martin, D. J., 1974, Pinon seed assessment by the Pinon Jay, *Gymnorhinus cyanocephalus*, *Anim. Behav.* **22**:421–429.

Lorenz, K., and von St. Paul, U., 1968, Die Entwicklung des Spiessens und Klemmens beiden drei Würtgerarten *Lanius collurio*, *L. senator*, und *L. excubitor*, *J. Ornithol.* **109**:137–156.

MacArthur, R. H., 1972, *Geographical Ecology*, Harper & Row, New York.

Maccarone, A. D., 1987, Age-class differences in the use of food sources by European Starlings, *Wilson Bull.* **99**:699–704.

MacLean, A. A. E., 1986, Age-specific foraging ability and the evolution of deferred breeding in three species of gulls, *Wilson Bull.* **98**:267–279.

Magrath, R. D., and Lill, A., 1985, Age-related differences in behaviour and ecology of Crimson Rosellas, *Platycercus elegans*, during the non-breeding season, *Aust. Wildl. Res.* **12**:299–306.

Maron, J. L., 1982, Shell-dropping behavior of Western Gulls, *Auk* **99**:365–369.

Maynard Smith, J., 1982, Evolution and the Theory of Games, Cambridge University Press, Cambridge.

Maynard Smith, J., Burian, R., Kauffman, S., Alberch, P., Campbell, J., Goodwin, B., Lande, R., Raup, D., and Wolpert, L., 1985, Developmental constraints and evolution, *Quart. Rev. Biol.* **60**:265–287.

Moreno, J., 1984, Parental care of fledged young, division of labor, and the development of foraging techniques in the Northern Wheatear (*Oenanthe oenanthe* L.), *Auk* **101**:741–752.

Morrison, M. L., Slack, R. D., and Shanley, E., Jr., 1978, Age and foraging ability relationships of Olivaceous Cormorants, *Wilson Bull.* **90**:414–422.

Morse, D. H., 1980, *Behavioral Mechanisms in Ecology*, Harvard University Press, Cambridge.

Moyle, P., 1966, Feeding behavior of Glaucous-winged Gull on an Alaskan salmon stream, *Wilson Bull.* **78**:175–190.

Mueller, H. C., 1974, The development of prey recognition and predatory behaviour in the American Kestral *Falco sparverius, Behaviour* **49**:313–324.

Mueller, H. C., Berger, D. D., 1970, Prey preferences in the Sharp-shinned Hawk: The roles of sex, experience, and motivation, *Auk* **87**:452–457.

Murton, R. K., 1971, The significance of a specific search image in the feeding behaviour of the Wood Pigeon, *Behaviour* **40**:10–42.

Newton, I., 1967, The adaptive radiation and feeding ecology of some British finches, *Ibis* **107**:33–98.

Norton-Griffiths, M., 1967, Some ecological aspects of the feeding behaviour of the oystercatcher *Haematopus ostralegus* on the edible mussel *Mytilus edulis, Ibis* **109**:412–424.

Norton-Griffiths, M., 1968, *The feeding behaviour of the oystercatcher*, Ph.D. thesis, Oxford University, Oxford, England.

Norton-Griffiths, M., 1969, The organization, control and development of parental feeding in the oystercatcher (*Haematopus ostralegus*), *Behaviour* **34**:55–114.

Olive, C. W., 1982, Behavioral response of a sit-and-wait predator to spatial variation in foraging gain, *Ecology* **63**:912–920.

Olton, D. S., Handelmann, G. E., and Walker, J. A., 1981, Spatial memory and food-searching strategy, in: *Foraging Behavior: Ecological, Ethological, and Psychological Approaches* (A. C. Kamil and T. D. Sargent, eds.), Garland STPM Press, New York, pp. 333–355.

Orians, G. H., 1969, Age and hunting success in the Brown Pelican (*Pelecanus occidentalis*), *Anim. Behav.* **17**:316–319.

Orians, G. H., 1981, Foraging behavior and the evolution of discriminatory abilities, in: *Foraging Behavior: Ecological, Ethological, and Psychological Approaches* (A. C. Kamil and T. D. Sargent, eds.), Garland STPM Press, New York, pp. 389–405.

Palameta, B., and Lefebvre, L., 1985, The social transmission of a food-finding technique in pigeons: What is learned? *Anim. Behav.* **33**:892–896.

Partridge, L., 1976, Individual differences in feeding efficiencies and feeding preferences of captive Great Tits, *Anim. Behav.* **24**:230–240.

Partridge, L., 1979, Differences in behaviour between Blue and Coal Tits reared under identical conditions, *Anim. Behav.* **27**:120–125.

Partridge, L., and Green, P., 1985, Intraspecific feeding specializations and population dynamics, in: *Behavioral Ecology: the Ecological Consequences of Adaptive Behaviour* (R. M. Sibley and R. H. Smith, eds.), Blackwell Scientific Publishing, Oxford, pp. 207–226.

Partridge, L., and Green, P., 1987, An advantage for specialist feeding in Jackdaws, *Corvus monedula, Anim. Behav.* **35**:982–990.

Paszkowski, C. A., and Moermond, T. C., 1984, Prey handling relationships in captive Ovenbirds, *Condor* **86**:410–415.

Perrins, C. M., 1980, Survival of young Great Tits, *Parus major,* in: *Acta XVII Congress Internationalis Ornitholoigici* 159–174.

Pienkowski, M. W., 1983, Development of feeding and foraging behaviour in young Ringed Plovers *Charadrius hiaticula* in Greenland and Britain, *Dansk Orn. Foren. Tidsskr.* **77**:133–147.

Porter, J. M., and Sealy, S. G., 1982, Dynamics of seabird multispecies feeding flocks: Age-related feeding behaviour, *Behaviour* **81**:91–109.

Pulliam, H. R., 1980, Do Chipping Sparrows forage optimally? *Ardea* **68**:75–82.

Puttick, G. M., 1978, The diet of the Curlew Sandpiper at Langebaan Lagoon, South Africa, *Ostrich* **49**:158–167.

Puttick, G. M., 1979, Foraging behaviour and activity budgets of Curlew Sandpipers, *Ardea* **67**:111–122.

Pyke, G. H., 1984, Optimal foraging theory: A critical review, *Ann. Rev. Ecol. Syst.* **15**:523–575.

Pyke, G. H., Pulliam, H. R., Charnov, E. L., 1977, Optimal foraging: A selective review of theory and tests, *Quart. Rev. Biol.* **52**:137–154.

Quinney, T. E., and Smith, P. C., 1980, Comparative foraging behaviour and efficiency of adult and juvenile Great Blue Herons, *Can. J. Zool.* **58**:1168–1173.

Rabinowitch, V., 1968, The role of experience in the development and retention of food preferences in gull chicks, *Anim. Behav.* **16**:425–428.

Rabinowitch, V., 1969, The role of experience in the development and retention of seed preferences in Zebra Finches, *Behaviour* **33**:222–236.

Recher, H. F., and Recher, J. A., 1969, Comparative foraging efficiency of adult and immature Little Blue Herons (*Florida caerulea*), *Anim. Behav.* **17**:320–322.

Richardson, H., and Verbeek, N. A. M., 1987, Diet selection by yearling Northwestern Crows (*Corvus caurinus*) feeding on littleneck clams (*Venerupis japonica*), *Auk* **104**:263–269.

Royama, T., 1970, Factors governing the hunting behaviour and selection of food by the Great Tit (*Parus major* L.), *J. Anim. Ecol.* **39**:619–668.

Ryan, M. R., and Dinsmore, J. J., 1980, The behavioral ecology of breeding American Coots in relation to age, *Condor* **82**:320–327.

Ryder, J. P., 1980, The influence of age on the breeding biology of colonially nesting seabirds, in: *Behavior of Marine Animals, Vol. 4: Marine Birds* (J. Burger, B. Olla, and H. Winn, eds.), Plenum Press, New York, pp. 153–168.

Savory, C. J., 1977, The food of Red Grouse chicks, *Ibis* **119**:1–9.

Schneider, K. J., 1984, Dominance, predation and optimal foraging in White-throated Sparrow flocks, *Ecology* **65**:1820–1827.

Schnell, G. D., Woods, B. L., and Ploger, B. J., 1983, Brown Pelican foraging success and kleptoparasitism by Laughing Gulls, *Auk* **100**:636–644.

Schoener, T. W., 1971, Theory of feeding strategies, *Ann. Rev. Ecol. Syst.* **2**:269–404.

Schreiber, R. W., 1968, Seasonal population fluctuations of Herring Gulls in central Maine, Bird-Banding.

Searcy, W. A., 1978, Foraging success in three age classes of Glaucous-winged Gulls, *Auk* **95**:586–588.

Sherry, T. W., and McDade, L. A., 1982, Prey selection and handling in two neotropical hover-gleaning birds, *Ecology* **63**:1016–1028.

Shettleworth, S. J., 1984, Learning and behavioural ecology, in: *Behavioural Ecology: An Evolutionary Approach* (J. R. Krebs and N. B. Davies, eds.), Sinauer Associates, Sunderland, Massachusetts, pp. 170–194.

Siegfried, W. R., 1972, Aspects of the feeding ecology of Cattle Egrets (*Ardeola ibis*) in South Africa, *J. Anim. Ecol.* **41**:71–78.

Siegfried, W. R., 1977, Mussel dropping behaviour of Kelp Gulls, *South African J. Sci.* **73**:337–341.

Slater, P. J. B., 1983, The development of animal behavior, in: *Animal Behaviour: Genes, Development, and Learning* (T. R. Halliday and P. J. B. Slater, eds.), Freeman, New York.

Smith, J. N. M., 1974a, The food searching behaviours of two European thrushes. I: Description and analysis of search paths, *Behaviour* **48**:276–302.

Smith, J. N. M., 1974b, The food searching behaviours of two European thrushes. II: The adaptiveness of the search patterns, *Behaviour* **49**:1–61.

Smith, J. N. M., and Dawkins, R., 1971, The hunting behaviour of individual Great Tits in relation to spatial variations in their food density, *Anim. Behav.* **19**:695–706.

Smith, J. N. M., and Sweatman, H. P. A., 1974, Food searching behaviour of titmice in patchy environments, *Ecol.* **55**:1216–1232.

Smith, S. M., 1972, The ontogeny of impaling behaviour in the Loggerhead Shrike, *Lanius ludovicianus* L., *Behaviour* **42**:232–247.

Smith, S. M., 1973, Factors directing prey-attack by the young of three passerine species, *Living Bird* **12**:7–67.

Smith, S. M., 1983, The ontogeny of avian behaviour, in: *Avian Biology, Vol. VII,* (D. S. Farner and J. R. King, eds.), Academic Press, New York, pp. 85–160.

Southern, H. N., 1970, The natural control of a population of Tawny Owls (*Strix aluco*), *J. Zool.* (London) **162**:197–285.

Spaans, A. L., 1971, On the feeding ecology of the Herring Gull *Larus argentatus* Pont. in the northern part of the Netherlands, *Ardea* **59**:1–188.

Stevens, J., 1985, Foraging success of adult and juvenile Starlings *Sturnus vulgaris:* A tentative explanation for the preference of juveniles for cherries, *Ibis* **127**:341–347.

Stokkan, K. A., and Steen, J. B., 1980, Age-determined feeding behaviour in Willow Ptarmigan chicks *Lagopus lagopus lagopus, Ornis Scandinavica* **11**:75–76.

Sullivan, K. A., 1988a, Ontogeny of time budgets in Yellow-eyed juncos: Adaptation to ecological constraints, *Ecology* **69**:118–124.

Sullivan, K. A., 1988b, Age-specific profitability and prey choice, *Anim. Behav.* **36**:613–615.

Sullivan, K. A., 1989, Predation and starvation: Age-specific mortality in juvenile juncos (*Junco phaenotus*), *J. Anim. Ecol.* **58**:275–286.

Sutherland, W. J., Jones, D. W. F., and Hadfield, R. W., 1986, Age differences in the feeding ability of Moorhens *Gallinula chloropus, Ibis* **128**:414–418.

Thorpe, W. H., 1963, *Learning and Instinct in Animals,* Methuen, London.

Tinbergen, N., Impekoven, M., and Franck, D., 1967, An experiment on spacing out as a defence against predation, *Behaviour* **28**:307–320.

Turner, E. R. A., 1964, Social feeding in birds, *Behaviour* **24**:1–46.

Ulfstrand, S., 1979, Age and plumage associated differences of behaviour among Black-headed Gulls *Larus ridibundus:* Foraging success, conflict victoriousness, and reaction to disturbance, *Oikos* **33**:160–166.

Verbeek, N. A. M., 1977a, Age differences in the digging frequency of Herring Gulls on a dump, *Condor* **79**:123–125.

Verbeek, N. A. M., 1977b, Comparative feeding behavior of immature and adult Herring Gulls, *Wilson Bull.* **89**:415–421.

Vince, M. A., 1960, Developmental changes in responsiveness in the Great Tit (*Parus major*), *Behaviour* **15**:219–243.

Wakeley, J. S., 1978, Factors affecting the use of hunting sites by Ferruginous Hawks, *Condor* **80**:316–326.

Walton, K. C., 1979, Diet of Meadow Pipits *Anthus pratensis* on mountain grassland in Snowdonia, *Ibis* **121**:325–329.

Weathers, W. W., and Sullivan, K. A., 1989, Juvenile foraging proficiency, parental effort, and avian reproductive success, *Ecol. Mongr.* **59**:223–246.

Wemmer, C., 1969, Impaling behavior of the Loggerhead Shrike, *Lanius ludovicianus, Z. Tierpsychol.* **26**:208–224.

Werner, E. E., 1976, Niche shift in sunfishes: Experimental evidence and significance, *Science* **191**:404–406.

Wunderle, J. M., and Lodge, D. J., 1988, The effect of age and visual cues on floral patch use by bananaquits (Aves: Emberizidae), *Anim. Behav.* **36**:44–55.

Wunderle, J. M., and Soto-Martinez, J., 1987, Spatial learning in the nectarivorous bananaquits: Juveniles versus adults, *Anim. Behav.* **35**:652–658.

Wunderle, J. M., and O'Brien, T. G., 1985, Risk aversion in hand-reared bananaquits, *Behav. Ecol. Sociobiol.* **17**:371–380.

Zach, R., 1979, Shell dropping: Decisionmaking and optimal foraging in Northwestern Crows, *Behaviour* **68**:106–117.

Zach, R., and Smith, J. N. M., 1981, Optimal foraging in wild birds? in: *Foraging Behavior: Ecological, Ethological, and Psychological Approaches* (A. C. Kamil and T. D. Sargent, eds.), Garland STPM Press, New York, pp. 95–109.

INDEX

Acanthis flammea: see Redpoll
Acanthisitta chloris: see Rifleman
Acanthisittidae, 2, 8–9, 29
Acanthizidae, 228
Accipiter nisus: see Sparrowhawk, European
Accipiter striatus: see Hawk, Sharp-shinned
Accipitridae, 3
Acedinidae, 3
Acridotheres tristis: see Myna, Common; Myna, Indian
Acrocephalus
arundinaceus: see Warbler, Great Reed
scirpaceus: see Warbler, Reed
Africa, 9, 124, 125, 129
Agelaius phoeniceus: see Blackbird, Red-winged
Age, 162–163
breeding, 57
Aggression, 173
Air sac, 85
Aix sponsa: see Duck, Wood
Alarm calls, 174–176, 179, 188–191, 194
Alauda arvensis: see Skylark
Alaudidae, 46
Albatross, 3, 5
Buller's, 51
Royal, 6
Alcedinidae, 46
Alectoris chukar: see Chukar
Allelic diversity, 45
Allen's rule, 23
Allometry, 33
Allozyme, 56
Altricial species, 92, 294
Altruism, reciprocal, 176, 182, 192–195
Amblyornis macgregoriae, 224, 225
Anarhynchus: see Plover
frontalis: see Plover, Wrybill
Anas
aucklandica: see Teal, Brown
clypeata: see Shoveler, Northern
crecca: see Teal, Green-winged
discors: see Teal, Blue-winged
platyrhynchos: see Mallard
Anatidae, 3, 29, 46, 104, 105
Anatinae, 104
Anhingidae, 3
Anomalopteryx
didiformis, 30, 33
oweni, 30
Antarctica, 1, 9, 56
Anthornis melanura: see Bellbird, New Zealand
Anthus pratensis: see Pipit, Meadow
Apodidae, 3
Aposematic coloration, 287
Apterygidae, 3, 29
Apterygiformes, 2
Apteryx
australis: see Kiwi, Common
haasti: see Kiwi, Great Spotted
oweni: see Kiwi, Little Spotted
Aptornis, 6
otidiformis, 35
Archaic avifauna, 5, 6
Archey, G., 51
Ardea
cinerea: see Heron, Gray
herodias: see Heron, Great Blue
Ardeidae, 3

Arenaria interpres: see Turnstone, Ruddy
Asia, 46
Athene nocua: see Owl, Little
Atlantic Flyway, 105
Australasian species, 197
Austral avifauna, 5, 6
Australia, 4ff, 46–47
Avocet
American, 310
Red-necked, 4
Aythinae, 105
Aythya
affinis: see Scaup, Lesser
australis: see Duck, White-eyed
collaris: see Duck, Ring-necked
fuligula: see Duck, Tufted
marila: see Scaup, Greater
novaeseelandiae: see Scaup, New Zealand

Banaquit, 284, 285, 314
Begging calls, 201–203
Bellbird, 38
New Zealand, 8, 16, 20–22, 36, 194
Bergmann's rule, 13, 17
Bill size, 297
Binomial test, 148–157, 223ff
Biogeography, 4–9
Biomagnification, 108, 110, 133
Biotic transport, 122ff
Bird of Paradise, 217
Bittern, 3
Brown, 37
Blackbird, 45, 46, 48, 193, 194
Red-winged, 161–162, 259, 285, 287, 314
Yellow-headed, 160–161
Bluebird, Eastern, 87, 89, 92, 96, 97, 145, 148, 153–155, 313
Bobwhite, Northern, 46, 77
Bone, 75, 81, 85, 108
marrow, 85
Botaurus poiciloptilus: see Bittern, Brown
Bowdleria punctata: see Fernbird
Brain receptor, 175
Branta canadensis: see Goose, Canada
Breeding biology, 55
Breeding success, 54–58; *see also* Reproductive success
British Isles, 124
Broadbill, 8
Brood parasitism, 176ff, 197–205
Brood size, 148–157, 190
Bubulcus ibis: see Egret, Cattle
Bucephala albeola: see Bufflehead
Budgerigar, 144
Bufflehead, 105, 110
Bulbul, Red-vented, 46
Bullfinch, Eurasian, 314
Bull, Sir Walter, 50
Bunting, Cirl, 46
Butorides striatus: see Heron, Green-backed

Cacatua galerita: see Cockatoo, Sulphur-crested
Cacatuidae, 3, 46
Caffinch, Blue, 7
Calidris
canutus: see Knot
ferruginea: see Sandpiper, Curlew
Callaeas cinerea: see Kokako
Callaeidae, 2, 8, 29
Carduelis
cannabina: see Linnet, Eurasian
carduelis: see Goldfinch
chloris: see Greenfinch; Greenfinch, European
flammea: see Redpoll, Common
Carnivore, 298
Cassowary, 31
Casuariaus casuarius: see Cassowary
Catharacta
maccormicki: see Skua, South Polar
skua: see Skua, Southern Great
Cat, feral, 43–44, 55, 193
Cenozoic era, 4, 33, 34
Cereopsis novaehollandiae: see Goose, Cape Barren
Chaffinch, 7, 45, 46, 48, 192–194
Blue, 7
Common, 290
Charadriiformes, 277
Charadrii, 3
Charadrius, 6
hiaticula: see Plover, Ringed
Chen caerulescens: see Goose, Snow
Chernobyl, 121–132
Cherry, 281, 289
Chick development: *see* Growth
Chick recognition, 173
Chicken, Domestic, 77, 80, 86–88, 90–96, 114–116, 175, 286
China, 5
Chlidonias niger: see Tern, Black
Chromosome evolution, 146
Chrysococcyx lucidus: see Cuckoo, Shining Bronze

Chukar, 46
Ciconiiformes, 277
Circus
approximans: see Harrier, Pacific Marsh
cyaneus: see Harrier, Northern
Clam, 284, 299, 300
Clutch size, 95, 157, 190; *see also* Brood size
Cnemiornis, 28: *see also* Goose, New Zealand
calcitrans, 35
Cnemophilus loriae, 224, 225
Cockatoo, Sulphur-crested, 46
Coenocorypha: see Snipe
aucklandica: see Snipe, Sub-Antarctic
Coerba flaveola: see Banaquit
Cognitive processing, 175
Cognitive recognition model, 176–180, 185
Coleoptera, 290
Colinus virginianus: see Bobwhite, Northern
Colonial breeding, 174
Colonization, 9–12
Color, 173
Columba livia: see Dove, Rock
Columbidae, 3, 46, 217
Communal breeding, 57–58
Competition, 196, 273ff
intraspecific, 22
Conservation, 38–45
biology, 196
genetics, 44–45
Contamination, 70ff
Cook, Captain, 49
Cooperative breeding, 152, 163–165
Coot, American, 77–79, 105, 107, 108, 110, 112–120, 127, 309
Copulation, 59
Coraciidae, 3
Cormorant, 3, 6, 28, 279
Common, 49
King, 40
Olivaceous, 281, 308
Rough-faced, 49
Corvidae, 47, 297
Corvus
brachyrhynchus: see Crow, Common
caurinus: see Crow, Northwestern
frugilegus: see Rook
monedula: see Jackdaw
ossifragus: see Crow, Fish
Coturnix novaezealandiae; see Quail, New Zealand
Cracticidae, 47
Cradion caruncululatus: see Saddleback
Craig, J. L., 58–59
Crane, 3
Creeper, New Zealand Brown, 8, 42, 192, 194
Cretaceous period, 4, 8, 33
Crockett, David, 50
Crow, 299
Common, 109
Fish, 109
Northwestern, 284, 290, 300, 313
Crypsis, 173
Cuckoldry, 248ff
Cuckoo, 3
Long-tailed, 42
Shining Bronze, 176, 193, 197–203
Cuculidae, 3
Culmen, 91, 93
Cultural transmission, 60, 177, 182, 195–97
Cyanoramphus
auriceps: see Parakeet, Yellow-fronted
Novaezelandiae: see Parakeet, Red-fronted
malherbi: see Parakeet, Orange-fronted
Cygnus
atratus: see Swan, Black
olor: see Swan, Mute
sumnerensis, 35

Dacelo gigas: see Kookabura
Darter, 3
Deception hypothesis, 247–269
Deer, red, 38
Delayed breeding hypothesis, 276
Denham, W. M. C., 51
Denmark, 46
Dicaeidae, 229
Dieffenbach, E., 51
Dinornis maximus, 30, 32–33
Dinornithiformes, 2
Diomedea
bulleri: see Albatross, Buller's
epomophora: see Albatross, Royal
Diomedeidae, 3
Diphorapteryx, 6
Diptera, 290
Disease resistance, 98
Dispersal, 58, 59
Distraction display, 181
Distress calls: *see* Alarm calls
DNA
fingerprinting, 57

DNA (*cont.*)
  hybridization, 8, 9, 231
  mitochondrial, 5, 7
Dotterel, Banded, 40
Dove, 3, 217
  Rock, 46
  Spotted, 46
Drepanididae, 30
Dromaius novaehoolandiae: see Emu
Duck
  Blue, 6, 8, 40, 45
  Grey, 45
  Ring-necked, 105, 110
  Ruddy, 105, 110
  Tufted, 5
  White-eyed, 4
  Wood, 75, 76, 78, 79, 98, 99, 104, 112, 114, 115, 120
Dumont, D'Urville, 49
Dunnock, 46, 48, 194

Eagle
  Bald, 294, 166–168
  Chatham Island, 5
  Giant, 35
  White-bellied Sea, 5
Earthworm, 295
Edward Grey Institute, 53
Egg
  laying date, 54
  production, 95
  recognition, 198ff
Egretta caerulea: see Heron, Little Blue
Egret, 295
  Cattle, 290, 308
Elaphe obsoleta: see Snake, rat
Electromagnetic waves, 73ff
Emberiza
  cirlus: see Bunting, Cirl
  citrinella: see Yellowhammer
Emberizidae, 46
Embryo, 92
Emeus, 32
Emu, 31
Endangered species, 35–45
Enemy recognition, 173–205
England, 47, 50
Eocene epoch, 5, 8, 35
Eopsaltriidae, 224
Epimachus meyeri, 225
Erithacus rubecula: see Robin, European
Erythrura papuana, 224
Eudocimus albus: see Ibis, White
Eudyptula minor: see Penguin, Little Blue
Europe, 8, 47, 125, 129, 130
Euryapteryx curtus, 30, 32
Evolutionarily stable strategy, 142ff
Extinction, 30–38

Facultative manipulation, 141–169
Falco
  novaeseelandiae: see Falcon, New Zealand
  sparverius: see Kestrel, American
Falconiformes, 277
Falcon, 3
  New Zealand, 193
Falla, R. A., 51
Fantail, Grey, 28, 192, 193
Fat, 91
Feedback hypothesis, 182, 185–192
Feeding, 273–315
Fernbird, 16, 29, 36, 68, 40
Ficedula
  hypoleuca: see Flycatcher, Pied
  parva: see Flycatcher, Red-breasted
Fieldfare, 188
Fiji, 46
Finch, 297
  Darwin's, 30, 32–33
  Zebra, 144, 158–160, 169, 176
Finschia novaeseelandiae: see Creeper, New Zealand Brown
Fisher, R. A., 141ff
Fish, 277, 290, 304
Fitness, 141ff, 174ff, 248ff
  genetic, 214ff
Fledging success, 90, 95, 262, 263; *see also* Reproductive success
Fleming, C. A., 50
Flightlessness, 27–28
Fly, biting, 237
Flycatcher
  Pied, 248–254, 259–268
  Red-breasted, 193, 194
  Spotted, 313
Flyway
  Europe, 125
  European Russia, 125
Food
  birds as, 124ff
  recognition, 266–291
  selection, 266–291
  web, 70
Food-handling time, 297–303
Foot, 92
Foraging
  proficiency, 273–315
  site, 278–283
  site fidelity, 107, 108

Forest reserves, 42–43
Fregata magnificens: see Frigatebird, Magnificent
Fregatidae, 3
Frigatebird, 3
  Magnificent, 296, 308
Fringilla
  coelebs: see Chaffinch; Chaffinch, Common
  teydea: see Chaffinch, Blue
Fringilliade, 46
Frugivore, 219, 230
Fulica americana: see Coot, American

Gaimard, P., 49
Galapagos, 30, 33
Glacial refugia, 17
Gallinula chloropus: see Gallinule, Common; Moorhen, Common
Gallinule, Common, 110
Gallirallus: see Rail
  australis: see Weka
Gallus domesticus: see Chicken, Domestic
Gambusia affinis: see Mosquitofish
Gannet, 3
Gene flow, 24, 27
Genetic disequilibrium, 216
Genetic drift, 16
Gens, 198
Geographic variation, 12–27
Geospizinae, 30–32
Germany, 46
Gerygone igata: see Gerygone, Grey; Gerygone, New Zealand
Gerygone
  Grey, 49, 176, 192, 193, 197–203
  New Zealand, 16
Gigantism, 98
Glacial refugia, 33
Godwit, Bar-tailed, 9
Goldfinch, 46, 48
Gondwanaland, 4, 8–9
Goose, 28, 35
  Canada, 7, 46
  New Zealand, 8
  Snow, 148, 155–157
Gray, G. R., 50–51
Grebe, 3
  Horned, 110
  Pied-billed, 110
Greenfinch, 46, 48, 198–201
  European, 290
Grouping, 173
Growth, 90–93, 98, 133, 186
Growth rate, 166–168
Gruidae, 3
Gruiformes, 277
Gull, 3, 178, 277, 288, 297–299, 306, 310
  Black-billed, 5, 6, 50
  Black-headed, 5, 6
  Bonaparte's, 305, 311
  California, 283, 312
  Common Black-headed, 282, 294–296, 311
  Glaucous-winged, 279, 299, 300, 312
  Herring, 284, 294, 299, 305, 312
  Laughing, 281, 288, 292, 310
  Red-billed: *see* Gull, Silver
  Ring-billed, 296, 305, 311
  Southern Black-backed, 6
Guthrie-Smith, H., 50
Gymnorhina tibicen: see Magpie, Black-billed
Gymnorhinus cyanocephalus: see Jay, Pinyon
Gypaetus barbatus: see Lammergeyer

Habitat selection, 278
Haematopus
  fuliginosus: see Oystercatcher, Black
  longirostris: see Oystercatcher, Pied
  ostralegus: see Oystercatcher, European
  ostralegus finschi: see Oystercatcher, South Island Pied
  unicolor: see Oystercatcher, Variable
Haemoproteus, 217ff
Halcyon sancta: see Kingfisher
Haliaeetus
  australis: see Eagle, Chatham Island
  leucocephalus: see Eagle, Bald
  leucogaster: see Eagle, White-bellied Sea
Haplodiploidy, 146, 147
Harpagornis moorei: see Eagle, Giant
Harrier
  Northern, 309
  Pacific Marsh, 193
Hatching
  asynchrony, 165, 166
  success, 90, 95
Hawaii, 8
Hawk, 3
  Harris, 165–166
  Sharp-shinned, 287, 309
Hector, Sir James, 50
Helpers, 59; *see also* Cooperative breeding
Hemiphaga novaeseelandiae: see Pigeon, New Zealand

Heritability, 54, 56
Heron, 3, 279
  Gray: *see* Heron, Gray
  Great Blue, 290, 296, 302, 308
  Green-backed, 287, 308
  Grey, 287, 290, 304, 308
  Little, Blue, 308
Heteralocha acutirostris: see Huia
Heterozygosity, 56
Hierarchy, 59
Himantopus
  himantopus: see Stilt, Black-winged
  leucocephalus: see Stilt, Pied
  mexicanus: see Stilt, Black-necked
  novaezealandiae: see Stilt, Black
History, New Zealand ornithology, 49–53
Holarctic, 5, 9–10
Holocene epoch, 33
Homeotherm, 76, 78
Honeycreeper, Hawaiian, 30
Honeyeater, 217
  Australian, 8
Hormesis, 98–100
Huia, 29
Human, 179, 279
  predation by, 33
Hunger, 297
Hutton, F. W., 50
Hybrid, 5
Hybridization, 10–11, 45
Hydrobatidae, 3
Hymenolaimus: see Duck, Blue
Hymenoptera, 290

Ibis, 3
  White, 296, 308
Impaling, 298, 299
Imprinting, 195, 278, 298
Inbreeding, 58–61
  depression, 45
Incest, 59
Incubation
  period, 201
  time, 95
India, 47
Innate releasing mechanism, 173
Insectivore, 298
Introduced species, 45–49, 192
Island
  endemic species, 196
  refuge, 42–43

Jackdaw, 305
Jaeger, Parasitic, 9
Jay, Pinyon, 287
Jenkins Effect, 110
Jenkins, P. F., 60–61
Junco
  hyemalis: see Junco, Dark-eyed
  phaeonotus: see Junco, Yellow-eyed
Junco
  Dark-eyed, 89
  Yellow-eyed, 290, 297, 301, 314
Juvenile, 273–315

Kakapo, 28, 29, 37, 39, 42–44, 51
Kaka, 24–27, 42
Kea, 24–27
Kestrel, American, 286
Kingfisher, 3, 193
Kite, 279
  Snail, 290, 302, 308
Kiwi, 2, 3, 6, 8, 31, 35, 42, 49
  Brown, 45
  Common, 29
  Great Spotted, 29, 50
  Little Spotted, 29, 37, 39, 42, 45
Kleptoparasitism, 282, 287, 294, 305
Knot, 9
Kokako, 29, 37, 39, 42
Kookabura, 46

Lack, David, 53
Lammergeyer, 299
Lanius
  collurio: see Shrike, Red-backed
  excubitor: see Shrike, Northern
  ludovicianus: see Shrike, Loggerhead
  senator: see Shrike, Woodchat
Lapwing
  Masked, 309
  Northern, 282, 294, 295, 309
Laridae, 3
Larus
  argentatus: see Gull, Herring
  atricilla: see Gull, Laughing
  bulleri: see Gull, Black-billed
  californicus: see Gull, California
  delawarensis: see Gull, Ring-billed
  dominicanus: see Gull, Southern Black-backed
  glaucescens: see Gull, Glaucous-winged
  novaehollandiae: see Gull, Silver
  philadelphia: see Gull, Bonaparte's
  ridibundus: see Gull, Common Black-headed
Learning, 176–180, 274ff
Leatherjacket, 289
Lemur, 35
Lesson, R. P., 49

Leucocytozoon, 217ff
Limosa lapponica: see Godwit, Bar-tailed
Lindsey, C. J., 51
Linnet, Eurasian, 290
Lion, 148, 150–152
Lithornis, 8
Liver, 75
Longevity, 98
Lophorina superba, 224, 225
Lophortyx californica: see Quail, California
Lorenz, K., 173
Lung, 75, 85

Madagascar, 35
Magpie, 189
  Black-backed, 47
  White-backed, 47
Mallard, 45, 46, 77, 88, 104, 108, 127, 130
Mammal
  as competitor, 36
  as predator, 27, 36, 42, 43, 47
  introduced, 53
Maori, 35
Mat fidelity, 54
Mating status, 59, 247–269
Mealworm, 301, 302
Mediterranean region, 129
Megaegotheles, 28
Megalapteryx
  benhami, 30
  didinus, 30, 33
Melanerpes formicivorous: see Woodpecker, Acorn
Melanism, 10, 13, 17, 21, 27–30
Melanocharis
  striativentris, 224
  versteri, 224
Meliphaga: see Honeyeater, Australian
Meliphagidae, 217, 228
Melipotes fumigatus, 224
Melopsittacus undulatus: see Budgerigar
Melopiza melodia: see Sparrow, Song
Merganser, 51
  Auckland Island, 5, 38
  Chinese, 5
Mergus
  australis: see Merganser, Auckland Island
  squamatus: see Merganser, Chinese
Metabolic rate, 76, 78
Microfilaria, 217ff, 221
Mills, J. A., 55
Mimicry, 173, 176, 198ff
Mimus polyglottos: see Mockingbird, Northern
Miocene epoch, 33
Miro, 12
Mixed species flocks, 192–195
Moa, 2, 6, 8, 28, 30–35, 50
Mobbing, 174ff, 180–197
Mockingbird, Northern, 289, 290, 291, 314
Mohoua
  albicilla: see Whitehead
  ochrocephala: see Yellowhead
Mollusc eaters, 298
Monogamy, 249ff
Moorhen, Common, 303, 309
Mortality, 54, 55, 57, 92, 133, 273ff
  juvenile, 144
Mosquitofish, 114
Muscicapa striata: see Flycatcher, Spotted
Muscicapidae, 29, 46
Muscle, 75, 76, 80, 124, 127, 132
Mussel, 280, 281, 301
Myna
  Common, 17
  Indian, 47

Natural selection, 274ff
Nectivore, 219, 285
Nest
  defense, 190
  predation, 260, 261
  site fidelity, 54
Nesting success, 181
Nestling, 92
Nestling period, 95
Nestor
  meridionalis: see Kaka
  notabilis: see Kea
Nestoridae, 3
Nest, 198ff
New Caledonia, 9
New Guinea, 213–242
New Zealand, 1–62, 196–197
Niche width, 22
Ninox novaeseelandiae: see Owl, Morepork
North America, 8, 47, 130, 160
Nothofagus, 52
Notiomystis cincta: see Stitchbird
Notornis, 8; *see also* Rail
  mantelli: see Takahe
Nuthatch, European, 298, 313

Oenanthe oenanthe: see Wheatear, Northern

Oligocene epoch, 35
Oliver, W. R. B., 50
Omnivore, 214, 230
Ontogeny, 274ff
Opossum, 145, 146, 158
Optimality theory, 174ff, 182–185, 189
Orbell, G. B., 51
Ornithological Society of New Zealand, 52
Osprey, 277, 279, 282, 290, 292, 297, 306, 308
Ostrich, 31
Otago Museum, 55
Owen, Richard, 50
Owlet-nightjar, 28
Owl-Parrot; *see* Kakapo
Owl, 3
  Laughing, 8, 37, 39
  Little, 46, 193
  Morepork, 193
Oxyura jamaicensis: see Duck, Ruddy
Oystercatcher, 279, 297
  Black, 10–12
  Chatham Island, 40
  Eurasian, 280, 281, 290, 301, 306, 310
  Pied, 10–12
  Sooty, 10–12
  South Island Pied, 5, 10
  Variable, 7, 10–12, 22–23, 28

Pachycephala soror, 225
Pachycephalidae, 229
Pachyornis elephantopus, 30, 33
Palaearctic, 5
Paleocene epoch, 8
Pandion haliaetus: see Osprey
Panthera leo: see Lion
Paradisaeidae, 217, 228
Parakeet
  Forbes', 39, 45
  Orange-fronted, 39, 45
  Red-fronted, 36, 45
  Yellow-fronted, 16, 36, 42, 45
Parasite, 213–242
Parasite resistance: *see* Parasite
Parent recognition, 173
Parental care, 273
Parental investment, 160
Parental investment theory, 174ff
Parotia lawesii, 224, 225
Parrot, 3, 6, 9, 217, 233, 279, 297
Partridge, 3
  Grey, 46
Parus
  ater: see Tit, Coal
  atricapillus: see Tit, Willow
Parus (cont.)
  bicolor: see Titmouse, Tufted
  major: see Tit, Great
Passer domesticus: see Sparrow, House
Passeres, 3
Paternal care, 256–260
Pavo cristatus: see Peafowl, Common
Peafowl, Common, 46
Peale, T., 50
Pelecanidae, 3
Pelecaniformes, 277
Pelecanoididae, 3
Pelecanus occidentalis: see Pelican, Brown
Pelican, 3, 277, 279, 288
  Brown, 281, 282, 292, 293, 306, 308
Penguin, 3–5, 35
  Little Blue, 44
  Yellow-eyed, 51, 54–55
Perdix perdix: see Partridge, Grey
Pesticide, 109
Petrel, 5, 58, 59
  Chatham Island, 40
  Common Diving, 6
  Storm: *see* Storm Petrel
  Westland, 40
Petroica
  australis: see Robin, New Zealand
  macrocephala: see Tit, New Zealand; Tomtit
  multicolor: see Robin, Scarlet
Phaetontidae, 3
Phalacrocorax
  carbo: see Cormorant, Common
  olivaceus: see Cormorant, Olivaceous
Phasianidae, 3, 46
Phasianus colchicus: see Pheasant, Ring-necked
Pheasant, 3
  Ring-necked, 46
Philopatry, 165
Phylidonyris: see Honeyeater, Australian
Phylloscopus sibilatrix: see Warbler, Wood
Pica pica: see Magpie
Picoides borealis: see Woodpecker, Red-cockaded
Pigeon, 3, 180, 217
  New Zealand, 8, 39
Pipilo erythrophthalmus: see Towhee, Red-eyed
Pipit, Meadow, 290
Piracy, 294–297
Piscivore, 298, 305
Pitta, 8

Pituitary, 100
Plasmodium, 217ff
Platycercidae, 3, 46
Platycercus
  elegans: see Rosella, Crimson
  eximius: see Rosella, Eastern
Pleiotropy, 216
Pleistocene, 7, 12, 17, 21, 27, 33, 38
Pliocene epoch, 5, 12
Ploceidae, 46
Plover, 6
  Golden, 9
  Ringed, 290, 309
  Shore, 8, 40
  Wrybill, 8
Plumage, 277ff
  brightness, 214ff
Pluvialis dominica: see Plover, Golden
Podiceps auritus: see Grebe, Horned
Podicipedididae, 3
Podilymbus podiceps: see Grebe, Pied-billed
Poecilodryas albispecularis, 224
Poephila guttata: see Finch, Zebra
Poikilotherm, 76
Polygamy, 249ff
Polygyandry, 58
Polygyny, 161ff, 248, 254ff
Polymorphism, 10, 28
Polynesia, 33, 35
Polyterritoriality, 248–250, 256ff
Population
  bottleneck, 44
  decline, 36, 55
  turnover, 54
Potts, T. H., 50
Precocial species, 92
Predation, 247ff
Predator, 181ff, 214/m
  avoidance, 273ff
Prey capture, 291–297
Prosthemadera novaeseelandiae: see Tui
Protein electrophoresis, 24
Prunella modularis: see Dunnock
Prunellidae, 46
Pseudemys scripta: see Turtle, Yellow-bellied
Psittacidae, 29, 217
Pterodroma magentae: see Taiko
Ptiloprora guisea, 224
Pukeko, 58–59, 61
Pycnonotidae, 46
Pycnonotus cafer: see Bulbul, Red-vented
Pyrrhula pyrrhula: see Bullfinch, Eurasian

Quail, 3
  Brown, 46
  California, 46
  New Zealand, 37, 49
Quaternary, 35
Quelea quelea: see Quelea, Red-billed
Quelea, Red-billed, 89, 96, 98
Quoy, J. R. C., 49

Radiation
  atomic, 69–140
  ionizing, 72–85
Radioecology, 69–140
Radioisotope, 71ff
Radionuclides, 70ff
Rail, 3, 6, 35
  Banded, 38
  Chatham Island, 38
Rallidae, 3, 29
Rallus
  modestus: see Rail, Chatham Island
  philippensis: see Rail, Banded
Raptor, 297, 305
Ratite, 8, 31
Rat, 193, 196
Recurvirostra
  americana: see Avocet, American
  novaehollandiae: see Avocet, Red-necked
Red blood cell, 217ff
Redpoll, 46, 48
  Common, 192, 194
Reproduction, 93–98
Reproductive failure, 131
Reproductive investment, 56
Reproductive rate, 33, 54–56
Reproductive success, 145ff, 162, 247ff
Respiratory system, 85; *see also* Lung, Air sac
Rhagologus leucostigma, 223
Rhea americana: see Rhea
Rhea, 31
Rhipidura fuliginosa: see Fantail, Grey
Richdale, L. E., 51, 54–55
Rifleman, 42, 192
Ripiduridae, 228
Robin
  American, 283, 290, 313
  Chatham Island, 17, 18, 39, 43
  European, 188
  New Zealand, 12–20, 36, 42, 49
  Scarlet, 12, 18
Roller, 3
Rook, 47, 299, 313

Rosella, 298
  Crimson, 46, 280, 296, 313
  Eastern, 46
Rostrhamus sociabilis: see Kite, Snail
Royal Forest and Bird Protection Society, 53

Saddleback, 29, 36, 37, 39, 40, 42, 60–61
Salmon, 279
Sandpiper, Curlew, 290, 310
Savannah River Plant, 100–121, 128
Saxicola torquata: see Stonechat
Scandinavia, 124
Scaup
  Greater, 5
  Lesser, 5, 105, 110
  New Zealand, 4
Scavenger, 298
Sceloglaux albifacies: see Owl, Laughing
Secondary sexual characteristics, 213ff
Selection, 141
  sex specific, 144–169
Sewall Wright effect, 16
Sex ratio, 54, 118, 141–169
Sexual dimorphism, 144
Sexual selection, 213–242
Shearwater, Hutton's, 40
Shelduck, New Zealand, 7
Shell-dropping, 299, 300
Shore Plover: *see* Plover, Shore
Shorebird, 9, 28
Shoveler, Northern, 10, 89
Shrike, 297
  Loggerhead, 286, 298
  Northern, 299
  Red-backed, 299
  Woodchat, 299
Sialia sialis: see Bluebird, Eastern
Silvereye, 192, 193, 314
Sitta europaea: see Nuthatch, European
Skua, 3
  South Polar, 56
  Southern Great, 56–58
Skylark, 46, 48
Snail, 302
Snake, rat, 76, 78, 79
Snipe, 6
  Sub-Antarctic, 8, 36, 37, 40, 51
Social facilitation, 288
Social organization, 58
Song
  development, 60
  dialects, 60–61
Songbird, 3
South Africa, 6
South America, 6, 9
Soviet Union, 69, 100, 121ff
Sparrowhawk, European, 55, 309
Sparrow
  House, 17, 45, 46, 48
  Song, 89, 189
Speciation, 10, 12–27, 33
Species
  diversity, 33
  rare, 35–45
Spheniscidae, 3
Spoonbill, 3
Starfish, 294
Starling, European, 45, 47–48, 89, 279, 281, 289, 292, 296, 314
Stead, E., 50
Stercorariidae, 4
Stercorarius parasiticus: see Jaeger, Parasitic
Sterna
  hirundo: see Tern, Common
  maxima: see Tern, Royal
  sanvicensis: see Tern, Sandwich
Sternidae, 3
Stilt
  Black, 7, 10, 39, 196
  Black-necked, 296, 310
  Black-winged, 288, 310
  Pied, 10
Stitchbird, 8, 37, 40, 42
Stomach contents analysis, 105
Stonechat, 188
Storm Petrel, Kermadec, 40
Streptopelia chinensis: see Dove, Spotted
Strigidae, 3, 46
Strigops: see Parrot
  habroptilus: see Kakapo
Struthio camelus: see Ostrich
Sturnidae, 47
Sturnus vulgaris: see Starling, European
Sulidae, 3
Survival, long-term, 54
Swallow
  Tree, 87, 90–92, 95, 96
  Wood: *see* Wood Swallow
Swan, 35
  Black, 46
  Mute, 46
Sweden, 248
Swift, 3
Synoicus ypsilophorus: see Quail, Brown

Tachycineta bicolor: see Swallow, Tree
Tadorna variegata: see Shelduck, New Zealand
Taeniopygia guttata: see Finch, Zebra

Taiko, 39, 43
Takahe, 29, 36, 38, 39, 41, 51, 52
Tanzania, 148
Tarsus, 91, 93
Teal
  Blue-winged, 89
  Brown, 7, 28, 29, 37, 40
  Green-winged, 89, 119
  New Zealand, 45
Tern, 3, 277, 306
  Black, 110
  Common, 110, 301
  Royal, 287, 288, 293, 296, 312
  Sandwich, 292, 313
Territorial defense, 59
Territory, 247ff
Tertiary, 8, 26, 33
Testes degradation, 100
Thermoregulation, 30
Thinornis: see Plover; Plover, Shore
Thrasher, Brown, 97
Threskiornithidae, 3
Thrush
  New Zealand, 2, 6, 37, 39
  Song, 45, 46, 48
Thyroid, 70, 75, 85
Tick, 219, 233
Tinamou, 28
Tinbergen, N, 173ff
Titmouse, 279, 297, 300
  Tufted, 96
Tit
  Coal, 280
  Great, 183
  New Zealand, 196; *see also* Tomtit
  Willow, 280
Tomtit, 7, 12–20, 28
  New Zealand, 18, 49
Towhee, Rufous-sided, 97
Toxorhamphus poliopterus, 224, 225
Toxostoma rufum: see Thrasher, Brown
Troglodytes aedon: see Wren, House
Tropicbird, 3
Trypanosoma, 217ff
Tui, 8
Turbott, E. G., 51
Turdus
  merula: see Blackbird
  migratorius: see Robin, American
  philomelas: see Thrush, Song
  pilaris: see Fieldfare
Turnara capensis: see Thrush, New Zealand
Turnagridae, 8
Turnstone, Ruddy, 9, 310

Turtle, 81
  yellow-bellied, 114
Tyrannides, 8
Tyranni, 3

United Kingdom, 46
United States, 46, 119

Vanellus
  miles: see Lapwing, Masked
  vanellus: see Lapwing, Northern
Variability, genetic, 56
Vocal mimicry, 201–203
von Haast, J., 50

Wader, 3
Warbler, 279
  Great Reed, 248–251, 254, 258–268
  Reed, 291, 292, 298, 313
  Wilson's, 131
  Wood, 248–254, 258–268
Waterfowl, 3
Wattlebird, New Zealand, 2, 6
Weight, 91, 92, 98
  nestling, 257
Weka, 28, 29, 37, 42
Wheatear, Northern, 291, 293, 294, 313
White, E. R., 51
Whitehead, 8, 42
Wilkes, Captain Charles, 50
Wilsonia pusilla: see Warbler, Wilson's
Wing, 91–93
  shape, 18
Wood Swallow, Dusky, 5
Woodpecker
  Acorn, 146
  Red-cockaded, 148, 152–154, 162–165
Wren
  Bush, 26, 29, 36, 37, 39
  House, 87, 90, 92, 95, 97
  New Zealand, 2, 6, 8
  South Island Rock, 26, 29
  Stephen Island, 29, 37
Wrybill, 49

Yellowhammer, 46, 48
Yellowhead, 8, 42
Young, E. C., 56

Xanthocephalus xanthocephalus: see Blackbird, Yellow-headed
Xenicus
  gilviventris: see Wren, South Island Rock
  longipes: see Wren, Bush
  lyalli: see Wren, Stephen Island

Zosterops lateralis: see Silvereye